Springer Series in Healthcare Management and Innovation

The Springer Series in Healthcare Management and Innovation presents a curated collection of research monographs, contributed volumes, and professional titles that illuminate emerging theories, empirical studies, practical applications, strategic frameworks, and real-world case studies in healthcare. Each volume addresses a broad spectrum of topics in healthcare management and innovation, including—but not limited to—leadership, financial management, ethics and corporate social responsibility (CSR), human resource management, managed care, workforce and patient diversity, research and development (R&D) of medical technologies, organizational strategy and planning, and healthcare information systems.

By bridging the gap between theory and practice, this series serves as an essential resource for academics, researchers, scholars, and practitioners. It equips readers with the insights and competencies needed to navigate and lead effectively in the dynamic and evolving landscape of healthcare.

Luigi Negri

De-Risking Investments in Medical Technology Innovation

Strengthening Decision-Making and Valuation Accuracy in Medical Startups

Springer

Luigi Negri
Thinkeen Global
Miami, FL, USA

ISSN 3091-3810 ISSN 3091-3829 (electronic)
Springer Series in Healthcare Management and Innovation
ISBN 978-3-032-15006-6 ISBN 978-3-032-15007-3 (eBook)
https://doi.org/10.1007/978-3-032-15007-3

This Springer imprint is published by the registered company Springer Nature Switzerland AG
The registered company address is: Gewerbestrasse 11, 6330 Cham, Switzerland

Foreword

It is an honor and a pleasure to write this preface to Luigi's book on *"De-Risking Investments in Medical Technology Innovation."* Luigi and I crossed paths at a very special juncture of our careers. He had started his Doctoral in Business Administration (DBA) at SDA-Bocconi, the business school of Bocconi University in Milan. Luigi was already a competent and expert executive and consultant in the medical technology business, who wanted to move to the next level by getting an advanced academic degree (not an easy task given his outstanding background!). I was, at the time, struggling with my colleague Arnaldo Camuffo, thinking of whether we could devise better ways to help executives and managers make better decisions for their organizations. As professors of firm strategy and organization, we were puzzled by the rhetoric of some academic literature and practitioner insights about the myth of "failing" in entrepreneurial ventures. To be sure, this "myth" underlies, in fact, a very good intuition: entrepreneurs should try and experiment, learn from failure, and move to the next step, which can provide a successful new spin. At the same time, we conjectured that while entrepreneurs learn from failing, we could think of a way, a method, to help them make better decisions anyhow, and possibly learn even more from failures.

Arnaldo and I mumbled. In the end, we landed on the idea that the best method we were aware of for understanding the world and making decisions about the future in a disciplined way was the "scientific method." We then asked: "What if we teach entrepreneurs to think like scientists?" This calls for teaching them to develop theories—that is, to make logical statements based on cause and effects relations, or more generally on logical links such as in Wittgenstein's truth tables. This is not an infallible method, as scientists do make mistakes. But on the one hand, it is the best method we know to make good decisions under uncertainty, and on the other hand, it is a disciplined method that forces decision-makers to think about what they do. Armed with these ideas, we run a randomized control trial with 116 early-stage entrepreneurs at Bocconi, documented in Camuffo et al. (2020). We trained a random sample of these entrepreneurs to think like scientists, by developing and testing theories about their business, that is, by developing causal links, and by testing their theories. We trained the control group with a standard entrepreneurial training, like the ones that we had taught for many years in several master or executive classes. The results were striking: the entrepreneurs treated to act like scientists showed a higher rate of termination, but performed better conditional on staying in business.

Further research and interviews with our sample entrepreneurs showed that the earlier termination was motivated by an early recognition that the business was going to be unsuccessful. We concluded that the entrepreneurs treated to act like scientists were better able to see opportunities and threats, and more generally they were better able to discern good and bad options.

These encouraging results prompted us to undertake a new step. We run a new randomized control trial, this time with 759 early-stage entrepreneurs (Camuffo et al., 2024). This is an unprecedented scale of a randomized control trial in the academic field of strategic management, and, as a matter of fact, the paper won a prize for the best field experiment paper in 2024 issued by the Innovation Growth Lab, a London-based institute supporting field experiments in entrepreneurship and management. The new randomized control trial confirmed the results of the previous smaller scale study.

Luigi started his DBA in 2018, the year in which I met him as a student of the class that I taught in the first year of the program. In my class, I taught the participants, all experienced executives, managers, or consultants, to make decisions using a scientific approach. At the time, Arnaldo Camuffo and I, along with PhD students, were working on our first randomized control trial, and ideas were still in a flux. Luigi became immediately passionate about the idea, and after several deep brainstorming discussions, we decided to try to test the theory in a quasi-experimental setting in which he taught a group of his clients to make decisions using a scientific approach, while working with other clients using the traditional methods that his company, ThinKeen Global, had used up to that point. In fact, Luigi went further. He successfully revisited our theory, largely based on academic language and concepts, and translated it into practical applications mixed with his own experience, methods, and backgrounds. The end result has been a concrete framework to make real decisions in real companies. Not only were Luigi's results consistent with our randomized control trials, but he was also able to produce details about the process and the results that we could not highlight with our broad average and quantitative effects. These initial results, the additional data he gathered later, through the routine application of the theory in real consulting life, and more generally, Luigi's detailed insights about this decision-making process are the object of this book. What makes the book valuable is indeed his effort and ability to move a framework from its academic basis to a practical tool. We are also trying to work out ways to process what we have done to get closer to our executive and master students, and to the concrete use by firms. In this respect, we look at Luigi's book as an exemplary attempt to take this step building on the insights of our framework at the outset of the thinking and brainstorming process. In Camuffo, Gambardella, and Pignataro 2024, we have moved a step further, and outlined a general model to apply the scientific approach—which we now call a theory-driven approach to strategic management decisions—not just to early-stage entrepreneurs but to managers of larger companies. Luigi has pioneered this generalization by applying the scientific approach that we had worked out for early-stage entrepreneurs to managers in the medical technology business. As a matter of fact, this book is not about early-stage entrepreneurs, but about full-fledged managers, and, in this respect, the book covers an important extra-mile with respect to the context of our initial randomized control trials.

The focus on the medical technology business is another plus of the book. A better understanding of the applications cannot be context-free. Applying it to a specific context makes the framework more realistic, usable, and practical. The book is a journey. The reader will navigate the approach, understand how it works, and comprehend how they can apply it to their context—whether by replicating what Luigi did, or by understanding what Luigi did to adapt it to the reader's own context and goal. As one of the pioneering attempts to lay out this framework in practice, this book sets the ground for building a disciplined process to managerial decision-making. For me, this has not only been an opportunity to learn from Luigi's expertise in business and from our own discussion, but also to meet a new friend with whom I have walked together a path to improve decision-making in organizations. By reading this book, the reader will be able to experience what I experienced and learn about the fascinating opportunity to make our organizations more creative, efficient, and capable of generating more values for our societies.

Milan, Italy Alfonso Gambardella

References

Camuffo, A., Cordova, A., Gambardella, A., & Spina, C. (2020). A scientific approach to entrepreneurial decision making: Evidence from a randomized control trial. *Management Science, 66*(2), 564–586.

Camuffo, A., Gambardella, A., Messinese, D., Novelli, E., Paolucci, E., & Spina, C. (2024). A scientific approach to entrepreneurial decision-making: Large-scale replication and extension. *Strategic Management Journal, 45*(6), 1209–1237.

Camuffo, A., Gambardella, A., & Pignataro, A. (2024). Theory-driven strategic management decisions. *Strategic Science, 9*(4), 382–396.

Preface

Similar to other executives, during the last 25 years of my corporate career, I have often been surprised by how BioTech and MedTech corporations could survive such a high frequency of poor business decisions. Global BioTech and MedTech corporations are well-organized establishments with well-defined leadership teams, functional leaders, general managers, boards of directors, established policies and procedures, and highly specialized, highly paid decision-makers. My colleagues and I were best positioned to make and contribute to the most brilliant and rational decisions. Still, today, we can all remember the numerous bad critical decisions we witnessed and, sometimes, sadly, collaborated to make.

Later, when, with some colleagues, we decided to focus on innovative medical startups as consultants, the fresh corporate memories triggered a natural doubt: if big, established, and well-organized corporations can give origin to so many memorable decision-making mistakes, what about early-stage MedTech, Health Tech, and BioTech startups?

What is the decision-making accuracy of the two enthusiastic and lonely scientists hoping to bring their new medical idea to the market? How do they decide when their medical invention is "good enough" to form a proper company and start risking investors' cash? How do they convince investors about the quality of their market assumptions in a global context of market stakeholders holding different and often conflicting priorities? How do they predict the time and money required to conclude clinical trials and reach market entry milestones? How do they make and validate hypotheses about the economic value of their medical innovations?

Due to these questions, we were not truly surprised by the decision-making and predictive biases we observed while working with early-stage medical ventures. For instance, we noticed how frequently the estimation of future revenues was based on untested assumptions, often disconnected from actual market constraints. We observed how initial biases in making business assumptions are sometimes followed by Microsoft Excel exercises in which, in a revenue forecast attempt, an analyst mouse-pulls untested growth formulas from year to year.

In some other projects, we observed medical ventures underestimating the complex issue of predicting adoption. For instance, we met startups that confused the idea of invention with the idea of innovation and, consequently, assumed that a patent-protected technical improvement over the current standard of care would be sufficient to justify the expectations of broad adoption or even changes in global clinical practices.

We also observed many medical ventures overlooking the financial expectations of the most sophisticated investors, such as venture capitalists, forgetting that an early-stage medical startup represents a high-risk investment and, therefore, must be able to promise highly rewarding returns credibly. These returns can be supported only by genuine medical innovations driving quantum-leap improvements—for instance, in patients' lives, not by marginal clinical benefits, marginal usability, or technical improvements that are too often mistakenly equated to innovations.

Not surprisingly, literature and specialized websites report that MedTech startups fail at a rate of 80–87%—rates that may discourage potential investors and, therefore, deprive this sector of precious economic resources. A large amount of literature highlights decision-making as the most frequent cause of startup failure across industries. On the other hand, the tight relationship between the quality of decision-making and the probability of the venture's success is apparent to most long-course practitioners of this field.

Recent cross-industry research comes in very handy while looking for a systematic solution to address decision-making in startups. Extensive research proves that, in early-stage startups, a scientific approach to decision-making enables entrepreneurs to make better decisions and predictions while testing and launching their business ideas (Camuffo et al., 2020).

My colleagues and I relied on these findings and invested significant time and effort in testing their validity in the real-life practice of medical startups. The undebatable results of the ventures adopting this approach made us commit to deploying the most practical and usable tools to help medical entrepreneurs adopt a scientific approach to decision-making. These field-testing activities enabled the elaboration of hands-on prescriptions for medical entrepreneurs, embracing the application of the scientific approach to decision-making, representing the main result of our efforts.

This approach presents proven processes and tools that help protect medical ventures from typical decision-making biases, such as those rooted in our love for our ideas, our dream of their survival, and our desire for a rewarding economic payback. It helps medical ventures to rationally and analytically support their claims about the value of an innovation and about the capital required for its development, encouraging both entrepreneurs and investors to shift the emphasis of their interactions from relational to rational. Moreover, the approach provides several practical and proven tools that help medical entrepreneurs move further away from the dangerous philosophy of "fake it until you make it" and embrace the realism of "let's think it rationally first, so we are more likely to make it later." It also suggests that, in medical ventures, the scientific approach to decision-making should not remain confined to clinical trials and clinical decisions but must be adopted in the broader field of managerial and business decisions.

High complexity and contextual uncertainty, which are typical of medical innovation, can impulsively discourage the use of rigorous and disciplined decision-making methods. They can be perceived as overly analytical, intellectually burdensome, time-consuming, and, therefore, unfit for under-resourced and time-constrained ventures. Consequently, ventures may be inclined to resort to more casual, offhand decision-making processes, exposing their business assumptions and decisions to critical flaws or to the dangerous charm of self-stabilizing predictions.

Still today, in complex decision-making settings, we observe some medical entrepreneurs or even executives of large BioTech and MedTech multinationals employing decision-making processes that resemble that of those movie cowboys who, given the difficulty of shooting and hitting a can thrown in the air, decide to close their eyes while shooting as they prefer to rely on their "feelings" than on observing reality.

Our field experience suggests that, even in complex environments, a scientific approach to decision-making does not necessarily call for heavy, overcomplicated decision-making processes, nor years-long experiments such as those of large phase III clinical trials. It is more of a mental discipline and a managerial approach to carefully selecting, collecting, refining, and rationally processing market information.

Milan, Italy — Alfonso Gambardella
Trieste, Italy — Nicola Redi

Reference

Camuffo, A., Cordova, A., Gambardella, A., & Spina, C. (2020). A scientific approach to entrepreneurial decision making: Evidence from a randomized control trial. *Management Science, 66*(2), 564–586.

Acknowledgments

I'd like to sincerely thank Prof. Alfonso Gambardella for his support and mentorship in helping me understand and translate a complex theoretical framework into a practical approach to managing medical technology innovation. I also wish to thank Dr. Nicola Redi, who not only contributed to the content of this book but also generously shared his valuable business insights and actively helped recruit some of the innovative organizations where the methodology has been tested. My gratitude also goes to the senior editor, Nitza Jones, who provided me with patient support throughout the publishing process. Finally, a special thank you to my wife and my mom, who understood and supported my extended absence and mental distance while I wrote this book, continued my consulting work, and fulfilled my teaching duties.

Contents

List of Authors

Alfonso Gambardella Department of Management and Technology & ION Management Science Lab, Bocconi University, Milan, Italy

Nicola Redi Obloo Capital, Milan, Italy

MIB Trieste School of Management, Trieste, Italy

Introduction

1

Abstract

High failure rates in medical technology startups have cooled investors' sentiment and consequently hindered innovation in this space, despite rapid scientific and technological progress. This book introduces a practical and evidence-based methodology that applies the scientific approach to decision-making in medical technology ventures. Drawing on research and field testing with real startups, it provides a structured approach for identifying, mapping, and prioritizing the complex decision-related problems that determine success or failure. By applying a scientific approach not just to clinical trials but also to the most critical business problems at each stage of a venture's development, entrepreneurs and investors can improve prediction accuracy, mitigate cognitive biases, and thereby implement faster paths to market. The proposed methodology combines conceptual rigor with visual maps, making complex decisions more manageable and transparent. Entrepreneurs and investors alike will find in this book a pragmatic framework for navigating uncertainty, making better and timelier choices, and ultimately increasing the probability of successful value creation in the demanding landscape of medical innovation.

High failure rates of medical startups tend to discourage many potential investors, depriving the sector of precious financial resources. In the last two decades, managing early-stage medical technology ventures hasn't become easier: It still involves dealing with considerable complexity due to various factors, including the difficulty of predicting the behavior of numerous stakeholders with conflicting priorities, the stretched breadth of competencies required of medical entrepreneurs, and the ever-accelerating rate of growth of global medical knowledge.

L. Negri, *De-Risking Investments in Medical Technology Innovation*, Springer Series in Healthcare Management and Innovation,
https://doi.org/10.1007/978-3-032-15007-3_1

Medical entrepreneurs and innovators are required to make precise predictions and decisions in a highly complex, uncertain environment, which poses specific challenges to the accuracy of decision-making and, ultimately, to the prediction of a medical innovation's economic value and the actual investment risk. These decision-related challenges strongly threaten entrepreneurial accountability, particularly the ability to deliver unbiased predictions of future adoption dynamics and, in turn, the capacity to craft and execute adequate monetization strategies.

This book offers practical solutions to entrepreneurs and investors in the medical field. It proposes an efficient, pragmatic approach to decision problem identification and solution based on essential research findings that have been tested in real life with medical innovation startups. Research has shown that a scientific approach to decision-making can enable entrepreneurs to make better predictions while testing and launching their business ideas (Camuffo et al., 2020). It has also proven that a scientific approach can not only mitigate the cognitive biases that influence entrepreneurs' ability to efficiently collect, prioritize, and process the critical market information required for estimating the actual value of their product ideas (Cohen et al., 2019), but also provide them rational support while making important decisions (Camuffo et al., 2020; Camuffo, Gambardella, et al., 2024; Camuffo, Gambardella, Messinese, 2024).

The author tested the use of this scientific approach to decision-making with early-stage medical innovation startups in the real world, verifying and validating the outcomes. Based on the knowledge gained through this fieldwork experience and the necessary adaptation of the methodology to industry specificities, this book provides practical tools to both medical entrepreneurs and investors seeking to maximize their probability of success by employing a scientific approach to decision-making.

While presenting the fundamentals of applying the scientific method to decision-making, the book also cites examples. It outlines ideas on how medical entrepreneurs can field-test some of the most common assumptions and decisions.

Centered on the application of the scientific approach, the methodology recommends the use of a map that visually frames and connects the numerous decision-related issues typical of medical innovation. This map is a valuable visual tool that guides the identification of all critical decision-related problems, affording efficiency and real-life usability to the methodology.

Furthermore, the book suggests and exemplifies ways to implement selective prioritization and identification of the mapped decision-related challenges. Indeed, the author's fieldwork experience revealed that the application of the scientific approach to decision-making exhibits greater efficiency when it is selectively applied to a few decision-related issues in each phase of a startup's development.

The reader of this book will find practical tools that help navigate the map and rapidly identify critical problems—those inhabiting the critical path that begins with an innovative idea and ends with the highly desired monetization. Addressing these problems with high-quality decisions can make the difference between successful value creation and expensive failure.

Any medical venture that applies the suggested approach can expect to witness what both our field tests and the broader research have demonstrated. Compared to ventures that do not employ the scientific approach to decision-making, early-stage medical startups using the proposed methodology are

- Less likely to linger on projects that should be terminated.
- More likely to make a timely exit.
- More likely to pivot than not.
- More likely to focus on only a small, manageable number of critical pivots.
- More likely to enjoy higher and earlier revenues.

In summary, this book presents medical innovators and investors with an entirely new, efficient, and guided approach to decision-making in medical innovation: a methodology that is based on the latest scientific evidence and successfully tested in real life.

It addresses the need for a "deliberate" approach to decision-making that encompasses all the investigations and decisions faced by the medical technology entrepreneur, whose absence remains one of the most important reasons for failure in medical innovation (Behrns et al., 2021).

Bibliography

Behrns, K., Gingles, B., & Sarr, M. (2021). *Medical innovation.*

Camuffo, A., Cordova, A., Gambardella, A., & Spina, C. (2020). A scientific approach to entrepreneurial decision making: Evidence from a randomized control trial. *Management Science, 66*(2), 564–586.

Camuffo, A., Gambardella, A., Messinese, D., Novelli, E., Paolucci, E., & Spina, C. (2024). A scientific approach to entrepreneurial decision-making: Large-scale replication and extension. *Strategic Management Journal, 45*(6), 1209–1237.

Camuffo, A., Gambardella, A., & Pignataro, A. (2024). Theory-driven strategic management decisions. *Strategic Science, 9*(4), 382–396.

Cohen, S. L., Bingham, C. B., & Hallen, B. L. (2019). The role of accelerator designs in mitigating bounded rationality in new ventures. *Administrative Science Quarterly, 64*(4), 810–854.

The Vocabulary of Medical Ventures: The Anatomy of Success and Failure

2

Abstract

This chapter defines a shared vocabulary for understanding medical technology ventures. It clarifies medical innovation across medical devices, in vitro diagnostics, and digital health solutions. The notion of startup failure is refined, distinguishing true failure—when expected returns cannot be achieved—from strategic pivots that enable learning. The chapter explains the differences between precision and accuracy in a company's valuation, highlighting how both entrepreneurial optimism and decision biases might distort these delicate estimations. It differentiates marginal improvements from innovative ideas, underlining that true innovation drives clinical and economic transformation. Finally, it links investigation and decision-making within the Stage-Gate Model and introduces scenario decisions as assumption-based choices guiding strategic direction. A shared vocabulary supports the alignment of understanding among entrepreneurial stakeholders, which is particularly important in business environments characterized by high degrees of uncertainty, where environmental complexities and fast-changing assumptions often lead to misunderstandings and inefficiencies. It is therefore essential to establish a shared understanding of the fundamental concepts and their corresponding names.

2.1 Medical Innovation

Before commencing, it is essential to agree on the meaning of certain terms. Medical technologies include products and services that are used to save and improve lives. These products and services are generally grouped into three main categories: medical devices, in vitro diagnostics (IVDs), and digital health solutions. A medical device can be an instrument, apparatus, appliance, software, implant, reagent, or

L. Negri, *De-Risking Investments in Medical Technology Innovation*, Springer Series in Healthcare Management and Innovation,
https://doi.org/10.1007/978-3-032-15007-3_2

material, among others. From syringes and wheelchairs to cardiac pacemakers and medical imaging technology (such as MRI, CT, and X-ray machines), medical devices can play a range of roles in maintaining and restoring health (MedTechEurope, 2025). Medical devices are used, either alone or in combination, for one or more medical purposes:

- Diagnosis, prevention, monitoring, treatment, or alleviation of disease (Czekster et al., 2023).
- Diagnosis, monitoring, treatment, alleviation, or compensation for an injury or a handicap.
- Investigation, replacement, or modification of the anatomy or a physiological process (Brönneke et al., 2021).

IVDs never come into direct contact with an individual; they provide information based on a sample. Their value resides in the quality of the information they provide. IVDs do not treat patients; their role is to provide information relevant to healthcare decisions. In vitro diagnostics provides critical information at every step of the patient's journey, from prognosis, screening, and diagnosis to monitoring of disease progression and prediction of treatment responses.

Digital health and care, in turn, refer to tools and services that use information and communication technologies (ICTs) to improve the prevention, diagnosis, treatment, monitoring, and management of health and lifestyle (Dolezel & Smutny, 2021). Digital health and care hold the potential to innovate and improve people's access to and quality of care and to amplify the overall efficiency of the healthcare sector. Digital tools will play a central role in some of the most promising areas of healthcare innovation. Data will be critical as we advance toward more personalized, predictive, and precision healthcare. In the future, healthcare will be more integrated, value-based, and focused on patient outcomes.

2.2 Failure of a Medical Startup

It is essential to clarify what this book means by "startup failure" and to demystify some common beliefs, including those that attribute most strategic changes to the category of failures. While refining the definition of startup failure, we want to answer some key questions:

- Can a market-stage medical startup that is financially sustainable but unable to deliver the promised returns be considered a failure?
- Is an early, deliberate exit synonymous with failure?
- Can pivoting be considered a failure?
- Is a startup with several failed attempts to enter the market a failure?

According to our guiding principle, a venture is considered successful when it can deliver the promised returns to investors through a solid pattern of adoption and monetization and in a timeframe consistent with the business plan. One of the worst cases of failure is characterized by the burning of all invested cash and subsequent forced exit due to a venture's inability to enter the market or its failure to reach a financial breakeven.

While it is relatively easy to classify the extreme cases of failure and success, labeling intermediate conditions of the venture may be less straightforward. Hence, specificity is crucial.

In this book, a venture that, despite its ability to sustainably commercialize its product, chronically and significantly misses the promised financial returns is considered a failure.

A pivot is the strategic change of a business concept, product, or different elements of a business model. This book does not regard an early, deliberate pivoting or exit as a failure, as it doesn't typically allow the time to burn a significant amount of cash. Instead, when triggered promptly, it enables investors and entrepreneurs to reassign the remaining funds to more promising ventures or pivots. While high pivoting frequency is a dangerous sign of strategic weakness, a limited number of pivots must be actively embraced and considered crucial as the fruit of newly validated learning that can lead to brilliant ideas (Bajwa et al., 2017).

Figure 2.1 visually represents the confines of this book's working definition of actual startup failure. The dark green box labeled "Stagnant Pre-Market Development" represents the variegated conditions in which a venture struggles to reach the commercialization milestone. For instance, the startup may struggle to secure the financing necessary to support the clinical validation and, in turn, the

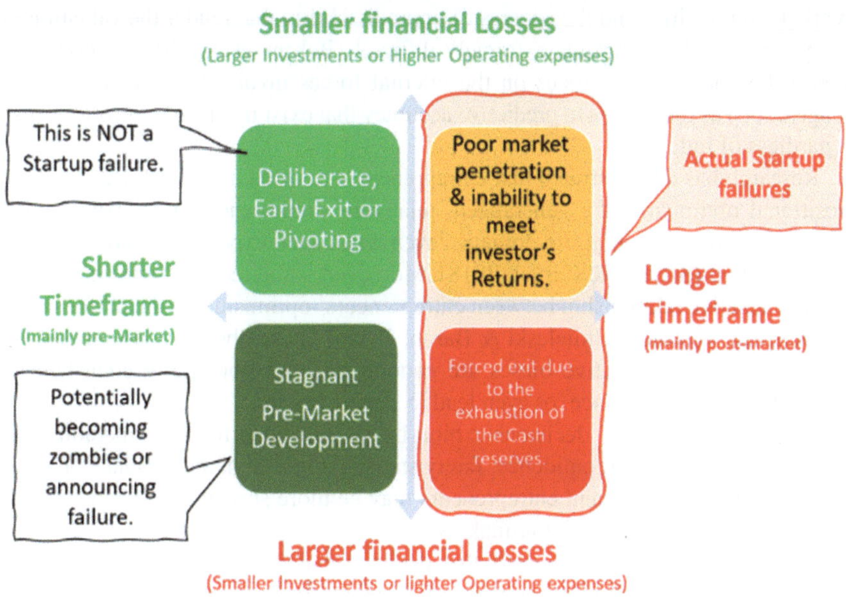

Fig. 2.1 Defining failure of a medical startup

efficacy claims necessitated by regulatory bodies. In these cases, it becomes difficult even to predict a market entry milestone or a rough penetration target. While this situation is not predictive of a venture's ability to deliver the promised returns, the venture cannot yet be considered a failure. The venture may later move toward the failure boxes, or it may change speed and finally deliver. This condition can't be regarded as a failure until the uncertainty surrounding the commercialization date becomes a realistic threat to financial survival. In summary, and as shown in Fig. 2.1, we consider that a medical startup fails when:

- It is forced to exit due to cash issues linked to long-standing market entrance or adoption issues.
- Poor market penetration or limited profitability makes the delivery of returns at the level of those promised to investors impossible.

2.3 Threats to Valuation Precision and Accuracy

This book constantly addresses the central issue of entrepreneurial accountability based on ventures' ability to generate the financial returns promised to investors. The book will consider this challenge the ultimate entrepreneurial mission. It spotlights valuation efforts as the most difficult exercise for a medical entrepreneur. It involves comprehending the uncertainties surrounding, for instance, the post-market adoption rates, the likelihood of obtaining a public or private reimbursement, and the post-market revenues.

Let us temporarily postpone the analysis of the distinctive environmental and market complexities and the sources of unpredictability that render the valuation of early-stage medical startups extremely difficult (Behrns et al., 2021; Herzlinger, 2006). Instead, we now focus on the internal forces involved in predictive challenges, such as the threats to predictive accuracy that exist in all early-stage startups, regardless of industry.

Research has demonstrated that entrepreneurs are often highly affected by "dispositional optimism": the generalized, relatively stable tendency to expect good outcomes across essential life domains, even when such expectations are not rationally justified (Scheier & Carver, 2018). Moreover, some recent studies have indicated a negative relationship between entrepreneurs' optimism and the performance of their new ventures (Hmieleski & Baron, 2017). These observations support the idea that certain predictive biases are endogenous and potentially contribute to unsatisfactory performance, or even lead to startup failure.

Often, entrepreneurs' decisions exhibit biases and imperfections, including the illusion of control, overconfidence, faulty statistical intuition, and reliance on decision-making heuristics, and entrepreneurs may be more susceptible to biased decision-making than other decision-makers.

For instance, research suggests that decision-making errors related to market entry may be the result of entrepreneurs' focus on themselves (their expectations, wishes, and competencies) at the expense of understanding the competition (Moore et al., 2007). Another common challenge comes from the particular forecasting necessary for all early-stage ventures. In this case forecasting constitutes predicting the future based on either statistical elaboration of historical data or other predictive methods when no history exists (Mun, 2004).

However, pre-market startups lack historical revenue data. Hence, the venture's forecasting strength strongly depends on its ability to set unbiased assumptions and predict the most probable future scenarios. Another endogenous source of predictive bias often comes from high-level methodological myopia, which narrows the focus mainly to predictive precision while neglecting the requirement for accuracy while selecting, feeding, and building predictive models. However, is it just precision that entrepreneurs need? How can the quality of a prediction be measured?

It is essential to clarify that while precision is a critical component of prediction, accuracy is an equally or perhaps more precious element. Precision is generally defined as the degree to which a score obtained by a person or a model on one occasion is repeated on a second occasion. The measure of precision is expressed as the standard deviation of the test results: a more significant standard deviation reflects a lower degree of precision (Streiner & Norman, 2006).

Accuracy is the degree of approximation of a prediction with respect to a specific value. It refers to the quality of the result in predicting a particular value (Hofer et al., 2005). Figure 2.2 graphically exemplifies different cases of accuracy and precision in estimating, for instance, the value of peak sales of a new rehabilitation device worth US$50 million. The outputs of precise models are always tightly grouped (low standard deviation) around a mid-value that is not necessarily close to the value that should be predicted (Cases A and B), while the outputs of accurate models tend to be placed much closer to the value sought independently by their standard deviation.

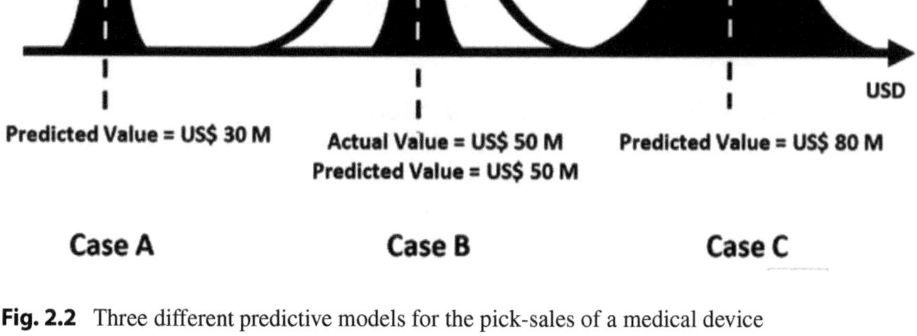

Fig. 2.2 Three different predictive models for the pick-sales of a medical device

There are many, very sophisticated and precise models employed by early-stage medical startups, from the most complex Microsoft Excel-based models to tools such as Crystal Ball Professional, Predictor, OptQuest, and Real Options Analysis Toolkit, which are immensely popular among financial analysts and CFOs. With such tools, a new user can pick up software products and rush immediately into predictions.

As mentioned, management usually places much emphasis and self-referential confidence on the precision of the predictive model, losing sight of its accuracy. This approach promotes the use of intrinsically precise models that could soon become inaccurate, as they are often fed biased assumptions. In these cases, most of the efforts and trust are put into the model's completeness, its ability to relate to several aspects of the complexity of reality, its mathematical coherence, its ability to perform statistical analyses, and even the technical skills of the startup's CFO. It is rather common to witness startup management over-trusting the analyst's technical skills, believing that the analyst's understanding of the model's complexity is sufficient for generating reliable predictions. This attitude may often become a managerial excuse to avoid devoting significant resources to the rugged process of setting the unbiased, core business assumptions that are essential for any predictive models.

In reality, to perform reliable prediction using sophisticated modeling, the analyst must be provided with a clear understanding of most of the underlying assumptions. It is important to remember that, from an accuracy standpoint, using an evolved predictive software tool without any sedimented knowledge of the nature and impact of key hypotheses and strategic decisions is akin to giving a three-year-old child a loaded machine gun (Mun, 2004). The risks of setting unrealistic financial expectations and failing to deliver the promised returns are embedded in almost every aspect of innovative ventures. However, instead of passively accepting this reality and the related failure rates, we can develop a framework to better understand how the sources of risk can be mitigated and, therefore, our predictive abilities be improved.

Entrepreneurs must be able to demonstrate that their valuation is the result of a compelling process in which precision is supported by the predictive model and accuracy is supported by unbiased, tested assumptions. Satisfying this condition increases the probability of raising capital from rational investors. These methodologies help manage both the contextual complexities of medical innovation and the impactful, contagious effects of entrepreneurial dispositional optimism.

2.4 New vs. Innovative Medical Ideas

What is the difference between a new medical technology idea and an innovative medical technology? All professionals of the medical technology industry know that the probability of a needleless blood testing device becoming a true innovation is much higher than the one of the newest and fastest lancet-based glucose meters. How can this example be generalized? Joseph Schumpeter (1942) distinguishes between "inventions" as new creations and "innovations" as inventions that become economically successful and facilitate the earning of significant profits (Ariew, 2009). It is one reason the methodology suggested by this book constantly seeks to make explicit the impact of any decision or assumption made and the economic value of the new medical idea.

Early-stage medical startups must understand whether the medical idea being funded is more likely to be a true innovation or not, because investors in medical technology innovation expect investment multipliers of 10 or 20. It is implausible for non-truly innovative medical technology or product improvements to generate such returns. As consultants in the field of medical technology innovation, we frequently come across startups that qualify their products as "innovative" when, de facto, they represent marginal improvements over existing solutions that constitute the standard of care.

These products enter a competitive market with the comfort of some measurable competitive advantages but without any significant chance of driving such changes in medical practices that would qualify them as "innovation." Non-disruptive innovations do not fundamentally alter the existing market structure or displace established products or services. In contrast to disruptive innovations, they generally improve existing offerings without causing significant turbulence in market dynamics. While disruptive innovations often require the willingness to cannibalize existing sales, non-disruptive innovations tend to be more likely to coexist with existing products and services, moderating the risk of cannibalization (Govindarajan et al., 2011).

Predicting which idea will become a true medical innovation is as critical as challenging. Providing such a straightforward prediction is not within the scope of this book. Still, in Sect. 6.3.2, several elements of the proposed methodology encourage disenchanted, science-based assessments of the medical idea vs. the current standard of care vs. existing competitors (if any), and the ability to truly satisfy critical unmet medical needs.

2.5 Investigations vs. Decisions

While assessing a medical innovation, we may start by browsing large sets of publicly available information. For instance, in an attempt to roughly quantify the size of a specific market, we may consult the database and statistics published by the US Center for Disease Control and Prevention (https://www.cdc.gov).

If we are wondering whether Medicare may reimburse our innovative technology, we can consult the list of existing reimbursed codes (https://perma.cc/2YLL-GBDQ) and verify if similar therapies or procedures are already codified and eligible for reimbursement.

Access to these large sets of information tends to be cheap, if not completely free of charge, but identifying and organizing the relevant information is likely to require significant time and resources. We'll refer to these preparatory activities as the investigation.

Decision-making typically starts with an **investigation** phase in which managers explore the environmental aspects relevant to their decision and identify the relations and patterns among them. The investigation phase concludes when the available information, its retrieval, and analysis are exhausted. At this point, decision-makers are in a better position to make actual **decisions** (Gambardella, 2021).

We, therefore, define **investigation** as an earlier stage in the preparation of the actual decision-making through the systematic collection of the relevant information that is publicly available or accessible with negligible costs, and making sense of the collected data. **Decision-making**, instead, is the cognitive process that results in the selection of a course of action among several possible alternatives.

Solving a **decision-related problem** can, therefore, be defined as the managerial process that leads to decision-making, starting with investigations. This sequential division of activities is coherent with and supported by the Stage-Gate Model: a structured framework for managing innovation, product development, and new project initiatives, introduced by Dr. Robert G. Cooper in the 1980s and has since become one of the most widely used models for guiding new product development in industries such as pharmaceuticals, MedTech, software, and consumer goods. As more extensively described in Sect. 4.5, the Stage-Gate Model divides the innovation journey into a series of stages, where data collection and data analysis work are performed, followed by gates, where key decisions are made. Each stage encompasses a cross-functional set of activities aimed at gathering information, reducing uncertainty, and advancing the project toward commercialization: the definition of stage is therefore in line with the definition of "investigation" given in Table 2.1. At each gate, management makes decisions about whether to proceed (Go), terminate (Kill), pause (Hold), or modify and resubmit (Recycle) the project: this description of a gate matches the above definition of "decision-making." In simple terms, the unit formed by one stage and one gate—introduced by the Stage-Gate Model—corresponds to the investigation-decision sequence that this book identifies with the solution of a "decision-related problem."

Table 2.1 Investigation vs. decision-making

Investigation	Decision-making
The investigation is an earlier stage in preparation for the actual decision-making through: • The systematic collection of relevant information that is publicly available or accessible with negligible costs. • Making sense of the collected information.	A decision-related problem is, thus, defined as the managerial process that leads to decision-making, starting with investigations

Table 2.2 The difference between investment and scenario decisions

Investment decisions	Scenario decisions
Investment decisions are choices among actions that may involve significant investments of resources	Scenario decisions are choices among actions that involve no significant investment of resources. These choices consist of the determination (election) of present or future scenarios. The by-product of a scenario decision is the election of a business assumption

2.6 Decisions vs. Scenario Decisions

All decision-related problems lead to a decision, but not all decision-related problems involve the commitment to an investment. Some decision-related problems are instead solely focused on the election of important business assumptions: an election that does not necessarily imply any immediate investment. For instance, if we are deciding whether to start a clinical trial that costs half a million dollars or to exit the project, we are confronted with a decision that involves an explicit commitment to an investment: this is an investment decision.

Let's suppose, instead, that we are simply deciding whether to continue or to leave ("stay vs. exit") a project involving a new home diagnostic device based on the expected (predicted) level of market adoption. In such a case, neither of the two possible decisions represents an explicit commitment to an investment. "Staying" means only continuing to investigate and assess the project's attractiveness and development plans. Exiting, by definition, does not require resources.

It is not an investment decision but rather a decision about which future scenario to adopt as a working assumption. For this reason, this book labels this type of decision as a "scenario decision." In other words, the scenario decision consists of the formal acceptance or refusal of a business assumption (Table 2.2). It is important to emphasize that the act of forming or electing a business assumption corresponds to making a decision and that, consequently, requires the discipline of the scientific approach presented in this book.

We'll see that the application of the scientific approach to both investment and scenario decisions requires very similar processes. Complementarily, an investment decision is a cognitive process that results in selecting among courses of action that potentially lead to significant economic investments.

Bibliography

Ariew, A. (2009). Innovation and invention from a logical point of view. *Innovation in Cultural Systems*, October, 21–36.

Bajwa, S. S., Wang, X., Duc, A. N., & Abrahamsson, P. (2017). 'Failures' to be celebrated: An analysis of major pivots of software startups. *Empirical Software Engineering, 22*(5), 2373–2408.

Behrns, K., Gingles, B., & Sarr, M. (2021). *Medical innovation.*

Brönneke, J. B., Müller, J., Mouratis, K., Hagen, J., & Stern, A. D. (2021). Regulatory, legal, and market aspects of smart wearables for cardiac monitoring. *Sensors, 21*(14), 4937.

Czekster, R. M., Grace, P., Marcon, C., Hessel, F., & Cazella, S. C. (2023). Challenges and opportunities for conducting dynamic risk assessments in medical IoT. *Applied Sciences, 13*(13), 7406.

Dolezel, M., & Smutny, Z. (2021). Usage of EHealth/MHealth services among Young Czech Adults and the impact of COVID-19: An explorative survey. *International Journal of Environmental Research and Public Health, 18*(13), 7147.

Gambardella, A. (2021). *Making low frequency/high impact managerial decisions*.

Govindarajan, V., Kopalle, P. K., & Danneels, E. (2011). The effects of mainstream and emerging customer orientations on radical and disruptive innovations. *Journal of Product Innovation Management, 28*(s1), 121–132.

Herzlinger, R. E. (2006). Why innovation in health care is so hard. *Harvard Business Review, 1*(May).

Hmieleski, K., & Baron, R. (2017). Entrepreneurs' optimism and new venture performance: A social cognitive perspective. *Academy of Management Journal, 52*(3), 473–488.

Hofer, M., Strauß, G., Koulechov, K., & Dietz, A. (2005). Definition of accuracy and precision-evaluating CAS-systems. *International Congress Series, 1281*(May), 548–552.

MedTechEurope. (2025). *What is medical technology? – MedTech Europe*. https://perma.cc/5SFL-FF89

Moore, D. A., Oesch, J. M., & Zietsma, C. (2007). What competition? Myopic self-focus in market-entry decisions. *Organization Science, 18*(3), 440–454. https://doi.org/10.1287/Orsc.1060.0243

Mun, J. (2004). *Applied risk analysis: Moving beyond uncertainty in business*. Wiley.

Scheier, M. F., & Carver, C. S. (2018). Dispositional optimism and physical health: A long look back, a quick look forward. *The American Psychologist, 73*(9), 1082.

Streiner, D. L., & Norman, G. R. (2006). 'Precision' and 'accuracy': Two terms that are neither. *Journal of Clinical Epidemiology, 59*(4), 327–330.

3 Why Medical Innovation Is So Difficult

Abstract

This chapter explores how the environmental complexities typical of medical innovation put an entrepreneur's decision-making capabilities under heavy pressure. Innovation in the healthcare industry is a challenging discipline since—together with medical and technological knowledge—it requires a deep understanding of the managerial implications of five complex forces: industry players, funding, public policy, customer needs and constraints, and innovation accountability (Herzlinger, Why innovation in health care is so hard. *Harvard Business Review*, Vol. 1, 2006).

3.1 The Complexities of Medical Innovation

A core component of the industry's complexity comes from demonstrating the value of a new medical technology. While judging the benefits of a new medical technology, various stakeholders focus on and prioritize different yet overlapping sets of value drivers. To succeed, medical innovators must understand and address each stakeholder's unique set of value drivers (Deloitte/AdvaMed, 2017). The medical innovator must observe, investigate, validate, and critique every aspect of each stakeholder's needs. This observation must include inputs from physicians, patients, nurses, technicians, hospital buyers, public payers, insurance, and other staff members (Behrns et al., 2021). This characterization is not a simple, linear process; it is iterative and must be re-validated frequently since stakeholder interactions change rapidly. Moreover, these observations must be performed on a global scale; local observations cannot satisfy the scope of these interactions, as the key stakeholders and their priorities change tremendously from region to region.

L. Negri, *De-Risking Investments in Medical Technology Innovation*, Springer Series in Healthcare Management and Innovation,
https://doi.org/10.1007/978-3-032-15007-3_3

The context of medical innovation is also challenging because medical knowledge has been growing exponentially: While the doubling time of the volume of medical knowledge was estimated at 50 years in 1950, it accelerated to 7 years by 1980, 3.5 years by 2010, and then estimated in 73 days by 2020, according to a 2011 study (Densen, 2011). Moreover, in today's society, medical technology is often blamed for deliberately increasing the costs associated with public healthcare, even if, on a net basis, it efficiently improves the capabilities of medicine (Richard A. Rettig, 1994). Indeed, the value of medical innovation must be cost-effective in bringing substantial and measurable improvements to patients' health and healthcare delivery. Consequently, new risk-sharing and shared-savings systems have compelled insurers, providers, and payers to constantly reassess their performance levels, including the selection and employment of new medical technologies. Additionally, for many years, the responsibility for the selection of medical technology has gone beyond individual physicians. Currently, it encompasses the entire healthcare system, patient advocates, and—very importantly—healthcare technology assessment (Deloitte, 2021).

Moreover, the liability risk associated with the commercialization of medical products has historically impacted the entire innovation chain, demonstrating a significant slowing effect on downstream innovation (Aghion et al., 2018).

Finally, the core components of the mentioned complexity can be summarized as follows:

1. Highly regulated globally but with country-level specificities.
2. High social impact, imposing strict control on healthcare expenditure.
3. A high number of relevant stakeholders.
4. Highly asymmetric knowledge-sharing (patients vs. physicians).
5. New technology's requirement of disciplined clinical and economic proof.
6. Innovation's necessity for very high initial capital due to the construction of clinical evidence.
7. A lengthy and expensive process of clinical validation.
8. Uncertainty due to the high speed of medical innovation.
9. Significant legal liability implications.

These are the main reasons why discovering, developing, and funding new therapies, pharmaceuticals, diagnostic systems, medical devices, and technologies is a highly complex, risky, lengthy, and expensive process. In short, medical innovation is impacted by specific complexities that increase the difficulties encountered by entrepreneurs, to address which a more disciplined decision-making approach may be highly beneficial.

3.2 The Cost of Failure in Medical Innovation

Despite the tremendous financial risk, many life sciences companies have accepted the challenge and spend over US$100 billion annually in their quest to alleviate human suffering (McDonough, 2024). Nevertheless, it is essential to remember that the magnitude of the social benefits of medical innovation far exceeds the related investments; this is why high failure rates in medical innovation imply a tremendous social impact.

The social cost of the failure of medical innovation is proportionate to the considerable social value created by such innovation. Historically, medical innovation has consistently responded to numerous health-related challenges in times of crisis; more recently, it was on the frontline of the battle against the COVID-19 pandemic.

Economists Kevin Murphy and Robert Topel evaluated the social benefits of medical innovation by valuing aggregate improvements in human longevity (Murphy & Topel, 2006). Over the period 1970–1990, the increase in the lifespan of the average American was significant. Using age-dependent values for an additional life-year and the increases in life expectancy over this period, the authors attributed a value of roughly US$57 trillion (or about US$2.8 trillion per year) to the increased life expectancy. To put these figures in perspective, improvements in life expectancy over the period 1970–1990 contributed about as much to overall welfare as to improvements in material wealth (National Academy of Sciences, 2002).

However, due to the mentioned increasing market complexities in today's MedTech industry, many medical startups struggle to survive the period between the initial investment and the creation of commercially viable products. Indeed, some corporate and venture capitalist (VC) investors hesitate to invest in MedTech's early-stage innovation (AdvaMed and Deloitte, 2016).

For HealthTech, the situation is no better. In the first quarter of 2024, funding for digital health in the USA reached US$2.7 billion through 133 deals, with an average deal size of US$20.6 million. It marks the lowest first-quarter funding by sector since 2019, despite Q1 being the top-funded quarter in 2022 and 2023. However, the number of deals remains strong, with 133 fundraises for digital health companies, surpassing the numbers for the past six quarters. Hence, it appears that investors and founders have moved their focus to deals with smaller check sizes compared to previous quarters (Krasniansky & Somaiya, 2024).

3.3 Decision-Making and Medical Startup Failures

In the innovation context across industries, startups generally fail for well-recognized reasons established by post-mortem analyses. These failure causes range from "lack of market need" to "lack of customer understanding," passing through "exhaustion of cash flow" (Bednár & Tarišková, 2017). Different observers emphasize different items: Some stress the impact of biases affecting customer characterization and the understanding of customer needs (Yohn, 2019).

Such biases, which are related to the assessment of market needs as well as customers, have damaging implications for the entrepreneur's ability to design and sharpen efficacious value propositions and realistic pricing models. Indeed, other observers have pointed fingers at the practice of postponing marketing and pricing decisions to the end of long journeys of product development. Financial resources are, thus, burned for a long time under the untested assumption (hope) that there will be customers who are willing to pay for the new idea (Ramanujam & Tacke, 2016). Some practitioners and scholars openly recognize that the common denominator across the mentioned causes of failure is rooted in biased decision-making (Wagrell & Baraldi, 2019). To more compellingly assess the key role of biased decision-making in the failure of medical startups, it is critical to focus on the most reported, common reasons for startup failure in medical innovation. Table 3.1 summarize a broad list of "reasons for failure" from the literature. Any rational attempt to group all the issues listed in these tables under a common label highlights the apparent concept behind most failures: biased decision-making.

The following can serve as examples:

- "Weak assessment of actual customers and their needs" means making wrong assumptions about customer needs.
- "Accountability failure of the product" refers to making a decision to launch a product when it does not meet the core expectations.
- "Overly optimistic assumptions about market share gains" indicates that the fundamental choices of assumptions and the predictive process were wrong.
- "Inadequate distribution model" implies a wrong decision while choosing the distribution model.

Table 3.1 Reasons for startup failures in HealthTech and MedTech

Causes for Life Sciences Startup Failure: Review Table		
	HealthTech	MedTech
Financial Viability	Lack of proper funding (Herzlinger, 2006)	Overcompensation of executives (MDDI, 2013)
		The medical device tax (MDDI, 2013)
	Excessive spending prior to the identification of a market and product-market fit (Dietsche, 2018)	Lack of funding (MDDI, 2013)
		Excessive cash usage in pilot programs (Grand & Zavala, 2020)
Business Model and Organization	Underestimation of the barriers to the direct-to-consumer approach (Dietsche, 2018)	Unqualified sales reps or poor sales processes (MDDI, 2013)
		Inadequate distribution model (MDDI, 2013)
		Failure to invest in clinician training (MDDI., 2013)
		Lack of cultural support (Behrns, 2021)
		Inadequate or incomplete team (MDDI, 2013; Behrns et al., 2021) or wrong choice of CEO (Grand & Zavala, 2020)
		Underestimation of the barriers to the B2C approach (Grand & Zavala, 2020)

(continued)

Table 3.1 (continued)

Causes for Life Sciences Startup Failure: Review Table		
	HealthTech	MedTech
Stakeholders	Weak understanding of industry players (Herzlinger, 2006)	Lack of funding (MDDI., 2013)
	Weak understanding of public policy (Herzlinger, 2006)	Inferior reimbursement strategy (MDDI, 2013)
		Excessively long stay in stealth (Grand & Zavala, 2020)
		Product's inability to fit into existing (Grand & Zavala, 2020)
	Lack of knowledge of stakeholder complexity (Cole, 2020; Lux & Kempf, 2021)	Misunderstanding of the payment and reimbursement dynamics (Grand & Zavala, 2020)
Competitiveness	Overly optimistic competitive assumptions about timely market share gains (Dietsche, 2018)	Lack of a disruptive technology (Grand & Zavala, 2020)
	Longer-than-expected sales cycle (Dietsche, 2018)	
	Scarce knowledge of market unmet needs and requirements (Lux & Kempf, 2021)	
Value Proposition	Lack of value proposition-needs fit (Cole, 2020)	Lack of a disruptive technology (MDDI, 2013)
		The assumption that invention equals innovation (Behrns et al., 2021)
	Loss of track in terms of what matters to customers (Dietsche, 2018)	Choice of the wrong indication (Grand & Zavala, 2020)
		Lack of a deliberate process to assess the actual value of the new ideas (Behrns et al., 2021)
		Failure to properly articulate the value proposition (Grand & Zavala, 2020)
Product Strength	Insufficient focus on innovation accountability (Herzlinger, 2006)	Failure to establish a timeline for product development and launch (MDDI., 2013)
	Failure to identify product indications (Cole, 2020)	Issues with intellectual property (MDDI, 2013)
	Failure to generate value-supporting evidence and data (Cole, 2020)	Absence of an evidence-generation process (Grand & Zavala, 2020)
	Accountability failure of the product or the service (Cole, 2020)	
Market Attractiveness and Unmet medical Needs	Weak assessment of actual customers and their needs (Herzlinger, 2006)	Lack of an established market (MDDI, 2013)
	Overly optimistic environmental assumptions about timely market share gains (Dietsche, 2018)	
	Scarce knowledge of market needs and requirements (Lux & Kempf, 2021)	

Bibliography

AdvaMed and Deloitte. (2016). *Out of the valley of death: How can entrepreneurs, corporations, and investors reinvigorate early-stage MedTech innovation?*

Aghion, P., Dana, J., Frakes, M., Grennan, M., Hegde, D., Rai, A., Spier, K., Schmucki, R., Watzinger, M., & Ziedonis, R. (2018). *When does product liability risk chill innovation? Evidence.*

Bednár, I. R., & Tarišková, I. N. (2017). Indicators of startup failure. *Industry 4.0.*

Behrns, K. (2021). *Medical innovation.*

Behrns, K., Gingles, B., & Sarr, M. (2021). *Medical innovation.*

Cole, E. (2020). *Why healthcare startups fail – Top 5 reasons to consider.* MyTechMag.Com, 2020. https://perma.cc/B77S-CN7C

Deloitte. (2021). *Value-based care and medical technology: Assessing the value of MedTech products and offerings.* https://perma.cc/2PUG-E83Z

Deloitte & AdvaMed. (2017). *A framework for comprehensive assessment of medical technologies: Defining value in the new health care ecosystem.* https://perma.cc/T8DQ-Y3FAh

Dietsche, E. (2018). *Why healthcare startups fail and 5 that bit the dust in 2018 – MedCity News.* Medcitynews.Com. https://perma.cc/U2H6-BRQH

Grand, B. P., & Zavala, K. (2020). 10 *Reasons why health care startups fail.* https://perma.cc/D67W-M635

Herzlinger, R. E. (2006). Why innovation in health care is so hard. *Harvard Business Review, 1*(May).

Krasniansky, A., & Somaiya, M. (2024). *Q1 2024 digital health funding: Great (reset) expectations.* RockHealth.Org. https://perma.cc/9AXV-DW7W

Lux & Kempf. (2021). *Success factors for market entry of mobile health startups.* https://doi.org/10.1109/AICCSA53542.2021.9686849

McDonough, K. (2024). *United States life sciences 2024 financial environment.* www.gbreports.com

MDDI. (2013). *Why medical device start-ups fail: LinkedIn group weighs in*, 1–10. https://perma.cc/KUL8-8XQQ

Murphy, K. M., & Topel, R. H. (2006). The value of health and longevity. *Journal of Political Economy, 114.* https://perma.cc/6TFJ-GKUL

National Academy of Sciences. (2002). Medical innovation in the changing healthcare marketplace. In P. Aspden (Ed.), *Medical innovation in the changing healthcare marketplace.* National Academy Press.

Ramanujam, M., & Tacke, G. (2016). *Monetizing innovation: How smart companies design the product around the price.*

Rettig, R. A. (1994). Medical innovation duels cost containment. *Health Affairs.* Project HOPE – The People-to-People Health Foundation.

Wagrell, S., & Baraldi, E. (2019). The joys and sorrows of a start-up's interactions with the public sphere: A case from medical technology. *Journal of Business & Industrial Marketing, 34*(1), 267–283.

Yohn, D. L. (2019). *Why startups fail.* Forbes.

Minimizing Decision-Making Biases

4

Abstract

There are several well-known choices (decisions) that a medical entrepreneur must typically make: whether to bring a therapeutic or diagnostic idea to market, whether to keep the initial idea unchanged or to pivot to a more rewarding solution, and so on. This chapter examines how medical entrepreneurs can minimize decision-making biases when evaluating, developing, and launching innovations. It contrasts non-predictive approaches such as effectuation and bricolage—based on experimentation and improvisation—with predictive approaches grounded in structured hypothesis testing. Non-predictive methods support learning, but they also expose entrepreneurs to cognitive biases like overconfidence and selective attention. The chapter suggests adopting a scientific approach to decision-making: formulating hypotheses, testing them through field evidence, and interpreting results rationally. Supported by the Stage-Gate Model, the suggested approach provides a disciplined methodology for mapping, prioritizing, and addressing high-priority decision-related problems, improving accuracy, reducing risk, and enhancing venture performance.

4.1 Predictive vs. Non-Predictive Approaches

Both in real life and in the literature, two major approaches address the issue of entrepreneurial decision-making: the non-predictive approach and the predictive approach (Camuffo et al., 2020).

Considering the high degree of uncertainty that characterizes the creation of a new venture, a well-known stream of research in entrepreneurship suggests that entrepreneurs should rely on non-predictive techniques, such as effectuation (Sarasvathy, 2001) and bricolage (Baker & Nelson, 2005). Effectuation is the

L. Negri, *De-Risking Investments in Medical Technology Innovation*, Springer Series in Healthcare Management and Innovation,
https://doi.org/10.1007/978-3-032-15007-3_4

preferred approach for acquiring abilities, whereas bricolage is the preferred approach for capturing resources and knowledge about these resources (Vasconcelos Scazziota et al., 2020). According to these theories, entrepreneurs learn through a trial-and-error process, which makes the future value of ventures difficult to predict.

While reviewing this stream of research, it appears that entrepreneurs embracing non-predictive approaches exhibit two essential behaviors: a lack of focus on estimating long-run returns and a lack of clarity in building explicit long-range goals and plans (Fisher, 2012). These considerations trigger natural questions about the actual ability of non-predictive approaches to protect entrepreneurs and their ventures from failure.

Without a systematic approach to predictive efforts, several psychological factors become more likely to influence the odds of entrepreneurial success (Chatterjee & Das, 2015). For instance, an entrepreneur's ability to evaluate a venture's value is exposed to cognitive factors such as overconfidence, attention, learning, beliefs, memory, and problem-solving (Sharma, 2015). The adoption of a non-predictive approach may, therefore, better explain why entrepreneurs often fail to objectively support claims about the value of their ventures (Kuntze & Matulich, 2016) and why their plans fail to compellingly link product ideas and market dynamics with future expected value.

Cognitive biases affect the rationality of decision-making and, in turn, the reliability of the idea's claimed value, potentially leading to a failure to represent realistic future revenues.

A newer area of research proposes that the frequency of a fallible entrepreneur's judgment can be reduced by using a structured and disciplined predictive process of idea evaluation and development (Hogarth & Karelaia, 2012). Entrepreneurs embracing a predictive approach can apply a more scientific methodology to understanding and testing the mechanisms that affect the performance of their new products and ideas. Scholars and practitioners alike have begun to explore this issue in the last two decades (Blank, 2006; Felin & Zenger, 2009; Zenger, 2016).

4.2 The Role of the Scientific Approach

Research has already demonstrated that, across industries, a scientific approach mitigates the cognitive biases that influence entrepreneurs' ability to efficiently collect, prioritize, and process the key market information required for estimating the actual value of their product ideas (S. L. Cohen et al., 2019). By doing so, entrepreneurs can behave more rationally while accessing, gathering, organizing, and interpreting information from the marketplace and evaluating potential pivots of the original idea. In general, this approach provides entrepreneurs with a strong rational support while making important decisions (Kahneman et al., 2019; Camuffo et al., 2020). In short, a scientific approach to decision-making better positions entrepreneurs to assess the value of their innovative ideas and make better decisions about future steps necessary to deliver that value.

To the advantage of this approach, it is essential to observe that high-quality assumption-making and decision-making are key ingredients in preparing a compelling business plan, which is based on key assumptions about the market and decisions about how to do things in the future (Cambridge Dictionary of English, 1996). The value of diligent business-planning activities in the early stages of a startup comes from the added benefit of potentially making better ex ante decisions, including whether to launch a product, enter a particular market segment, or terminate a poor venture project before significant investments (Chwolka & Raith, 2012). Moreover, since the business plan is the first (and often the only) substantial point of communication and contact between a potential investor and the entrepreneur (Mason & Stark, 2004), it seems plausible that a business plan based on assumptions obtained using the scientific approach is more attractive to rational investors. It is even more critical when one considers that angel investors who emphasize the quality of predictive processes make significantly larger venture investments than investors stressing non-predictive approaches (Wiltbank et al., 2009).

The use of a scientific approach requires entrepreneurs to follow a series of steps:

- Precise formulation of a theory.
- Identification of falsifiable hypotheses derived from the theory.
- A disciplined test of these hypotheses.
- A rational interpretation of the test results.

With the promise of providing further clarifications and examples in Chap. 5, we can say that using the scientific approach for decision-making involves following the steps listed in Table 4.1.

4.3 The Scientific Approach Alone Is Not Enough

As already mentioned, before starting the production of this book, the employment of a scientific approach in decision-making was tested in real life in early-stage medical startups. One of the first and most prominent experimental results suggests that biased decision-making is only one of the sources of decision-related problems encountered by medical entrepreneurs. Indeed, medical entrepreneurs suffer additional difficulties, such as the following, in particular:

- Issues with overlooking, forgetting, or even ignoring the critical role of specific decision-related problems.
- Issues with prioritizing and, therefore, identifying the individual critical decisions to be approached.

Therefore, biased decision-making, when a decision-related problem has been correctly identified, is only one of a larger group of biases. An example of misprioritization can be observed when entrepreneurs develop their ventures for a long time without ever testing their price assumptions or the assumptions related to

Table 4.1 The six steps of the scientific method

Step	Description
1. Identify scenarios	Identifying which future scenarios may be encountered *For instance, an entrepreneur launching a wearable medical device identifies two possible future scenarios: high customer and low adoption.*
2. Identify actions	Defining the possible actions among which the decision-maker may choose *For instance, the entrepreneur decides whether to launch the wearable device or to exit.*
3. Map values in scenario-action pairs	Estimating the economic value of each possible action in each possible future scenario *For instance, the entrepreneur calculates the economic value of each scenario-action pair.*
4. Develop testable hypotheses	Developing testable hypotheses to rationally predict the likelihood of each scenario *For instance, the entrepreneur precisely defines the conditions under which the future adoption scenario can be defined as "high."*
5. Test hypotheses	Using field experiments to test these hypotheses *For instance, the entrepreneur performs "experiments," such as a series of future customer interviews designed to attempt to predict future adoption rates.*
6. Make decisions	Making a final decision *For instance, based on the results of the experiment, the entrepreneur decides whether to launch or exit, depending on the respective values and probabilities of the predicted scenario-action pairs.*

customers' willingness to pay, postponing the solution to a critical business problem. In many cases, entrepreneurs have a clear vision of the decision-related problems plaguing their respective areas of expertise; however, they may overlook the importance of certain business decisions. Engineers and MDs are perfect examples.

Figure 4.1 summarizes the three primary sources of entrepreneurial bias:

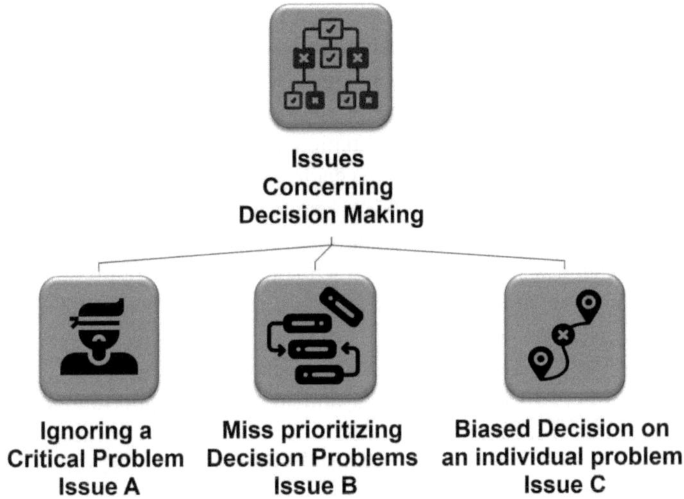

Fig. 4.1 The main challenges of decision-making

- Ignoring the existence of a certain problem (i.e., neglecting the importance of an early, sharp, and tested value proposition).
- Misprioritizing a problem (i.e., verifying pricing ranges only after the end of product development) and focusing first on less critical problems.
- Making biased decisions to address individual problems (i.e., setting the wrong price).

The scientific approach to decision-making supports the quality of individual decisions but does not necessarily help the entrepreneur mitigate the issues of overlooking, forgetting, ignoring, or misprioritizing the problems to focus on. With this in mind, we designed and tested a process that provides a sequential approach to the issue of grouping and ordering the complex and interconnected decision-related problems of innovation.

4.4 The Stage-Gate Model Suggests Using a Sequential Map

In what other ways can entrepreneurs avoid both issue A and issue B shown in Fig. 4.1? The Stage-Gate Model (Cooper, 2001) provides a strong basis for a solution.

The Stage-Gate Model, also known as the Phase-Gate process, is a structured framework for managing innovation, product development, and new project initiatives. Introduced by Dr. Robert G. Cooper in the 1980s, it soon became one of the most widely used models for guiding new product development across various industries, including pharmaceuticals and MedTech.

The Stage-Gate system acknowledges that product innovation is a process that can be effectively managed. As already mentioned, this model divides the innovation journey into a sequential series of stages, where work is performed, separated by gates, where key decisions are made. Each stage encompasses a cross-functional set of activities aimed at gathering information, reducing uncertainty, and advancing the project toward commercialization. At each gate, management evaluates the project and decides whether to proceed (Go), terminate (Kill), pause (Hold), or modify and resubmit (Recycle) the project. The relevant characteristic of the Stage-Gate system is that it suggests a predetermined sequence of stages for product innovation. For instance, a classic Stage-Gate process has six stages (Cooper, 2001):

1. Stage-Gate 0—The Discovery and Idea Generation (also called "pre-stage"), where innovators focus on scouting and identifying opportunities, collecting new ideas, or new technology insights.
2. Stage-Gate 1—The Idea Scoping and Feasibility, when technical feasibility, market potential, and strategic fit are preliminarily investigated.
3. Stage-Gate 2—The construction of a Business Case: where entrepreneurs organize all considerations about the financial sustainability of the venture and the attractiveness of the predicted future profits.
4. Stage-Gate 3—Product Development: the actual product design, development, and industrialization, which includes regulatory planning and initial marketing activities, such as cross-functional R&D and marketing activities.
5. Stage-Gate 4—Product Testing and Validation, which includes the validation of the product, the business case, and the launch strategy.
6. Stage-Gate 5—Launch and Commercialization, which refers to all the activities supporting full-scale market launch; sales, distribution, and production; and post-launch monitoring.

This predetermined sequence represents an invaluable sequential map for entrepreneurs who need to make decisions in a context where hundreds of technical, medical, legal, business, and managerial problems compete for the management's busy attention. Given the mentioned extreme complexities faced by the medical entrepreneur, this map-based approach fits perfectly with their needs.

For this reason, the process proposed in this book fully endorses the fundamentals of the Stage and Gate system and adapts it to the specific issues of medical technology innovation.

Chapter 6 focuses on the construction of a sequential map similar to the one illustrated above, provided by the Stages and Gates system: it's a map adapted to the issues of medical innovation and where the unit formed by a stage and a gate are called "clusters" to recall that each one of them contains an high number of lower level decision-related problems.

The most recent version of the Stages and Gates system (Cooper, 2017) adds several degrees of flexibility to the original version by allowing, for instance, decision-makers to focus only on specific stages.

In line with this enhanced, more flexible version of Cooper's system, this book recognizes that, in the field practice, it is unrealistic to believe that each stage-gate can be isolated and must be fully resolved before moving to the next. It is also unrealistic to imagine that an entrepreneur can ignore or dismiss the subsequent stages until final decisions about the preceding gates-stages are made.

In other words, this book proposes a process that encourages the use of a predetermined sequence of mapped clusters and concurrently proposes additional, non-sequential criteria to navigate the same map more dynamically. Chapter 8 focuses precisely on those criteria that can be employed to prioritize those clusters and the respective underlying decision problems routinely. In line with the above considerations, the following chapter provides a summary of the proposed decision-making process.

4.5 The Proposed Three-Steps Method

The proposed process suggests using three sequential steps (Fig. 4.2): mapping decision-related problems, prioritizing them, and using the scientific approach to address each prioritized problem. The following three paragraphs offer a slightly deeper view of the three-step methodology. Example: a medical entrepreneur has invented a diagnostic device for post-stroke patients. It is a skin patch that collects and shares a set of vital signs during the post-acute stroke phases. The device uploads the respective measures to a cloud-based AI system that identifies dangerous trends in cardiac activity and immediately sends an alert to the patient and the hospital.

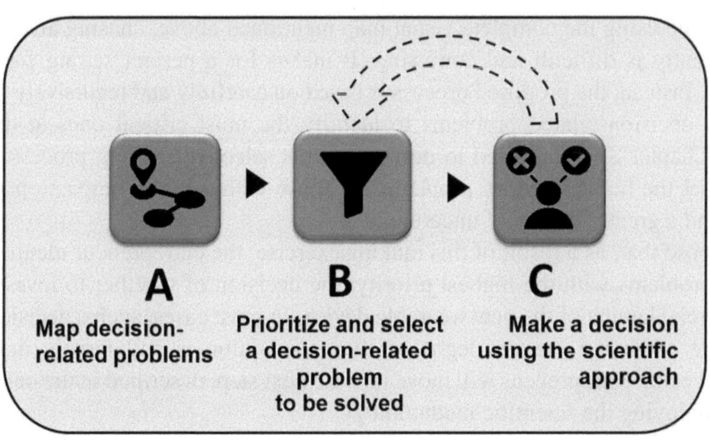

Fig. 4.2 The proposed three-steps process

4.5.1 Step A: Mapping Decision-Related Problems

Initially, the medical entrepreneur must address tens of complex decision-related problems, all of which are evidently highly important. For instance, the entrepreneur will face questions such as the following: Are there actual and compelling unmet medical needs that justify the investments in developing this new device? Is the invented device working and performing as promised? Is the promise to patients, users, and payers compelling and economically viable? Are there other products or existing standards of care that compete with the new product? If yes, is the invented product sustainably competitive? Who is the "client" for this innovation—the patient, the hospital, or the physician? Who are the most influential stakeholders in the always-complex buying-decision process? What is the exact value proposition of the new product, and how does it satisfy the most critical unmet medical needs? Does it make sense to prove the product's efficacy by investing in expensive clinical trials and usability tests?

We can continue listing the typically recurring problems for tens of pages. How can the medical entrepreneur safely navigate such a large number of decision-related problems? How can the entrepreneur avoid overlooking or completely ignoring some of these problems and questions?

The proposed approach frames and orders all these questions using mental (graphical) maps tailored to medical innovation. These maps can highlight the complete set of core decision-related problems and their relationships. Chapter 6 focuses on the proposed map and explores the principles that led to its structure.

4.5.2 Step B: Prioritizing and Selecting a Problem to Be Solved

Even while using the complete visual map mentioned above, chasing all problems concurrently is difficult and confusing: It makes for a perfect setting for biased choices. Instead, the proposed process is based on carefully and recursively ranking the key decision-related problems to identify the most critical ones at different times. Chapter 8 is dedicated to describing this selective ranking process. It will show that the higher-priority problems are those with a higher impact on venture value and a greater degree of uncertainty.

Suppose that, as a result of this ranking exercise, the entrepreneur identifies one of the problems with the highest priority: the decision of whether to invest in the commercial launch of the new wearable device. In most cases, such a decision must be made when the precise degree of future adoption is still largely unknown. Therefore, the entrepreneur will move into the next step, described in the below section, employing the scientific methodology.

4.5.3 Step C: Making a Decision Using the Scientific Approach

In short, and with the promise of providing more detailed clarifications in Chap. 5, using the scientific approach for decision-making involves the following actions:

- Defining the possible actions among which the decision-maker may choose (i.e., launch or exit).
- Identifying the future scenarios that may be encountered (i.e., high adoption vs. low adoption).
- Estimating the economic value of each possible action in each possible future scenario (launch or exit in a scenario of high or low adoption).
- Developing testable hypotheses to rationally predict the likelihood of each scenario.
- Using actual field experiments (as CMOs do for clinical trials) to test these hypotheses.
- Making a final decision based exclusively on the values of each scenario-action pair adjusted by the probability of each scenario.

Chapter 5 illustrates how these six steps can be practically implemented in the context of medical innovation.

4.5.4 What Happens After Step C?

After having approached a decision-related problem with a scientific method, the degree of uncertainty about that problem always drops significantly. Therefore, as more clearly described in Chap. 7, that specific problem's priority ranking is likely to fall at lower levels, changing the last prioritization revision set in Step B. In general, gathering market information, setting and testing hypotheses, and making decisions have an impact on the prioritization of the decision-related problems, which, therefore, need to be periodically reviewed.

4.5.5 Field-Test Results of the Three-Step Approach

The three-step process has been tested on actual medical technology startups with very encouraging results. My consulting partners and I have verified in the field that a scientific approach to decision-making actually enables entrepreneurs to make better predictions in testing and launching their business ideas, precisely as it was demonstrated by the literature across industries (Camuffo, Gambardella, Messinese, et al., 2024; Camuffo, Gambardella, & Pignataro, 2024). Compared to companies that are not exposed to the described three-step method, companies employing the guided scientific approach display common and superior performances in decision-making dynamics, managerial dynamics, and company performance. More

Table 4.2 Key indicators that depend on the employed decision-making approach

	Key Indicators, Key Events, and Other Observed Dimensions
Decision-making Dynamics	Higher propensity to actively search for relevant market information and quality signals
	Higher propensity and ability to use unsolicited, newly available information
	Better quality of information used for decision-making
	Stronger ability to connect financial valuations with a complete set of verified assumptions
Managerial Dynamics	Higher frequency of pivoting evaluations and pivoting decisions
	Higher propensity and ability to review decisions whenever new information becomes available
	Higher frequency and speed of exit decisions
	Higher strength and reliability of future revenue projections
Company Performance	Faster generation of proofs of concept and shorter times to go to market
	Better market entry performance
	Stronger cash position
	Greater likelihood of receiving investments from third parties

precisely, companies that employ the scientific approach display superior features compared to the other ventures (the placebo group), as described in Table 4.2.

When reading Table 4.2, it is important to notice that certain indicators of company performance (such as market performance and frequency of third-party investments) could only be measured for startups where the management did not decide to exit in the earlier stages.

Bibliography

Baker, T., & Nelson, R. E. (2005). Creating something from nothing: Resource construction through entrepreneurial bricolage. *Administrative Science Quarterly, 50*(3), 329–366.

Blank, S. G. (2006). *The four steps to the epiphany*. Lulu.com

Cambridge Dictionary of English. (1996). *Cambridge international dictionary of English*. Edited by Cambridge University Press, Vol. 24. Cambridge University Press.

Camuffo, A., Cordova, A., Gambardella, A., & Spina, C. (2020). A scientific approach to entrepreneurial decision making: Evidence from a randomized control trial. *Management Science, 66*(2), 564–586.

Camuffo, A., Gambardella, A., Messinese, D., Novelli, E., Paolucci, E., & Spina, C. (2024). A scientific approach to entrepreneurial decision-making: Large-scale replication and extension. *Strategic Management Journal, 45*(6), 1209–1237.

Camuffo, A., Gambardella, A., & Pignataro, A. (2024). Theory-driven strategic management decisions. *Strategic Science, 9*(4), 382–396.

Chatterjee, N., & Das, N. (2015). Key psychological factors as predictors of entrepreneurial success: A conceptual framework. *Academy of Entrepreneurship Journal, 21*(1), 105–117.

Chwolka, A., & Raith, M. G. (2012). The value of business planning before start-up – A decision-theoretical perspective. *Journal of Business Venturing, 27*(3), 385–399.

Cohen, S. L., Bingham, C. B., & Hallen, B. L. (2019). The role of accelerator designs in mitigating bounded rationality in new ventures. *Administrative Science Quarterly, 64*(4), 810–854.

Cooper, R. G. (2001). *Stage-gate systems: A new tool for managing new products.*

Cooper, R. G. (2017). Idea-to-launch gating systems. *Research Technology Management, 60*(1), 48–52.

Felin, T., & Zenger, T. R. (2009). Entrepreneurs as theorists: On the origins of collective beliefs and novel strategies. *Strategic Entrepreneurship Journal, 3*(2), 127–146.

Fisher, G. (2012). Effectuation, causation, and bricolage: A behavioral comparison of emerging theories in entrepreneurship research. *Entrepreneurship: Theory and Practice, 36*(5), 1019–1051.

Hogarth, R. M., & Karelaia, N. (2012). Entrepreneurial success and failure: Confidence and fallible judgment. *Organization Science, 23*(6), 1733–1747.

Kahneman, D., Lovallo, D., & Sibony, O. (2019). A structured approach to strategic decisions: Reducing errors in judgment requires a disciplined process. *MIT Sloan Management Review, 60*(3), 67–73.

Kuntze, R., & Matulich, E. (2016). Exploring cognitive bias in entrepreneurial startup failure. *Academy of Entrepreneurship Journal, 22*(2), 54–66.

Mason, C., & Stark, M. (2004). What do investors look for in a business plan? A comparison of the investment criteria of bankers, venture capitalists and business angels. *International Small Business Journal, 22*(3), 227–248.

Sarasvathy, S. D. (2001). Causation and effectuation: Toward a theoretical shift from economic inevitability to entrepreneurial contingency. *Academy of Management Review, 26*, 243–263.

Sharma, A. K. (2015). Venture capitalists' investment decision criteria for new ventures: A review. *Procedia – Social and Behavioral Sciences, 189*, 465–470.

Vasconcelos Scazziota, V., Guerrazzi, L., & Serra, F. A. R. (2020). Dynamics of the entrepreneurial process: The role of bricolage and effectuation. *Academy of Management Proceedings, 2020*(1), 17695.

Wiltbank, R., Read, S., Dew, N., & Sarasvathy, S. D. (2009). Prediction and control under uncertainty: Outcomes in angel investing. *Journal of Business Venturing, 24*(2), 116–133.

Zenger, T. R. (2016). *Beyond competitive advantage: How to solve the puzzle of sustaining growth while creating value.* Harvard Business Review Publishing.

Decision-Making Using a Scientific Approach

5

Abstract

This chapter presents an evidence-based framework for entrepreneurial decision-making based on the adoption of the scientific method for business decisions. It guides medical entrepreneurs through six sequential steps: identifying scenarios, defining actions, mapping their expected values, formulating testable hypotheses, conducting experiments, and making data-driven decisions. Using examples from real-life medical ventures—such as clinical trials, telehealth systems, and device competitiveness—the chapter demonstrates how hypotheses (e.g., adoption thresholds, efficacy rates) can be empirically tested to reduce uncertainty and bias. Because it combines quantitative approaches, pre-defined benchmarks, and continuous hypothesis testing, this approach transforms subjective entrepreneurial intuition into a rational, repeatable decision-making process that supports both pivoting and investment decisions.

5.1 The Scientific Approach to Decision-Making

It is finally time to dive deep into the third step of the approach shown in Fig. 5.1, which is the application of the scientific method to individual decision-related problems.

The six steps of decision-making using the scientific method consist of:

1. Identifying scenarios
2. Identifying actions
3. Mapping values in scenario–action pairs
4. Developing testable hypotheses

The original version of the chapter has been revised. A correction to this chapter can be found at https://doi.org/10.1007/978-3-032-15007-3_9

L. Negri, *De-Risking Investments in Medical Technology Innovation*, Springer Series in Healthcare Management and Innovation, https://doi.org/10.1007/978-3-032-15007-3_5

5. Testing hypotheses
6. Making a decision

Figure 5.2 visualizes this sequential list:

Fig. 5.1 The proposed three-step process with focus on the solution of individual decision problems

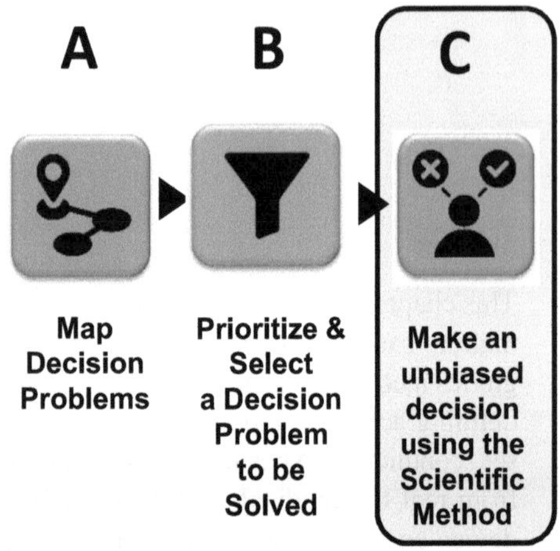

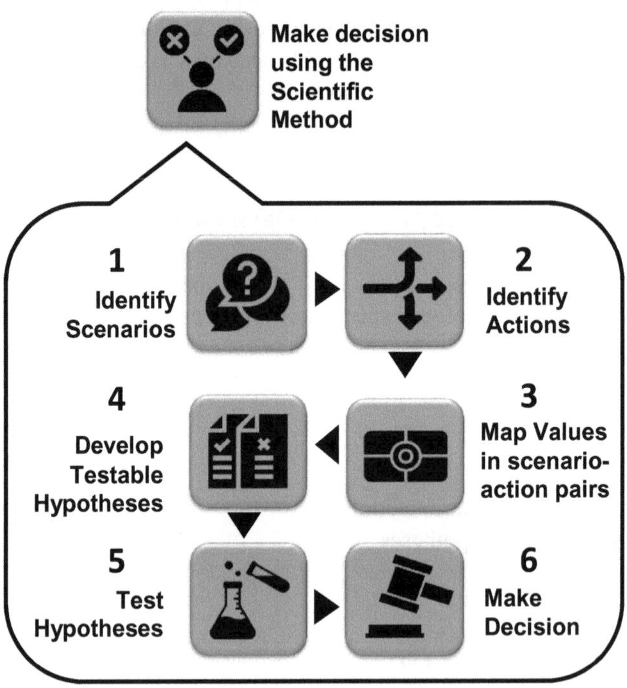

Fig. 5.2 A visual summary of the six steps of the scientific approach

The fundamentals of the scientific approach can be explained using a simplified real-life example: A medical entrepreneur has patented a wearable post-stroke monitoring device that allows an earlier and safer hospital dismissal for post-acute stroke patients, thanks to its ability to combine the real-time measurement, transmission, and cloud analysis of several vital signs, including a high-definition electrocardiogram.

After mapping and prioritizing all the relevant decision-related problems, the medical entrepreneur has identified a specific, relevant, impactful decision-related problem: Will the anesthesiologists of the Intensive Care Unit (ICU) recommend the new device to patients (and, therefore, to the hospital)? For simplicity, we assume that hospitals have already accepted the use of this device under the physician's recommendation and that the physician's recommendation alone represents an accurate predictor of adoption for the patient.

5.1.1 Step 1: Identifying the Possible Problem Scenarios

In order to apply the scientific approach, the entrepreneur must identify all possible future scenarios. In the simplest context, there will only be two possible scenarios. In the first scenario, the innovative product will encounter a high adoption rate. In the second scenario, the adoption rate will be low. The theory of the scientific approach states the following:

- The scenarios must be exhaustive and mutually exclusive.
- The scenarios must be uncertain.
- The decision-maker must not have any control over scenarios.

Indeed, the scenarios represent the source of uncertainty in decision-related problems, and decision-makers must attempt to predict the respective probabilities of occurrence. In our case, the two scenarios (a high adoption rate and a low adoption rate) met all the requirements: they are mutually exclusive, uncertain, and uncontrollable.

5.1.2 Step 2: Identifying the Possible Actions

As mentioned in 4.5, in contrast to scenarios, actions are investments, activities, and choices that the decision-maker controls. In this highly simplified example, the entrepreneur may identify two possible courses of action: to launch the innovative medical product or to exit. A rational entrepreneur wants to launch a new product only if there is an expectation of high product adoption, as it will potentially lead to high revenues and generate high net profits. However, how can the entrepreneur know how high the level of adoption must be in order to justify the decision to launch the product?

5.1.3 Step 3: Building the Table of the Scenario–Action Pairs

Following the prescriptions of the scientific approach, the entrepreneur must prepare a scenario–action table, like the one shown in Table 5.1, which represents the possible combinations of scenarios and actions and, for each pair, assigns the respective economic value.

The value of not launching and exiting equals zero, independent of the future scenarios. If the entrepreneur launches the innovation and, in the future, the scenario is one with a high adoption rate, the net value of the return is certainly greater than zero. This positive, best-scenario return value can be indicated with "r."

If, instead, the entrepreneur launches the product but later, the adoption scenario results in a low adoption rate, the market will not reward the investment made in launching. In this case, we indicate with "–z" the losses to be faced by the entrepreneur.

Following the scientific approach, decision-makers seek to predict the likelihood of each uncertain scenario; in other words, they want to attribute probabilities to the different scenarios. For this reason, Table 5.1 introduces the value "p," representing exactly the probability of high adoption; therefore, (1–p) is the probability of the other scenario, corresponding to low adoption.

5.1.4 Step 4: Developing Testable Hypotheses

We must develop a testable hypothesis to make the most accurate predictions about future scenarios and, therefore, about "p."

Developing hypotheses requires being very specific about the definitions of each scenario and clarifying what "high" and "low" precisely mean. For instance, the decision-maker may say—or decide—that the market is considered in a high scenario if, on a scale from 0% to 100%, ICU anesthesiologists recommend the wearable post-op sensor to at least 25% of post-op cardiac patients. This condition corresponds to setting a numeric benchmark of 25%, a percentage that we identify with the symbol "V*" here.

$$V^* = 25\%$$

Now that we have more precisely defined the meanings of "high" and "low," we can state the corresponding testable hypotheses.

The null hypothesis defines the low scenario; thus, it is as follows:

> If ICU anesthesiologists recommend the wearable sensor to less than 25% of post-op cardiac patients, the adoption rate is low (low scenario).

Table 5.1 The scenario–action table

Scenario / Action	H = High adoption rate P(H) = p	L = Low adoption rate P(L) = 1–p
Action 1: Develop	r	–z
Action 0: Exit	0	0

Therefore, the alternate hypothesis is the following:

If ICU anesthesiologists recommend the wearable sensor to 25% or more of post-op cardiac patients, the adoption rate is high (high scenario).

When the entrepreneur makes a decision, they always wish to choose the action that promises the highest economic value. Therefore, the decision will favor Action 1 only if the economic value of Action 1 is higher than the economic value of Action 0, which is equal to zero.

$$\text{Value}(\text{Action } 1) >= \text{Value}(\text{Action } 2)$$

Considering that the probability of Scenario 1 is equal to p and the probability of Scenario 2 is equal to (1–p), with simple calculations, the decision-makers can observe that the value of Action 1 is equal to.

$$\text{Value}(1) = rp - z(1 - p)$$

Since the above calculation shows that the value of Action 1 is a function of the probability p, the decision-maker can quickly calculate the minimum value of p that allows the value of Action 1 to be greater than the value of Action 0 which is equal to zero:

$$rp - z(1 - p) > 0$$

By isolating p the decision-maker will obtain the inequality below:

$$p > z / (r + z)$$

We call p* the minimum value of p that satisfies the above inequality and, therefore, allows Action 1 to hold an economic value greater than zero.

$$p^* = z / (r + z)$$

In other words, the entrepreneur will require the probability of the high scenario to be higher than p* to decide to go with Action 1. In order to provide a practical example we now make the assumption that, given the specific values or r and z, in this specific example, the calculated value of p* is:

$$p^* = 40\%$$

5.1.5 Step 5: Testing the Hypotheses

In order to test the hypothesis, the decision-maker is required to design and run some type of experiment. For instance, this experiment may constitute surveying a statistically significant number of ICU anesthesiologists and asking them about their inclination to recommend the new post-op sensor.

Suppose the sample of the interviewed anesthesiologists shows that 30% of them are inclined to recommend the new post-stroke monitor. In this case, one can say that, based on the experiment, our prediction of V (identified with the symbol "$\hat{V}$") is equal to 30%.

$$\hat{V} = 30\%$$

Earlier, in Step 4, we established a benchmark for V, identified by V* = 25%. Therefore, we can now affirm that the expected value of V ($\hat{V}$) is higher than its benchmark (V*):

$$\hat{V} > V*$$

This result favors the decision to launch. However, this information about the expected value of V alone is insufficient to make a final decision. There are two important questions that the decision-maker must still answer:

- Given that $\hat{V}$ is equal to 30%, what is the probability p that, in real life, V is greater than 25%?
- Is the probability p higher than the minimum benchmark value set as p* = 40%?

In summary, the decision-maker will have to verify the inequality below:

$$p\left(V > V^*\right) > p^*$$

This inequality check is performed to ensure that we can reasonably extend the result of the experiment to real life and predict V to be higher than V* not only for our test sample but also in the real world.

To analyze the sample distribution and verify whether p is greater than p*, the decision-maker can use the data obtained by surveying the anesthesiologists and process them with any software package with basic statistical abilities or with Microsoft Excel's statistical formulas. As anticipated above, the decision-maker set p* = 40%. Therefore, if, for instance, the statistical analysis of the surveyed sample indicates that p is equal to 55%, then the decision-maker's prediction is that p is greater than the benchmark p*:

$$p > p^*$$

5.1.6 Step 6: Making a Decision

Finally, we have verified the following:

$$\hat{V} > V * \text{and}\, p > p *$$

Therefore, we can reject the null hypothesis, accept the alternate hypothesis that supports a high adoption rate, and, thus, decide to launch our innovative medical product. This decision concludes the decision-making process in this highly simplified example.

5.2 Endpoints and Benchmarks of the Experiment

From the last example, we learned that the decision-maker must design experiments, elect the endpoints of the experiments, and establish benchmarks for the endpoints to test the decision hypotheses.

Using a parallel with the clinical trial design can help to better understand the critical function of the endpoints in business decisions. In clinical trials, an endpoint is, in many cases, a targeted outcome that is statistically analyzed to help determine the efficacy and safety of the therapy under study (NCATS, 2024). Among the several endpoints that a clinical trial may include, the primary endpoint is the main result that is measured at the end of a study to determine if a given treatment worked: It is a variable that is decided before the study begins (NCI. 2024).

Suppose we apply the scientific approach to business decisions. In this case, the endpoint of an experiment is the key variable that we study, and it is a variable that is chosen before the experiment is performed and on which we test our assumptions. In the previous simple example, V is the endpoint—the variable that the entrepreneur uses to identify future scenarios. V*, instead, is the threshold value used to set a specific, measurable distinction among scenarios.

In clinical trials, both the endpoint and (in most cases) the benchmark must be identified before the beginning of the experiment. Physicians designing clinical trials for new therapies can set their benchmarks by, for instance, examining the efficacies of existing therapies: These degrees of efficacy are hopefully lower than those of the latest, tested therapy.

In most cases, the medical entrepreneur, who must also establish benchmarks before starting the experiment, does not benefit from previous references such as those used by clinical trial designers. Later in this chapter, we explore how entrepreneurs can mitigate uncertainty related to the lack of pre-existing benchmarks. For now, it is important to emphasize the importance of linking a decision-related question to a specific experiment designed before starting any experiment, such as market research.

In daily practice, companies (even big corporations) tend to make significant investments in market research before having precisely identified the decisions they are trying to make. In such a case, companies start large market research endeavors without having identified either the research's primary endpoints or the respective benchmarks. In these contexts, organizations hire research-consulting companies, which return tick books full of beautiful charts that are difficult to translate into rational and useful business decisions. Therefore, these large market research efforts only serve as useful tools in earlier investigation phases, when the entrepreneur is making sense of the external medical or healthcare environment as well as of the potential impact of a certain medical innovation.

In my professional experience, for instance, in large biotech and medtech corporations in Europe, I have seen the opposite happening. I was given enormous market survey reports, usually designed by the global marketing team, and, on that basis, I was asked to make strategic business decisions for the European organizations. While those reports were valuable tools for any pre-decision, exploratory activity, they were almost always unfit to support the specific decision-related problems that I later decided to approach. In most cases, those tremendous marketing research investments exhausted the entire discretionary spending budget for the global marketing function; therefore, most of the decision-related problems that my regional team and I identified as critical were finally resolved without the support of a well-designed field-based test or experiment.

5.2.1 Setting V*

Setting the benchmark V* for the endpoint variable V is neither easy nor intuitive. In some cases, there are neither known general criteria nor results from previous similar experiments that can be used as references. The criteria and thinking processes leading to the choice of a benchmark depend on the type of endpoint. Therefore, the best approach to explaining this thinking process consists of analyzing the benchmark-setting process in real-life examples.

Example 1 The endpoint V is a medical device's degree of usability satisfaction.

In this case, the decision-maker could, for instance, establish a numeric scale that represents the degree of satisfaction, such as the one shown in Fig. 5.3.

One simple and instinctive way of setting the benchmark V* may be to choose any value between 7.5 and 10, as it would ensure a rating of "satisfied" or "more than satisfied."

While this common-sense-driven approach is intuitive, it lacks comparability; in other words, it relies on a supposedly common understanding of the term "satisfied." Instead, whenever possible, a safer way of setting V* consists of using external reference points: For instance, the decision-maker may set the benchmark using satisfaction ratings of similar products that have already been developed or similar experiments on earlier versions of the same product. Another solution can be to refer to the usability rating of similar medical devices: The decision-maker may design an experiment that also involves the collection of user feedback about competitive medical devices.

Users can be asked to assess the usability of the device under evaluation ($\hat{V}$) while concurrently assessing the usability of some of the "n" most relevant

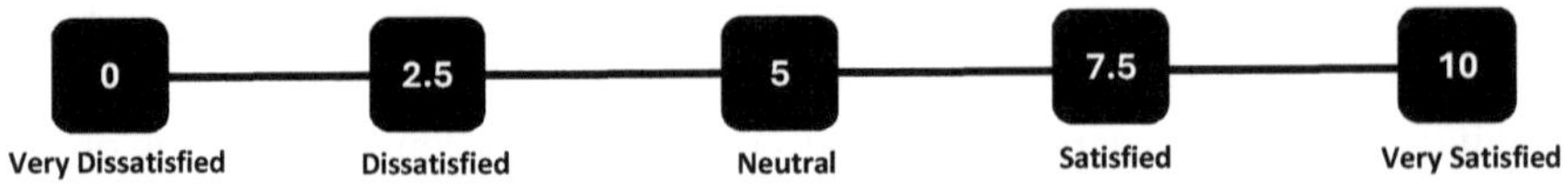

Fig. 5.3 A numeric scale that represents the degree of satisfaction

competitive devices ($\hat{V}_{Ci}$, $i = 1,.....,n$). In this case, the decision-maker can set the benchmark (V*) to ensure that the new device is likely to hold the best usability rating among its competitors:

$$V^* = \max\left(\hat{V}_{Ci}\right) i = 1,.....,n$$

Here, the value V* is not directly set before the experiment; rather, its mathematical formulation is defined prior.

Example 2 The endpoint V indicates the maximum percentage of adopting hospitals. It is also called "peak share" or "peak adoption," and it refers to the maximum market penetration or adoption rate that the new medical technology will ever reach during its life cycle. In this case, the entrepreneur predicts the market penetration dynamics and, in a highly simplified scenario, decides whether to commercialize. Here, the following question must be asked: What reference point of V can be used to set V*? An essential reference point comes directly from target revenues (T_R), which depend on the actual number of potentially adopting hospitals (N_H), the average selling price (P), and the ability to successfully drive the adoption rate (V). In summary,

Target revenues = T_R.
Price = P
# of potential hospitals = N_H
Adoption rate = V

T_R can be expressed as follows:

$$T_R = P * N_H * V$$

After the entrepreneur has defined the minimum target revenues (T_R *) that support the venture's financial strength till breakeven and then deliver the expected returns to investors, V* can be expressed as follows:

$$V^* = \frac{T^*_R}{P * N_H}$$

While setting V* using this approach may appear as a straight, safe, and easy way to find a benchmark, in reality, it requires additional discipline.

For instance, the number of potentially adopting hospitals must be carefully evaluated, ensuring that their healthcare delivery characteristics and their buying process actually allow for the acquisition of the proposed medical innovation.

Moreover, not all geographical regions of the world have hospitals with the same range of budget, the same buying process, and the same focus on innovation. Even the price at which adoption is supposed to take place must be carefully predicted by geographical region and needs to be market-tested as early as possible.

The adoption rate cannot be arbitrarily chosen by an optimistic entrepreneur. Otherwise, a significant bias may be introduced at the core of the "commercialize vs. exit" decision. It is, therefore, important to recall that using the scientific approach is not, by itself, a complete guarantee of bias avoidance. For instance, introducing benchmarks that are disconnected from market reality or based on wishful, non-tested assumptions can compromise the scientific method's ability to reduce biases.

5.2.2 Setting P*

Designing and performing an experiment to verify if $\hat{V}$ >V* is a crucial step, as it helps accept or refuse the set hypotheses. However, doing so without analyzing the distribution of the values of V observed in the experiment may be dangerous. Indeed, when the sample's variance is relatively high, measuring values of $\hat{V}$ "slightly higher than V* " may not be very meaningful to the final decision. For this reason, it is necessary to evaluate the probability p that V > V* qualitatively shown in Fig. 5.4 and expressed as follows:

$$p\left(V > V^*\right)$$

Otherwise, the result of the experiment may lose strength in its real-life applicability. In general, for a given value of V*, the decision-maker always needs to verify that the probability p(V > V*) is higher than the probability benchmark (p*).

Setting p* requires different approaches depending on the type of decision: investment decision vs. scenario decision.

5.2.2.1 Setting P* in an Investment Decision

We have already shown that the scenario–action table represents the possible combinations of scenarios and actions and that, in investment decisions, we assign the respective economic value to each combination (Table 5.2).

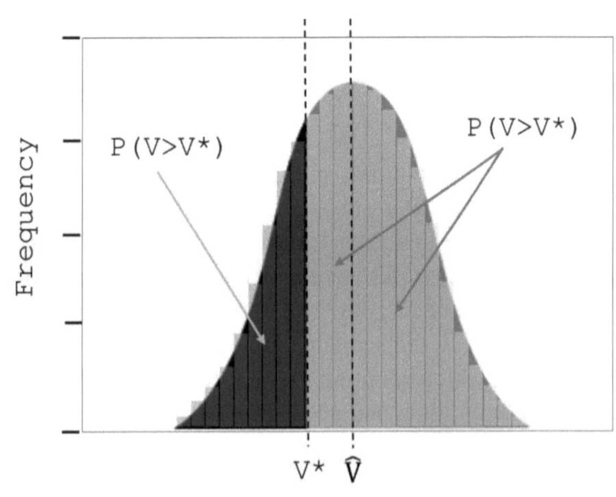

Fig. 5.4 How to graphically estimate the probability p that V > V*

Table 5.2 The scenario–action table

Action \ Scenario	H = High adoption rate P(H) = p	L = Low adoption rate P(L) = 1–p
Action 1: Develop	r	−z
Action 0: Exit	0	0

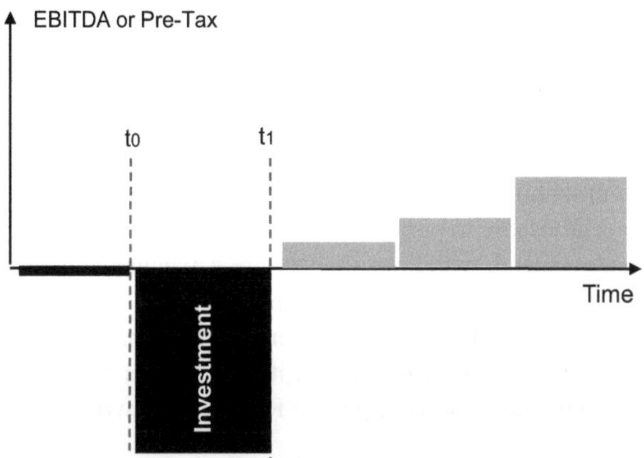

$$-z = -INV_{t0}$$

$$r = NPV_{t0} \sim NPV_{t1} - INV_{t0}$$

Fig. 5.5 The qualitative evolution of the Earnings Before Interests, Taxes, and Amortizations (EBITDA) of a startup before and after an important investment

We have also explained that, after having filled the scenario–action table, the decision-maker now knows the following:

$$p^* = z/(r+z)$$

It is essential to examine this calculation in greater depth and detail to understand how to calculate the values of r and z.

Figure 5.5 represents a typical and simplified evolution of the Earnings Before Interests, Taxes, and Amortizations (EBITDA) for a startup that must make an important investment in t_0 before it can start generating revenues and the resultant profits after t_1. The value of z corresponds to the value of the investment made in t_0, while the value of r is equal to the net present value of the entire cash flow starting at the time of the investment t_0. Assuming that the timeframe t_0– t_1 is relatively short, the value of r can be approximated by subtracting the net present value of the post-investment revenues from the value of z.

$$r = NPV_{t0} \sim NPV_{t1} - INV_{t0}$$

In the context of this simplification, p* can be expressed as follows:

$$\text{p}^* = \text{z}/(\text{r}+\text{z}) = \frac{INV_{t0}}{NPV_{t0} + INV_{t0}} \sim \frac{INV_{t0}}{NPV_{t1}}$$

This expression of p* allows for a very critical and practical consideration: In order to set p*, the decision-maker needs to be in the position to predict both INV_{t0} and NPV_{t1}. However, while the value of the investment INV_{t0} is relatively "easy" to predict, the present value of the post-market cash flow stream, instead, represents a significant predictive challenge. At the early stages of the startup, this prediction might include non-verified assumptions as well as several educated guesses.

This observation suggests the requirement to balance the efforts related to the sought precision used to set both V* and p*. Indeed, it makes little sense to spend an excessive amount of time and resources on the "perfect" definition of V* if the definition of the probability benchmark p* contains significant biases.

Therefore, it is advisable to periodically re-run and re-test this decision until the investment is made; indeed, with time, the venture's management gains a firmer grip on predicting future revenue and cash flow streams.

It is also worth noting that while it is challenging to manage the prediction of two separate variables, such as INV_{t0} and NPV_{t1}, it may be less difficult to predict their ratio. Indeed, using the definitions of p* and ROI, we can write the following:

$$\text{p}^* = \frac{INV_{t0}}{NPV_{t0} + INV_{t0}}$$

$$\text{Therefore,} \frac{1}{p^*} = 1 + \frac{NPV_{t0}}{INV_{t0}}$$

$$\text{Since ROI} = \frac{NPV_{t0}}{INV_{t0}}, \text{then} \frac{1}{p^*} = 1 + \text{ROI}$$

$$\text{And, finally: } p^* = \frac{1}{1 + \text{ROI}}$$

Using this last formula, we can calculate p* as a function of the target ROI. Conversely, if we set p* using other points of reference, we can reverse this formula to check if the set value of p* implies excessively low (unattractive) or excessively high (unrealistic) values of the predicted ROI:

$$\text{ROI} = \frac{1}{p^*} - 1$$

For instance, if the decision-maker picks a value of p* = 90%, the implication is that the expected ROI is as low as 11%, while a value of p* = 50% implies an ROI equal to 100%.

5.2.2.2 Setting P* in a Scenario Decision

In a non-investment decision (a scenario decision), the entrepreneur does not dispose of the values of r and z. In the scenario–action table presented below, the values of r and z are substituted by qualitative indicators such as "+++" and "– –." Therefore, the decision-maker cannot calculate p* using the formula described in the previous section (Table 5.3).

In these cases, the entrepreneur must find other references to help establish the value of p*. For instance, useful references can be the p* values used in other decisions on the same project. Other references to set p* can be found in the perception-of-probability table.

Setting p* Values Using the Perception-of-Probability Words

p* is the minimum probability of what we call a "high scenario" (V > V*). Then, what is a level of p* that is "high enough" for the entrepreneur or for potential investors? Entrepreneurs must be concerned about the degree of trust that different levels of p* generate in potential investors. For instance, if investors realize that the decision to pursue a critical medical innovation project is based on scenario predictions that have a 50% chance to become true in real life, they are likely to lose faith or even walk away. A degree of 50% uncertainty, in the absence of other quantitative indicators, is uncontroversial evidence that the entrepreneur does not actually know where that decision will take the venture in the future. For this reason, in the absence of other reference points, the entrepreneur may wish to use a scale called "perception-of-probability words" to translate the required level of confidence that investors are likely to require into numbers. Fig. 5.6 shows a typical example of a perception-of-probability scale that decision-makers can find in literature: for instance in the *Handbook of Decision Analysis* (Barclay et al., 1977), or in other more advanced and recent scales, such as the one developed by the University of Illinois Urbana-Champaign (Fagen-Ulmschneider, 2023).

Setting p* Values Consistently with V* Values

Another practical observation guiding a reasonable choice of p* consists in its statistical relationship with the value of V*. For instance, in a simplified setting (two scenarios and two actions), the statistical relationship between the variables V and p is qualitatively described by the charts presented in Fig. 5.7.

The higher the expectations set by the value V*, the lower the probability that these expectations will be met in reality. The higher the expected probability of a

Table 5.3 The qualitative version of the scenario–action table

Scenario / Action	H = High adoption rate P(H) = p	L = Low adoption rate P(L) = 1–p
Action 1: Develop	+++	--
Action 0: Exit	0	0

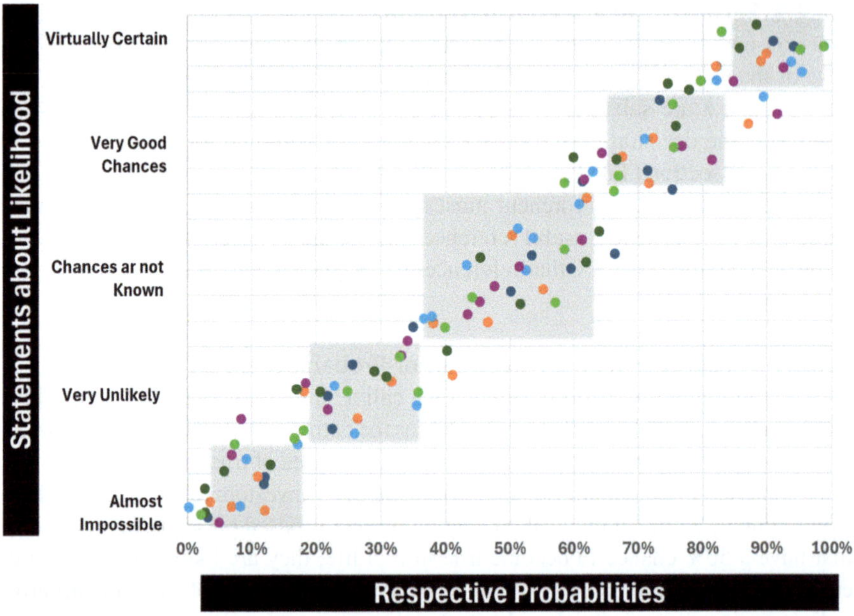

Fig. 5.6 A representative example that reflects the characteristics of a typical perception-of-probability scale: this was developed using datapoints from interviews about the expected probability of each statement of likelihood

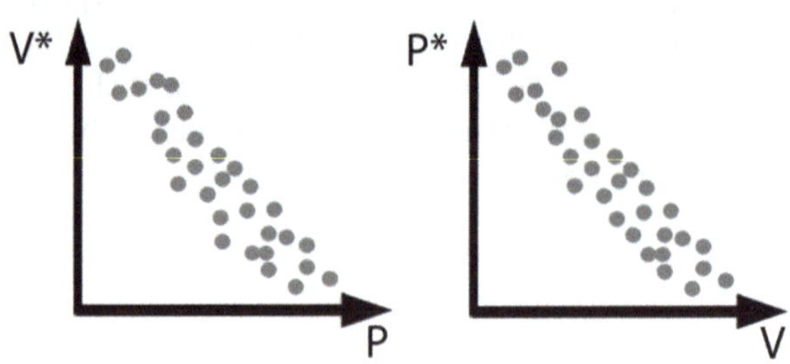

Fig. 5.7 A representation of the statistical relationship between p and V* and between V and p*

high scenario, the lower the benchmark value defining that scenario. Therefore, the decision-maker setting the values of V* and p* must keep in mind the nature of their statistical relationship, noting that thinking in terms of V means thinking in terms of expected values, while thinking in terms of p is like thinking in terms of the probability of the high scenario.

One Consideration on the Potential Biases Setting V* and p*

Many consulting this book will have the same question: Does it make sense to use the scientific approach with the knowledge that the decision-maker is highly exposed to biases in the choice of the benchmarks V* and p*? The answer is "yes," as the scientific method imposes tight boundaries on potential areas of bias. The goal of a medical entrepreneur is not to be 100% accurate; rather, it involves minimizing decision-making biases and, consequently, maximizing the probability of success for the medical innovation. Renouncing the opportunity to limit biases because one knows it is impossible to completely nullify them is not a rational thought.

I teach the approach of this book in different universities, usually, to bioengineering PhD candidates or to PhDs in various life science areas. When students ask questions about the advisability of the use of a scientific approach in such a complex and unpredictable environment, I provide common sense examples. Dismissing the scientific approach due to the fact that it still involves some elements of uncertainty would be similar to switching off the GPS in the desert because it is often inaccurate. Also, my high-school English teacher involuntarily offered a spontaneous, useful, practical example. She suffered from severe myopia, and on a gray, very foggy winter morning, she surprisingly drove to school without her usual thick glasses. When we asked her why she drove without glasses, she said, "There is so much fog this morning, it was impossible to see anything. So why would I even bother to put my glasses on?" My colleagues and I personally witnessed several failures of medical ventures guided by the idea that, in contexts of high uncertainty, individual instincts work better than thinking.

5.3 More Complex Decision Trees

There are more complex decision trees that hold more than two available actions or more than two possible scenarios. A French startup has come up with the idea of a therapeutic platform based on the ability to modify the human microbiota. While clinical research is still at an early stage, the startup's management must decide on which therapeutic areas future research investments should be oriented.

In this real case, the two most promising therapeutic areas are those of depression and obesity. Figure 5.8 visualizes the decision tree, including the options of investing in depression, investing in obesity, and exiting. Suppose the startup's management decides to base the decision on each branch's respective degree of market attractiveness. Since the estimation of the degree of attractiveness is based on a set of parameters, the decision-maker will decide to consolidate these parameters into a single numeric indicator (V) through weighted averages.

When focusing on the possible future scenarios, the assumption is that for each action, there are only two respective future scenarios: favorable ("high") and unfavorable ("low"). When market attractiveness is "high," the value of the variable V is "high," and vice versa, the value of the variable V is "low" when the attractiveness is "low." The scenario–action table, therefore, becomes more complex than the ones analyzed in the previous chapters: The number of possible future combined scenarios is indeed higher:

$$n_s = 2 * 2 = 4$$

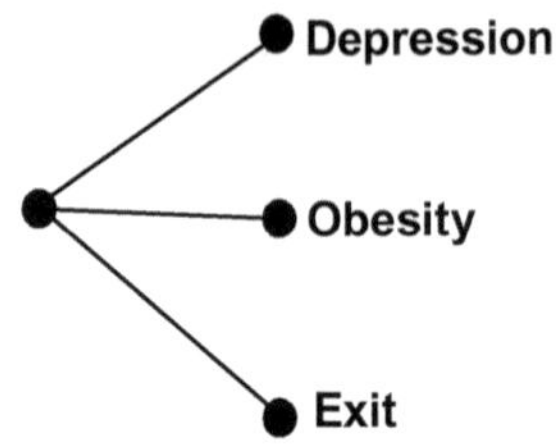

Fig. 5.8 An example of three-way decision-making tree in the case of a startup deciding which disease to target

Table 5.4 The scenario–action table for a decision with three possible actions and two possible scenarios

	Depression – Obesity Scenarios				
	HH	HL	LH	LL	
Depression	$V_{1,1}$	$V_{1,2}$	$V_{1,3}$	$V_{1,4}$	Value = V(D)
Obesity	$V_{2,1}$	$V_{2,2}$	$V_{2,3}$	$V_{2,4}$	Value = V(O)
Exit	0	0	0	0	Value = V(E)

Table 5.5 The two scenario–action tables that help simplify the solution of the three-action and two-scenario decision problem

	H	L
Depression	$V_{1,1}$	$V_{1,2}$
Exit	0	0

	H	L
Obesity	$W_{1,1}$	$W_{1,2}$
Exit	0	0

The number of possible decision options is higher:

$$n_d = 3$$

The number of scenario–actions pairs is therefore much higher:

$$n_{ds} = n_d * n_s = 4 * 3 = 12$$

In this case, the decision-maker must choose among three possible actions by evaluating 12 scenario–action pairs. In contrast to the case of the simpler decision trees discussed in the previous chapters, no straightforward formulas can alone support the resolution of the decision-related problem.

The best decision must support the highest economic value; therefore, using the variables indicated in Table 5.4, the final decision will pursue the following:

- "Depression" only if V(D) > Max(V(O), V(E))
- "Obesity" only if V(O) > Max(V(D), V(E))
- "Exiting" only if V(E) > Max(V(D), V(O)).

To simplify the resolution of this decision-related problem, instead of using a three-branch tree, the decision-maker may divide the problem into two simpler decision-related problems, each holding only two branches. The first decision compares the value of choosing "depression" with that of "exiting," while the second

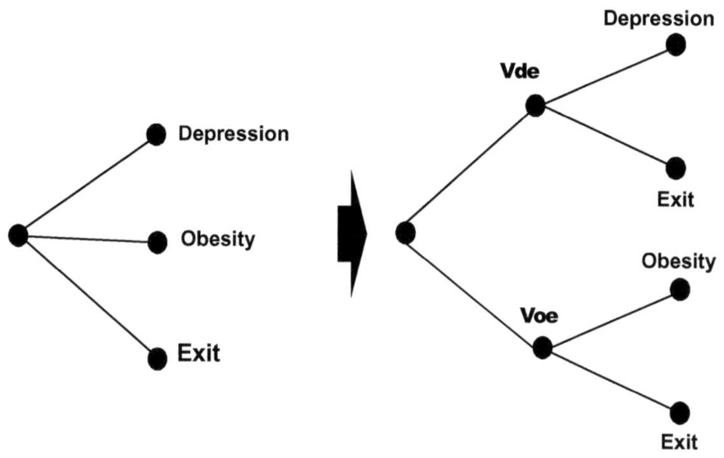

Fig. 5.9 How the two scenario–action trees that help simplifying the solution of the three-actions decision problem

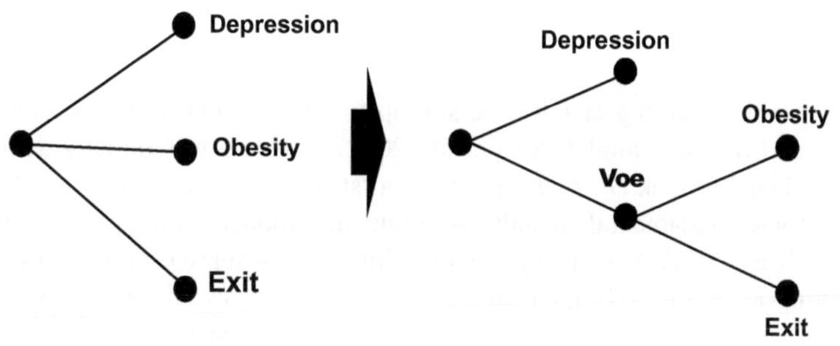

Fig. 5.10 How two sequential two-way decision trees help simplifying the solution of the three-action decision problem

compares the value of "obesity" with that of "exiting." As shown in Table 5.5, the two decisions can be approached simply using the more straightforward approach discussed in the previous chapter.

They will respectively deliver two expected values, Vde and Voe. The final decision must be made by comparing these two values and choosing the highest (Fig. 5.9).

There is another way to simplify this problem that involves 12 scenario–action pairs. Instead of transforming the decisions involving the three branches into two parallel branches, the decision-maker can divide the problem into two sequential decision-related problems (Fig. 5.10). For instance, first comparing the value of obesity with the value of exiting and, second, making a decision corresponding to the value Voe, comparing the value Voe with the value VD of depression. In other words, we first build a scenario–action matrix for the decision of obesity vs. exiting, selecting the action that holds the higher expected value: Voe. Later, we build the scenario–action matrix for depression vs. the value Voe. The final decision points to the greater of the two values between depression and Voe (Table 5.6).

Table 5.6 The two scenario–action tables that help simplifying the solution of the three-action and two-scenario decision problem using two sequential two-way decision trees

	H	L
Obesity	$V_{1,1}$	$V_{1,2}$
Exit	0	0

	H	L
Depression	$W_{1,1}$	$W_{1,2}$
V_{oe}	V_{oe}	V_{oe}

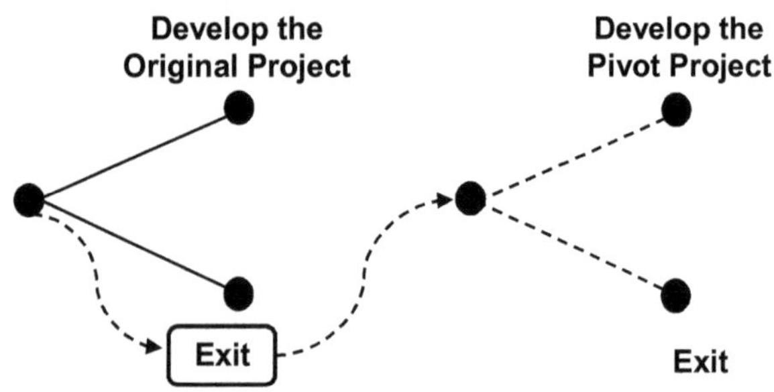

Fig. 5.11 An example of the very common shift between a decision-making process that leads to an exit decision and the discovery of a new pivotal idea

5.4 Pivoting

Entrepreneurial pivoting enables the startup to validate and refine the company's strategy and business model (Sala et al., 2022). Pivoting was initially defined by Ries (2011) in his book *The Lean Startup* as a "structured course correction designed to test a new fundamental hypothesis about the product, strategy, and engine of growth" (Ries, 2011). A science-based decision-making approach is often the source of pivoting ideas for early medical startups. Published research has demonstrated that early-stage startups employing the scientific approach for decision-making have a higher probability of performing at least a pivot and a higher likelihood of limiting the number of pivots within manageable proportions (Camuffo et al. 2020, 2024, 2024).

Based on the personal experience of my colleagues and me, a plausible explanation for these positive results is that the hard, unchallengeable "truth" of tested hypotheses offered by the scientific approach truly motivates (or forces) entrepreneurs to seek and test alternate ideas or even different applications of the same idea. Figure 5.11 exemplifies the very common shift between a decision-making process that leads to an exit decision and the often-sudden discovery and evaluation of a new (pivoted) idea that promises satisfying economic returns. A recent review of different types of startup pivots provides a synthetic grouping of various kinds of pivots (Sala et al., 2022), which is particularly interesting for medical innovation. The four suggested categories are product, market, strategy, and organizational pivots (Fig. 5.12).

Product Pivots

Zoom-in and zoom-out pivots, respectively, concern re-focusing the development efforts, emphasizing a specific "winning" and attractive product feature, and

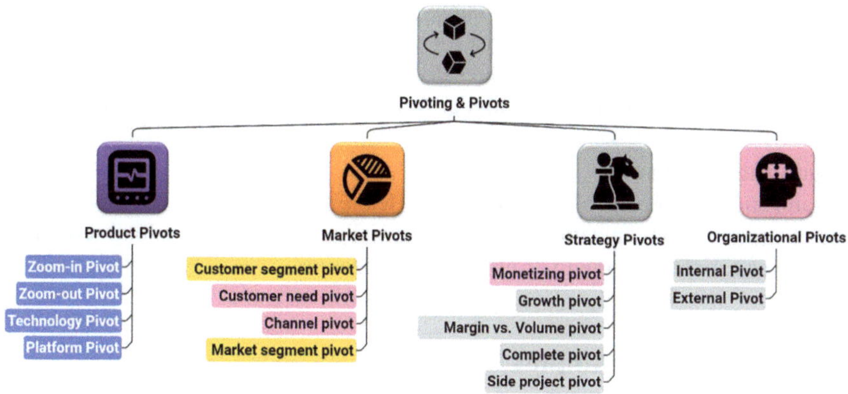

Fig. 5.12 The four main types of pivots

de-emphasizing particular product features in favor of a value proposition based on a broad set of product features. Technology pivots consist of delivering the same set of customer solutions but with different technologies. Platform pivots are related to the use of support software applications (platforms), which are very common in medical wearable technology and may change their role and even take the lead in the value proposition mix due to a pivot.

Market Pivots

Market segment pivots pertain to shifting the focus from one market segment to another. For instance, a medical device company may have better chances of success in the USA than in the EU as far as telemedicine is concerned and may, therefore, change its market focus. Customer segment pivot refers to changing customer segments within the same target market: In medical technology, for instance, a shift in customer segment may be represented by a shift in therapeutic indication. Customer needs pivots may occur when a startup realizes that its new medical innovation has a better fit with a different group of unmet medical needs than with the needs initially envisioned. Many startups focused on robotic rehabilitation have found that their product's diagnostic ability to measure a patient's progress has greater perceived value than the therapy itself. Channel pivot involves a change in the global commercial channels within the same target market and customer segment.

Strategy Pivots

Strategy pivots involve a change in at least one strategic choice. For instance, monetization pivots affect how a startup monetizes its innovation—for instance, moving from selling huge radiology equipment to charging for the equipment per diagnostic procedure. Growth pivots are strategic changes made to accelerate profitable growth and expand the customer base. Margin vs. volume pivots involve shifts between high margins with low volume and high volume with low margins. A complete pivot revolutionizes the entire startup, possibly concurrently changing the product, strategy, and target market. When an idea for a truly disrupting pivot emerges, the only alternative to a complete pivot is the side project pivot: the construction and management of an initiative that lives and grows in parallel to the original startup.

Organizational Pivots

Organizational pivots involve changes in the team, such as changes in individuals or specific roles. An organizational pivot is external when it involves a long-term partnership or collaboration with an external organization.

5.5 Real-Life Examples

This section offers some examples of actual cases from medical innovation startups, simplified to allow the reader to avoid processing a large number of practical details and focus on essential decision-making issues.

5.5.1 Example 1:Decision to Invest in a Large Clinical Trial

Let's focus on an example from the pacemaker market, where a young startup must make an essential decision about investment in a large clinical trial.

The Context

The global market size for pacemakers was valued at US$4.51 billion in 2022 and is estimated to grow at a CAGR of 3.6% from 2023 to 2030. Rising incidences of cardiovascular diseases (CVDs), the increasing cost of cardiac disease management, favorable government policies, and the launch of technologically advanced pacemakers, such as MRI-safe pacemakers and leadless pacemakers, are primarily driving the growth in the market (Ignite Healthwise, 2024). The rising burden of CVDs is one of the key factors behind the market's growth. Coronary heart disorder is one of the primary reasons for CVD deaths globally. The economic burden of these diseases is anticipated to be more than US$1044 billion by 2030, highlighting the tremendous need for cardiac-pacing devices.

The Decision-Making Problem

In this case, the decision-maker is the CEO of a startup planning to launch a new pacemaker, and they must decide whether to initiate an expensive (US$5 million) phase III trial for a leadless pacemaker (Fig. 5.13). A leadless pacemaker is a small, battery-powered device. It sends mild electrical signals to the heart to keep the heart beating normally. These signals are painless. The pacemaker can help stop the dizziness, fainting, and shortness of breath caused by a slow heart rate (Ignite Healthwise, 2024). The venture hopes that the results of the expensive clinical trial will confirm the sought-after high treatment efficacy, as, in this case, the CFO predicts accumulated profits above US$20 million. If this is not the case,

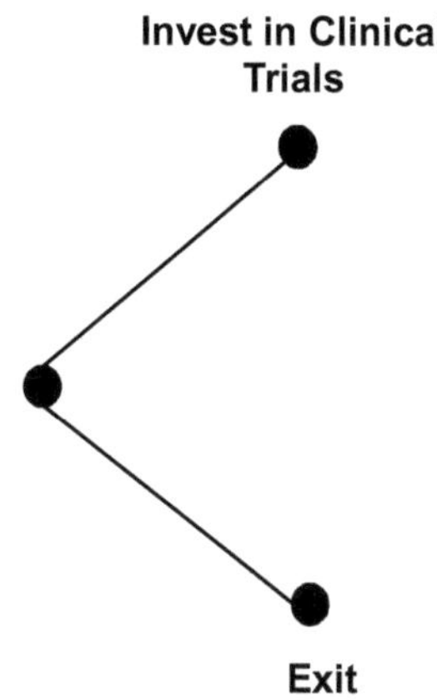

Fig. 5.13 The two-ways decision tree for the investment in clinical trials

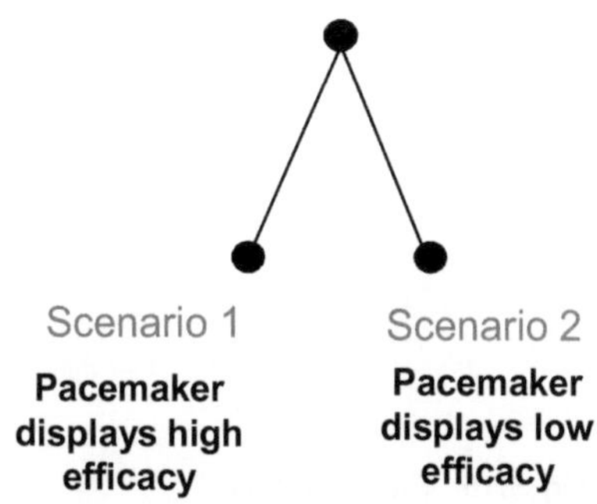

Fig. 5.14 The two scenarios related to the pacemaker performances

commercialization will not be possible, and the costs of the clinical trial will significantly contribute to the project's economic losses. Considering the monetary values at stake and the cornerstone problem in focus, it is feasible to use the scientific approach for this decision

Step 1: Identifying Possible Scenarios (Fig. 5.14)
In the simplest decision-making context, there are two possible future scenarios: The pacemaker proves to be highly efficacious, and the pacemaker's efficacy is low or unsatisfactory. The two working scenarios for the decision-making process are therefore: "high efficacy" and "low efficacy."

Step 2: Identifying Possible Actions (Fig. 5.15)
Now, the decision-maker focuses on the second step of the scientific approach, which constitutes identifying possible actions. The decision-maker must decide whether to initiate clinical trials to assess the pacemaker's efficacy. Let's suppose—for simplicity reasons—that exiting the project is the only viable alternative to investing in the clinical trial. In real life, in addition to investing and exiting, we would consider the option of pivoting the product features, functionalities, or therapeutic indications, which, in Fig. 5.15, would involve at least a third possible

decision arm. In our simplified context, there are only two possible actions in the decision-making process: investing in clinical trials or exiting. The investment required by a clinical trial is estimated to be close to US$5 million. This investment attracts the entrepreneur, as the CFO's risk-adjusted estimation of the net present value of the future cash stream indicates a value of US$20 million with an attractive internal rate of return (IRR).

Step 3: Mapping the Scenario–Actions Pairs (Fig. 5.16)
If the entrepreneur decides to invest and later the efficacy proves to be low, we talk about a "type 1 error." If, instead, the entrepreneur chooses to exit but the efficacy proves to be high, we are confronted with a "type 2 error." In all the other cases, the decisions are correct. The decision–action map is presented in Table 2, where the

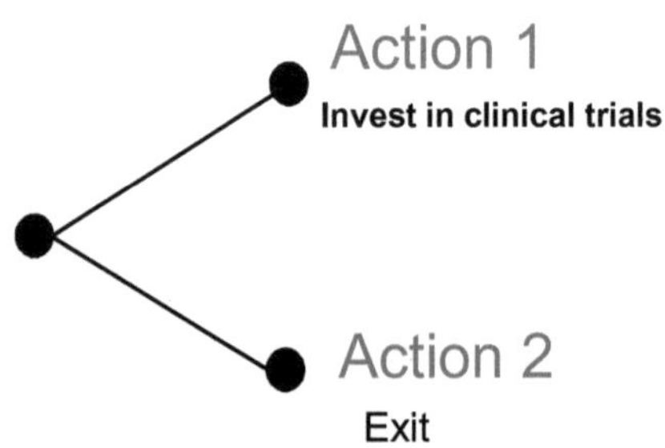

Fig. 5.15 The 2-action tree for the investment in clinical trials

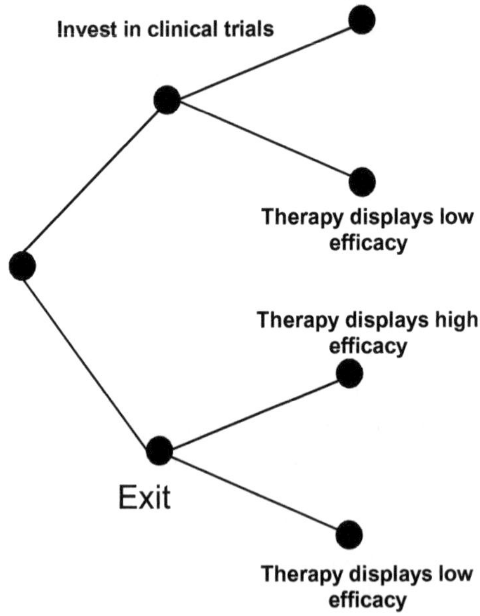

Fig. 5.16 The two-action and two-scenario tree for the investment in clinical trials

Table 5.7 The scenario–action table for the decision to invest in clinical trials in relation to the pacemaker

		Two Scenarios	
		Efficacy is high	Efficacy is low
		H(p)	L(1–p)
Two Possible Actions	Invest in clinical trials	r	–z
	Do not invest in clinical trials	0	0

values (r and –z) correspond to the economic values of the different scenario–action pairs we have described (Table 5.7).

As in previous cases, the letter "H" indicates the scenario of high efficacy, while the letter "L" indicates the scenario of low efficacy. Additionally, the letter "p" indicates the probability that the efficacy is high.

If the entrepreneur decides to invest in clinical trials, there are two possible outcomes, and they depend on the actual efficacy of the treatment. If the entrepreneur decides to invest, the high-efficacy scenario will likely reward the investor with a profitable return, indicated with the letter "r." If the treatment exhibits low efficacy at the end of the clinical trial, the entrepreneur is likely to lose the money spent on the clinical trial. The consequent loss is indicated with the symbol "–z." If, instead, the decision involves exiting, then the economic value in both scenarios is equal to zero.

Therefore, the decision-maker will choose to invest only if the predicted value corresponding to this decision is higher than the predicted value of exiting (equal to zero). In previous chapters, we explained that the minimum value of p required to support the decision to invest (p*) can be expressed as follows:

$$p^* = z/(r+z)$$

Here, r is equal to the net present value of the revenues minus the value of the investment:

$$r = NPV - INV$$

While z is equal to the value of the investment:

$$z = INV$$

Therefore, by substituting these values of r and z in the formula of p*, we can easily find:

$$p^* = INV/NPV = US\$5\,\text{million}/US\$20\,\text{million} = 25\%$$

Step 4: Developing Testable Hypotheses

In order to develop testable hypotheses, the decision-maker must precisely define the terms "high efficacy" and "low efficacy."

One of the indicators of a pacemaker's efficacy is the "pacing threshold." Pacing voltage thresholds are the minimum energy required for the pacemaker to cause the cardiac muscle to depolarize. In other words, the pacemaker needs that level of energy to take the lead in generating the heartbeat. The lower the energy required, the higher the efficacy.

With a strong simplification, we can suppose that the entire evaluation of therapeutic efficacy concerns the pacing threshold: The degree of "efficacy" can, therefore, be quantified using a numeric variable (V), which represents the actual measure of the voltage thresholds. To develop testable hypotheses, the decision-maker must be specific about the definition of high efficacy and low efficacy. It can be done by identifying a threshold value (V*) below which the pacemaker's efficacy is considered high. There are several ways to choose or calculate V*. For instance, the entrepreneur may use a comparative approach, deciding that, in order to achieve high efficacy, the pacemaker's voltage threshold must be the smallest among all the existing competitors. For example, we can suppose that the currently best-performing competitive pacemaker in the market holds a threshold of 1.35 volts. We can now set:

$$V^* = 1.35$$

Then, the null and alternative hypotheses can be expressed more precisely:

Null hypothesis:	If $V >= V^*$, the efficacy is low.
Alternate hypothesis:	If $V < V^*$, the efficacy is high.

In order to decide whether to accept or reject the null hypothesis, the decision-maker may decide to run an experiment that will provide refined signals to support this choice. Let's also keep in mind that even if the experiment proves that $V < V^*$, the decision-maker will also want to ensure that $p > p^*$.

Step 5: Testing the Hypotheses

The decision-maker, therefore, designs and implements an experiment that helps estimate the values of V and p. For instance, the experiment may consist of a simple, light, relatively cheap clinical trial involving a limited number of participating patients and with lighter procedural and administrative burdens. It is important to keep in mind that this experiment will support a business decision and not a medical decision or a peer-reviewed publication; therefore, even if it must comply with both patient safety and ethical standards, it does not need to comply with all the more complex and expensive standards of a large, randomized, controlled, multicenter clinical trial. With the data collected from this experiment, the decision-maker estimates the value of V. The expected value of V is calculated as the average of the values observed, and it is indicated by the symbol $\hat{V}$. Suppose that the decision-maker designs a test involving only 10 patients. Over the observation period, the performance of the pacemaker for each patient will be characterized by the respective average of the voltage thresholds, $\hat{v}i$ with $i = 1,\ldots,10$. Table 5.8 reports the average values $\hat{v}i$ obtained for each patient; therefore, the calculation of $\hat{V}$ can be expressed as follows:

$$\hat{V} = \sum_{i=1}^{10} \frac{\hat{v}i}{n}$$

When applying the actual experimental results the calculation delivers: $\hat{V} = 1.3 < V^* = 1.34$

With $\hat{V} < V^*$, the decision-maker can reject the null hypothesis, but to make the final decision to invest in a large clinical trial, the decision-maker must also verify whether $p(V < V^*) > p^*$, knowing that $p^* = 25\%$.

With a rapid statistical analysis applied to the collected sample, using a t-distribution with nine degrees of freedom, the calculated probability of $V < V^*$ is $p = 30.6\%$, which is greater than p^*.

Step 6: Making the Decision

The experiment provided the following signals:

$$\hat{V} > V^* \text{ and } p > p^*$$

With the support of these refined signals, the decision-maker can reject the null hypothesis and decide to invest in a large clinical trial.

5.5.2 Example 2: Telehealth and Patient Monitoring

The Context

Telehealth involves remotely distributing health-related services and information via electronic telecommunication technologies. It permits remote patient and clinician contact, care, advice, reminders, education, intervention, and even monitoring. Telehealth and telemedicine are sometimes used as synonyms. However, telemedicine is often used in a more limited sense to describe remote clinical services, such as diagnosis and monitoring. The size of the global telehealth market is anticipated to reach US$636.38 billion by 2028 and exhibit a CAGR of 32.1% during the forecast period. Fortune Business Insights™ presented this information in its report titled "Telehealth Market, 2021-2028." The market size stood at US$144.38 billion

Table 5.8 The measured average pacemaker's voltage thresholds

	Average voltage thresholds
Patient 1	1.35
Patient 2	1.5
Patient 3	1.3
Patient 4	1.1
Patient 5	1.28
Patient 6	1.3
Patient 7	1.34
Patient 8	1.1
Patient 9	1.4
Patient 10	1.5

in 2020. Telehealth can help bridge the gap and provide distance learning in circumstances where remote rural settings, lack of transport or mobility, epidemics, or pandemics restrict access to care. Telehealth may, for instance, involve different clinicians discussing a case over video conference; physical therapy can be done via digital monitoring instruments.

The venture is a medical device startup designing, developing, and commercializing a portable, all-in-one, cloud-connected monitoring system that can concurrently measure five vital signs. The product helps doctors, nurses, caretakers, and patients monitor and report the subject's current state of health by simultaneously collecting five key vital signs (body temperature, oxygen saturation, respiratory rate, heart rate, and blood pressure), besides providing an ECG and PPG medical-grade report, thereby offering a complete telemedicine system. The company has received US$200,000 in seed funds from a governmental institution. The product has been tested on humans, and a VC requires a feasibility study focused on the venture's ability to satisfy critical medical unmet needs in a competitive manner.

Decision-Related Problem 1: Value Proposition Fit to Needs
Problem Description

Question: Is the product capable of satisfying the current unmet medical needs?
Decision type: Scenario decision (no investment is involved)
Decision complexities:

- The definition of the satisfaction of needs is multidimensional: There are several needs that must be satisfied concurrently.
- The product needs to satisfy unmet needs while guaranteeing that the needs currently met by the current standard of care keep being satisfied.
- Not all needs are equally important. The satisfaction of important needs is a must, while the evaluation can be more flexible for other needs.

Step 1: Identifying Scenarios
There are two scenarios: high and low:

- H = The product is capable of responding to the most important stakeholders' medical unmet needs.
- L = The product is **not** capable of responding to the most important stakeholders' needs.

Step 2: Identifying Actions
There are two identified actions: continue and exit.

- Continue = The process of assessing and quantifying the business opportunity is continued.
- Exit = The project is dropped.

Table 5.9 Qualitative scenario–action table

	H	L
Continue	+++	---
Exit	0	0

Step 3: Building Scenario–Action Pairs

Since this is a scenario decision, the scenario–action matrix does not contain monetary values; therefore, Table 5.9 represents it well:

Step 4: Developing Testable Hypotheses

The decision-maker decides to refer to a one-dimensional scale (from 1 to 10) that mathematically represents the various degrees of satisfaction for each respective need. The minimum selected benchmark for satisfaction is set to 7.5.

H0 (null): *Key opinion leaders (KOLs) believe that the product does meet the list of unmet medical needs, assessing an average score higher than 7.5 on a 0–10 scale ($V^* = 7.5$) with a probability higher than 70% ($p^* = 70\%$).*

H1 (alt): *KOLs believe that the product does meet the list of unmet medical needs ($V < V^*$).*

In order to make the hypothesis testable, the entrepreneur decides to use and focus on the following variables:

1. For each of the 17 needs, the measured average satisfaction level is expressed as:

$$\hat{v}i \text{ with } i = 1,\ldots.,17$$

2. The average of $\hat{v}i$ across all needs is expressed as $\hat{\mathrm{v}}$:

$$\hat{\mathrm{v}} = \sum_{i=1}^{n} \frac{\hat{v}i}{n} > \boldsymbol{v}^* = 7.5$$

3. For each of the 17 needs, the calculated probability p_i that $vi > \boldsymbol{v}^* = 7.5$ is expressed as:

$$p_i \text{ with } i = 1,\ldots.,17$$

4. The average of $\hat{v}i$ across all needs is expressed as $\hat{\boldsymbol{p}}$

$$\hat{\boldsymbol{p}} = \sum_{i=0}^{n} \frac{p_i}{\mathrm{n}} > \boldsymbol{P}^* = 75\%$$

5. The decision-maker has decided to set the null hypothesis to be rejected if:

$$\hat{v} > v^* = 7.5 \text{ and } \hat{p} > P^* = 70\%.$$

In addition to the above, the decision-maker chooses to rank the relative importance of each need using an additional numeric scale. This approach is typically used to double-check if there are very unimportant needs (for instance, ranked below 2.5) that the experiment should exclude. It is also often used after the experiment to spot highly important needs that, regardless of the result of the experiment, show unacceptable dissatisfaction levels that need to be addressed individually. Table 5.10 shows both the scale used to assess the satisfaction of each need and the scale used to evaluate each need's relative importance.

Experiment Description

The designed experiment is consistent in the following steps:

Table 5.10 Scale used by stakeholders to assess the product's ability to satisfy each need and scale used by stakeholders to assess the needs' importance

Scale Used by Stakeholders to assess the Product's ability to satisfy each need.	
Quantitative Indicator	Description
0	Heavily negative compared to the current standard of care or existing competition
1	
2	
3	Significant drawbacks
4	
5	Neutral
6	
7	
8	Significant benefits
9	
10	Highly satisfying, a quantum leap improvement from existing solutions

Scale Used by Stakeholders to Assess the Needs' Importance		
Quantitative Indicator	Description	Notes
1	Not important at all	Needs that are not a real concern
2	Relevant but not Important	Noticeable needs with minor importance
3	Average importance	Needs that are worth being considered
4	Important	Critical unmet needs for at least one category among patients and physicians
5	Very important	Widely unmet needs that are well known to be a burden for patients, physicians, and hospitals

1. Developing a draft list of needs
2. Sharing and checking the list with KOLs for completeness
3. Adjusting and fine-tuning the list using KOLs' feedback
4. Creating a quantitative (1–10) scale of need satisfaction to ask KOLs about the product's ability to satisfy each need
5. Creating a quantitative (1–10) scale for need importance to ask KOLs to assess each need's importance
6. Interviewing at least 10 KOLs and assessing the product's ability to satisfy each need, as well as the relative importance of each need

The primary endpoint of the experiment is to make the decision of whether to continue or dismiss the project, and the secondary endpoint is to identify the most important needs with the weakest satisfaction levels.

Step 5: Testing the Hypotheses

The venture interviewed 10 KOLs and collected their quantitative feedback about the importance of each need as well as the product's ability to satisfy each need.

Data Collected from KOLs

Table 5.11 reports all the data collected by KOLs with regard to their opinion about the importance of each need. Table 5.12 reports all the data collected on KOLs' views about the product's ability to satisfy each need.

Step 6: Making a Decision

Figure 5.17 exhibits the mean satisfaction levels collected from the KOLs and shows that the cross-need's overall average satisfaction is higher than the benchmark set at 7.5. Observing the spider chart makes it easy to notice that no individual's satisfaction falls significantly below the benchmark.

Figure 5.18 shows the plotted values of p*i* (with i = 1,….,17) which represent the probability that the satisfaction of need "*i*" falls above the benchmark of 70%. Each of the 17 probabilities was calculated for each need using a t-student distribution with nine degrees of freedom. It is important to note that, in the spider chart, three needs bear a probability below the benchmark. These needs are the following: preventing diseases, avoiding the cost of visits, and saving time measuring. Despite these exceptions, the cross-needs' average $\hat{p}$ is higher than the benchmark.

In summary:

$$8.62 > v^* = 7.5 \text{ and } \mathbf{86.7\%} > P^* = 70\%$$

It means that:

$$\hat{v} > v^* = 7.5 \text{ and } \hat{p} > P^* = 70\%$$

Therefore, the decision-maker can accept the null hypothesis and continue the project.

Table 5.11 Data collected by KOLs with regard to their opinion about the importance of each need

Physician	Home availability	Continuous physician connection	Elderly alerts after measuring	Avoid cost of visit	Extend practice to family members	Portable device	Minimize clerical errors	Elderly remote checkups	Concurrent measures (5 vital signs)	Manage diseases	Ease of use	Measurement objectivity (data security)	Immediate data (make patient feel safe)	Saving time measuring	Saving time in data entry	Prevent diseases	Shorten hospital stay
1	5	5	2	4	3	3	4	2	4	5	3	5	4	3	3	2	5
2	5	4	3	3	3	3	5	3	4	5	4	5	5	5	5	2	5
3	5	4	3	2	4	4	5	4	4	5	4	5	5	3	5	3	5
4	5	5	4	5	3	3	5	4	3	4	4	5	5	4	4	4	2
5	2	3	4	4	3	5	5	2	4	4	4	5	5	5	5	1	5
6	5	5	4	1	2	3	1	5	3	4	5	5	5	5	4	1	5
7	5	5	3	2	2	3	5	3	4	5	5	4	5	2	5	2	1
8	4	5	4	2	2	4	4	4	4	5	5	4	4	5	5	5	5
9	4	5	5	5	4	5	5	5	5	5	5	5	5	4	5	4	5
10	4	4	5	2	3	4	4	4	4	5	5	4	4	5	5	4	5

Table 5.12 Data collected on KOLs' views about the product's ability to satisfy each need

Physician	Home availability	Continuous physician connection	Elderly alerts after measuring	Avoid cost of a visit	Extend practice to family member	Portable device	Minimize clerical errors	Elderly remote checkups	Concurrent measures (5 vital signs)	Manage diseases	Ease of use	Measurement objectivity (data security)	Immediate data (make patient feel safe)	Saving time measuring	Saving time in data entry	Prevent diseases	Shorten hospital stay
1	10	10	8	9	7	9	9	8	9	10	9	10	9	7	9	7	10
2	10	10	7	9	8	5	10	7	9	10	7	10	9	8	10	8	10
3	10	3	8	7	10	10	10	9	10	10	8	10	9	8	10	8	10
4	10	9	8	10	8	10	9	10	10	10	9	10	10	9	9	10	8
5	10	9	8	9	9	10	10	9	10	9	8	10	8	9	9	6	10
6	7	7	7	5	5	6	5	7	5	7	6	7	4	3	7	6	10
7	10	9	7	7	5	9	10	7	10	9	8	9	8	6	9	5	5
8	10	9	9	5	7	10	9	9	9	10	10	9	9	8	9	10	10
9	10	10	10	8	7	10	10	9	10	10	10	9	9	7	10	8	9
10	9	9	10	7	7	9	9	9	10	10	10	9	9	9	9	9	10

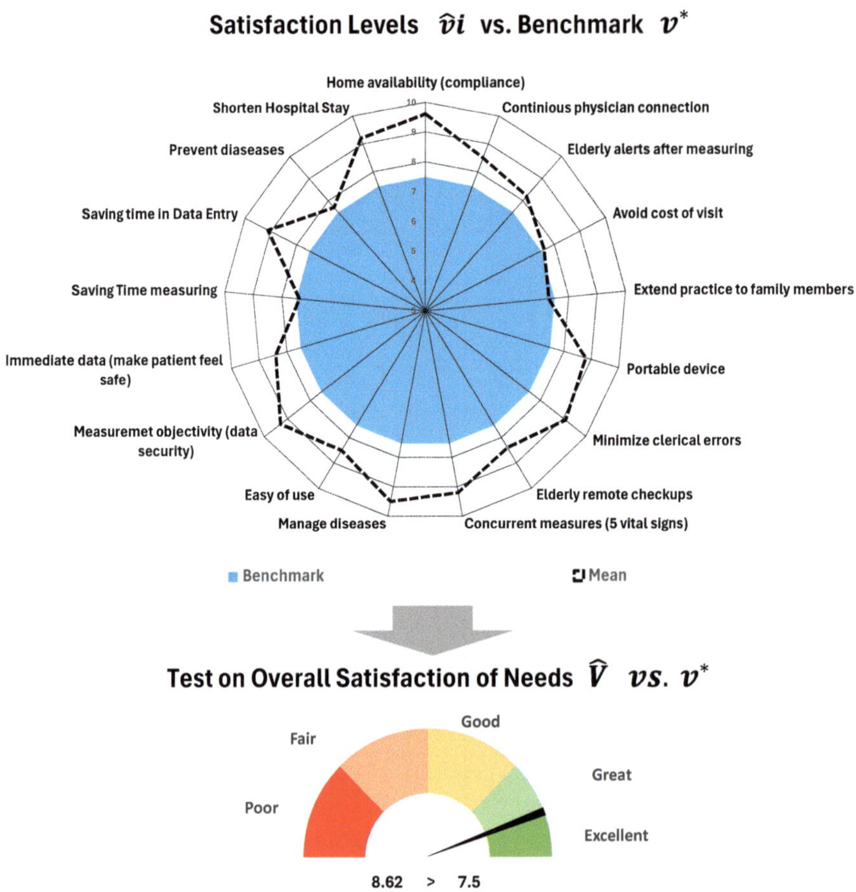

Fig. 5.17 Spider chart showing mean satisfaction levels collected from the KOLs, also showing that the overall average satisfaction is higher than the benchmark

Decision-Related Problem 2: Pivoting Evaluation

It is crucial to note that, in the previous problem, the decision-maker had ignored the ranking of needs by their importance; indeed, the focus was on average satisfaction across all needs. However, in this second case, the decision-maker draws additional information by carefully observing the chart presented in Fig. 5.18. In this figure, the circled areas indicate needs for which the respective p*i* are below the benchmark. In particular, this is the case for the need related to saving time in vital signs measurement, for which the experiment has shown a significant importance for most KOLs. Figure 5.19 shows the average degree of importance that KOLs have assigned to each need. This data-based observation led to a reconsideration of the previous decision and to the consideration of a product pivot that better responds to this highly ranked need.

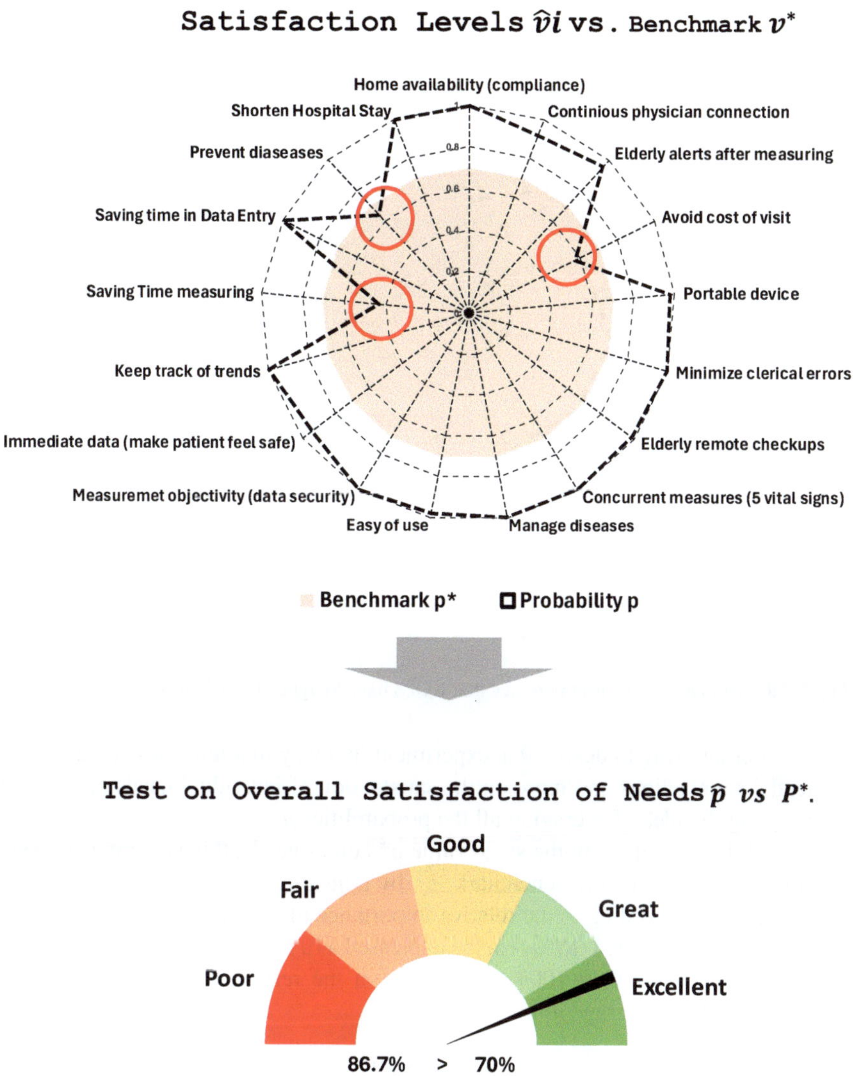

Fig. 5.18 Plotted values of p*i* representing the probability that the satisfaction of need "" falls above the benchmark of 70%. It also shows that the cross-needs average $\hat{p}$ is higher than the benchmark

Later, the venture decided to accept the pivoting inputs from the previous experiment for all the product needs that individually scored lower than the v^* with a probability lower than their respective p_i^* (not p*). This move led to the acceptance of one feature pivoting: the one related to the time required for each measurement of the vital signs. Product development activities were readdressed to prioritize the acceleration of the measuring process, and additional tools were put in place to minimize the required measuring time of the vital signs.

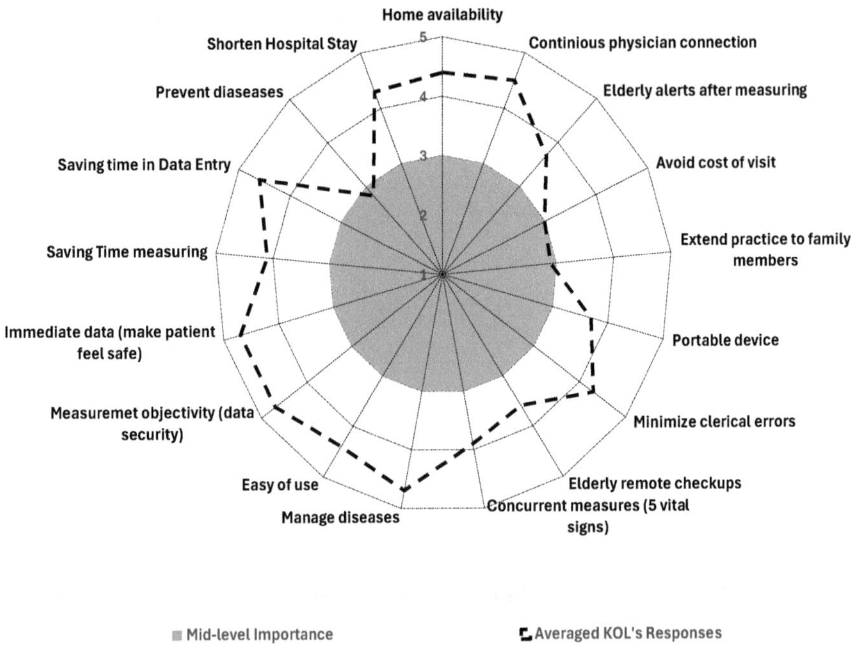

Fig. 5.19 Average degree of importance that KOLs have assigned to each need

An alternate way to design this experiment in a way that relies less on averages of needs' satisfactions and more on the satisfaction of individual needs consists of abandoning the idea of averaging all the probabilities p_i.

Indeed, instead of using the same value p* across needs, this venture can create need-specific levels of the benchmark p_i^* by adjusting (increasing or decreasing) the value of p* according to the relative importance of that specific need. The relative importance of a need ($RI\%_i$) can be expressed as a percentage that is proportionally higher or lower than 100% depending on the respective need's importance compared with the average importance.

$$p_i^* = \text{p} * \text{x}\, RI\%_i$$

Needs with an importance above average will demand a stricter probability benchmark than $\boldsymbol{P*} = 70\%$, while a more relaxed probability benchmark will be used for needs with lower-than-average importance.

After having created need-specific p_i, the old condition $\hat{\boldsymbol{p}} > \boldsymbol{P*} = 70\%$ must leave the place to the following one:

$$\Delta \boldsymbol{p} = \sum_{i=0}^{n} \frac{p_i - p_i^*}{\text{n}} > 0$$

Decision-Related Problem 3: Product Competitiveness

The same venture proposing new vital sign systems was challenged by the potential investor regarding its product's degree of competitiveness.

Problem Description

Question:

Is the product unique or superior and sustainably competitive in the market?

Decision type: Scenario decision

Decision Complexities:

Competitiveness is often assessed by focusing concurrently on a large set of product features. These features must be compared with those of all existing and emerging competition. Most of the competitive data must be collected from public records, which are not necessarily as detailed as needed, from users of competitive products who are not necessarily available, and from regional patent offices.

Step 1: Identifying Scenarios

Scenario H = The product has competitive features supporting an attractive market penetration level.

Scenario L = The product does not have any competitive features that support an attractive level of market penetration.

Step 2: Identifying Actions

- Continue = The process of assessing and quantifying the business opportunity is continued.
- Exit = The project is dropped.

Step 3: Building Scenario–Action Pairs

Since this is a scenario decision, the scenario–action matrix does not contain monetary values. Therefore, Table 5.9 can be used again to represent the qualitative scenario–action pairs.

Step 4: Developing Testable Hypotheses

An analysis performed on competition and on the key, meaningful competing dimensions helped with the identification of two primary ranking factors. One is the ease of use for the solution (EOU), and the second is the combination of all the measured vital signs, implying how many important vital-sign outputs each solution provides (VSS). These two ranking factors are quantified on a scale from 1 to 5.

$$\text{EOU} = \text{Ease of use } (0-5)$$

$$\text{VSS} = \text{Vital signs score } (0-5)$$

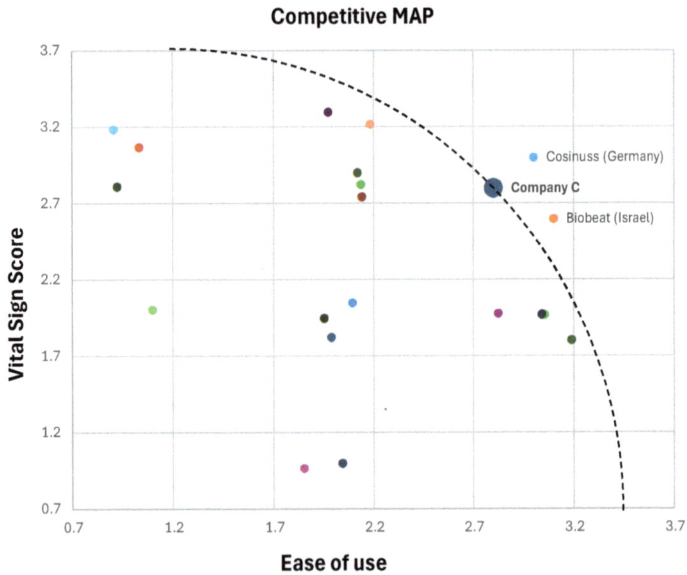

Fig. 5.20 The competitive map positions competitive products on the two key dimensions: ease of use and vital sign score

These same parameters are used as coordinates to place competitive products on a two-dimensional map that helps the decision-maker to formulate testable hypotheses consistent with the definition of a measurable competitive advantage.

H1 (alt): *When compared to all the competitive products on the* ***two-criteria ranking map,*** *this product's distance from the chart origin is at least 0.5 points higher than that of the best competitive alternatives.*

$$MAX_{i\neq 1}\left(\sqrt{EOU_{i\neq 1}^{\;\;2}+VSS_{i\neq 1}^{\;\;2}}\right)+0.5<\sqrt{EOU_{i=1}^{\;\;2}+VSS_{i=1}^{\;\;2}}$$

H0 (null): *When compared to all the competitive products on the* ***two-criteria ranking map,*** *this product's distance from the chart origin is* ***not*** *at least 0.5 points higher than that of the best competitive alternatives.*

$$MAX_{i\neq 1}\left(\sqrt{EOU_{i\neq 1}^{\;\;2}+VSS_{i\neq 1}^{\;\;2}}\right)+0.5\geq\sqrt{EOU_{i=1}^{\;\;2}+VSS_{i=1}^{\;\;2}}$$

Experiment Description

To make a final decision, the decision-maker designs an experiment that will help to test the two hypotheses. The experiment is composed of the following steps:

1. Identifying all products belonging to direct competition
2. Rating all competing products on the basis of the following:
   (a) Ease of use (EOU, scale of 0–5)
   (b) Vital sign score (VSS, scale of 0–5)

3. Positioning the concerned product and each competing product on a two-dimensional chart: EOU and VSS
4. Assessing the relative product positioning on the map

Step 5: Testing Hypotheses
The two-dimensional chart is shown in Fig. 5.20 together with all the positions of the competing products. The circle labeled "Company C" corresponds to the product and the venture tested against competition, and the dotted curve holds points of equal distance from the Cartesian chart's origin. The product identified with "Company C" is clearly inferior to at least two of the competitive products.

Step 6: Making a Decision

$$MAX_{i\neq 1}\left(\sqrt{EOU_{i\neq 1}^{2}+VSS_{i\neq 1}^{2}}\right)+0.5\geq \sqrt{EOU_{i=1}^{2}+VSS_{i=1}^{2}}$$

Thus, the experiment suggests we cannot reject the null hypothesis. The valid decision is to exit. Later, the company decided to challenge the two "superior" competitors through product pivots.

Decision-Related Problem 4: Product and Company Competitiveness
Despite the result of the previous scenario decision, the venture management is convinced that, if a product-based superiority cannot be demonstrated, a venture's superiority can be proven. In other words, management believes that a more comprehensive competitive test, including other features of the company beyond the product, would give a different result. In particular, management is considering including elements such as their venture's advancements in regulatory registration and in building a global commercial organization.

Problem Description
Question: Is my venture unique or superior and sustainably competitive in the market?

Decision type: Scenario decision

Decision complexities: Competitiveness must be assessed by focusing on a broader range of product and venture features. The decision about which additional features to add in the comparison is critical and cannot be completely arbitrary; otherwise, entrepreneurs are inclined to include only company features they feel strongly about.

Comparing a higher number of venture and product features requires the collection of a larger volume of information on all existing and emerging competition. Much of the related information is not easy to find and its accuracy is difficult to verify.

Step 1: Identifying Scenarios

Scenario H = The product holds competitive features that are able to support an attractive level of market penetration.

Scenario L = The product does not have any competitive features that are able to support an attractive level of market penetration.

Step 2: Identifying Actions

The simple, actionable context selected by the decision-maker consists of the two most common items:

- Continue = The process of assessing and quantifying the business opportunity is continued.
- Exit = The project is dropped.

Step 3: Building Scenario–Action Pairs

Since this is a scenario decision, the scenario–action matrix does not contain monetary values. Therefore, Table 5.9 can be again used to represent the qualitative scenario–action pairs.

Step 4: Developing Testable Hypotheses

The venture's management finally decides to extend the definition of competitive advantages to the following two dimensions:

(i) Regulatory status in the EU and the USA.
(ii) The company's development status and existing global organization.

The overall product-company ranking is finally built using four weighted numeric criteria: two strictly related to the product and two broader ones.

Numeric Criteria	Weights
1. EOU = Evolution of feedback generation (0–5)	30%
2. VSS = Compatibility with the current standard of care (0–5)	30%
3. Regulatory status in the EU and the USA (0–5)	20%
4. Company's development status (0–5)	20%

At this point, the decision-maker must formulate testable hypotheses. In this case, the decision-maker decides to consolidate the above four comparative parameters into a single one by averaging them accounting also for their individual weights.

In this way, one value, a competitive index (***C***i) included in the interval (0–5), corresponds to the degree of relative competitiveness of each respective company where $i = 1$ corresponds to the decision maker's venture of interest.

The final expression of the hypotheses results as follows:

H1 (alt): *When compared to all the other ventures, our competitiveness is the highest when:*

$$\boldsymbol{MAX}_{i \neq 1}\left(\boldsymbol{C}_{i \neq 1}\right) < \boldsymbol{C}_{i=1}$$

H0 (null): *When compared to all the other ventures, our competitiveness is NOT the highest when:*

$$\boldsymbol{MAX}_{i \neq 1}\left(\boldsymbol{C}_{i \neq 1}\right) \geq \boldsymbol{C}_{i=1}$$

Step 5: Testing the Hypotheses

The experiment consisted of in-depth research focusing on the quantification of competitive data pertaining to key competitive features, regulatory status, and the development stage of each competitor's commercial channels.

Table 5.13 summarizes the collected data and the calculated competitiveness index ***C***i in the column called "overall ranking."

Step 6: Making a Decision

The venture's positioning in the chart presented in Fig. 5.21 ("Company C") does not allow for rejecting the null hypothesis. The decision is, therefore, to exit.

As in many other real-life cases, the startup did not exit, but finally identified product pivots that were able to change the product positioning in the competitive ranking of both Figs. 5.20 and 5.21.

5.5.3 Example 3: Value Proposition—Epidural Safety

This section presents, again, a real-life example of a scenario decision—A decision that corresponds to the election of a vital business assumption about the strength of the product's value proposition.

The Context

Epidural anesthesia is a regional anesthesia that blocks pain in a particular region of the body (AADII Hospital, 2025). Epidural anesthesia systems include epidural catheters, epidural anesthesia needles, and other anesthesia accessories. These sets are ideal for use in chronic pain clinics for bolus injections into the epidural space. Epidural anesthesia involves the injection of a local anesthetic and an opioid into the lumbar epidural space, from which it is gradually diffused across the dura, into the subarachnoid space. Here, it acts primarily on the spinal nerve roots and, to a lesser extent, on the spinal cord and paravertebral nerves. Despite its widespread availability, this highly subjective technique depends on the anesthetist's experience and

Table 5.13 Summary of the collected competitive scores and the calculated competitiveness index Ci

Tier Segmentation	Product	Vital Sign Score	Ease-of- Use	Regulatory	Company Stage	Overall Ranking
Tier 1	**Company C**	3	3	1	1	2.2
	Biobeat (Israel)	3	3	2	1	2.4
	Caretaker (USA)	3	2	2	1	2.1
	Cosinuss (Germany)	3	3	1	2	2.4
	Winmedical (Italy)	3	2	1	1	1.9
	VitalConnect VitalPatch (UK)	3	2	2	2	2.3
	Vital Tracer (Canada)	3	3	0	1	2
Tier 2	Viatom Checkme Suit (China)	2	3	2	2	2.3
	Current Health (UK)	2	2	1	1	1.6
	Osler MD (USA)	2	2	0	1	1.4
	Hexoskin (Canada)	2	3	1	2	2.1
	Binah.ai (Israel)	2	3	0	1	1.7
	Bips Medical (Israel)	2	2	0	1	1.4
	drMAP watch (Hungary)	2	3	0	1	1.7
	Sensium (UK)	2	3	2	2	2.3
Tier 3	Apple watch (USA)	1	3	2	3	2.2
	AMD (USA)	2	1	1	3	1.7
	Omron Complete™ (Japan)	1	2	1	3	1.7
	Riester RVS-100 (Germany)	2	1	2	3	1.9
	Keenly Health (USA)	1	3	0	1	1.4
	Bodytrack (UK)	1	3	0	1	1.4

expertise. The failure rate for this procedure is significant. In addition to the suboptimal drug dose, more than 40% of epidural failures occur due to the incorrect positioning of the epidural catheter, and an additional 15% of the failures happen due to the incorrect identification of the epidural space (Arendt & Segal, 2008).

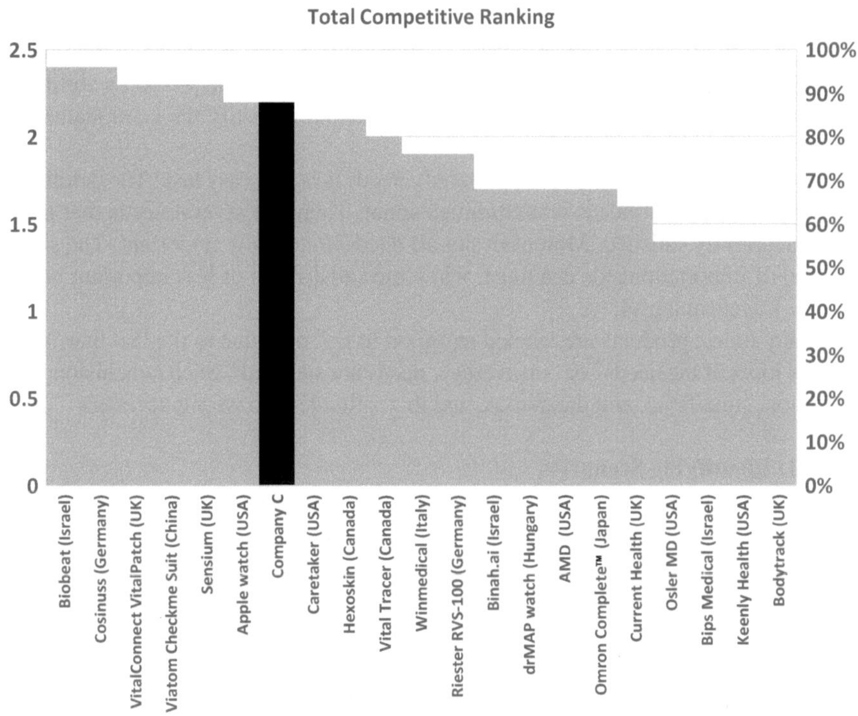

Fig. 5.21 Bar chart ranking competitors by ***C*i** score

Company A has developed and patented a new kit for epidural anesthesia, which can detect, in real time, the attainment of the epidural space and monitor the progress and correct positioning of the catheter inside the epidural canal, significantly reducing the risks associated with the epidural procedure.

The Decision-Related Problem

Suppose that the entrepreneur has ranked the multitude of decision-related problems that are relevant to this innovative idea and that this exercise has pointed the attention directly to the value proposition's ability to fit actual medical needs.

It is very common for wise entrepreneurs to prioritize problems related to value propositions at the beginning of the venture's life cycle. Indeed, an articulation of the value proposition, which is the value created for users by technology, is the key starting point for the evaluation of a new idea in medical technology (Davey et al., 2011).

The Decision-Related Problem

The decision-related problem may be stated as follows:

Does the product's ability to satisfy the currently unmet medical needs justify continuing to work on this project?

As described in Chap. 2, since this decision does not involve the commitment to an investment, it falls in the category of scenario decisions. Here, the entrepreneur decides whether to accept or reject a business assumption: The product's ability to meet unmet medical needs is sufficiently compelling to justify the continuation of the project.

Establishing a product's ability to satisfy needs is not an easy task. The definition of the satisfaction of needs is multidimensional: There are several needs that must be concurrently satisfied. Moreover, not all needs are equally important: The satisfaction of important needs is a must, while the satisfaction of less important needs can be less meaningful.

Many times, products are labeled as "good to go" only due to the fact that "they satisfy most of the needs" or "on average, needs are satisfied." Such conclusions are common, superficial, and dangerous, and they often lead to wrong decisions.

Step 1: Identifying Scenarios

As explained in Chap. 4, the first step is to identify scenarios. In this case, we can identify two scenarios with the letters "H" and "L":

- "H": The product is capable of responding to the most critical unmet medical needs (i.e., the value proposition has high efficacy).
- "L": The product is **not** capable of responding to the most important stakeholders' needs (i.e., the value proposition has low efficacy).

Step 2: Identifying Actions

For simplicity, we now exclude product pivoting options and assume that we have only two possible actions (decisions) to consider:

- Continue: The entrepreneur continues assessing and exploring the business opportunity.
- "Exit": The entrepreneur drops the project.

Table 5.14 Qualitative scenario–action table for the decision related to the continuation of the project based on the predicted efficacy of the product's value proposition

		Two Scenarios	
		Value proposition has high efficacy	Value proposition has low efficacy
		H(p)	L(1−p)
Two Possible Actions	Continue	+++	--
	Exit	0	0

Step 3: Building Scenario–Action Pairs

The scenario–action pairs are represented in Table 5.14: As discussed in Chap. 5, in the case of decision options that do not involve possible investments, it is difficult to fill these tables with actual numbers. Therefore, we use the symbols "+" and "-" to represent the positive and negative economic outcomes of the scenario–action pairs.

Step 4: Developing Testable Hypotheses

As explained in the previous sections, developing testable hypotheses involves the identification of the scenario-defining variable V and the definition of the benchmark value V*.

In this case, the decision-maker can quantify the degree of satisfaction of product users using a variable V, which ranges from 1 to 10.

0 = Very dissatisfied
2.5 = Dissatisfied
5 = Neutral
7.5 = Satisfied
10 = Very satisfied

The value V* must be arbitrarily set by the decision-maker: Let's suppose that the decision-maker chooses V* = 7.5 as a threshold, placing it exactly between a neutral and a very satisfied outcome.

In this case, "value proposition efficacy" is high only if V > V* = 7.5.

On the other hand, using only a single variable (V) to assess the ability of a medical product to satisfy multiple unmet needs is an excessively approximative approach. Indeed, the strength of the value proposition depends on the ability of the product to satisfy a significant number (n) of unmet needs concurrently.

Therefore, to be more precise, instead of defining variable V as the one that quantifies the overall assessment of product satisfaction, the entrepreneur seeks to have visibility into the product's ability to satisfy each of the "n" key unmet medical needs. In this way, variable V is defined as the average level of satisfaction related to all the "n" individual medical needs. In short, the decision-maker will first identify the "n" most relevant unmet medical needs and then estimate the respective level of satisfaction indicated as follows:

$$v_i \text{ with } i = 1, \ldots, \mathrm{n}$$

$$V = \sum_{i=1}^{n} \frac{v_i}{n} > \mathrm{V}*$$

Only after assessing the product performance for each need can the decision-maker calculate the average ($\hat{\mathrm{v}}$) and test it against the threshold V*:

This inequality corresponds to the first condition: condition A.

Averaging datasets that are characterized by unknown distributions is a very risky practice, as it can mask very important information (Savage, 2002). For instance, the average condition may prove to be satisfied owing to the high satisfaction of unimportant needs and a notable dissatisfaction with one very important need.

Therefore, using condition A alone to distinguish high scenarios from low scenarios and to build testable hypotheses appears to be a dangerous simplification.

For this reason, the decision-maker prefers to include in the definition of high and low scenarios the concurrent satisfaction evaluation of the two most important needs (need "y" and need "k"):

$$v_y \,\&\, v_k$$

Therefore, to define "high" and "low" scenarios, the decision-maker adds two new conditions that complement the more general condition A. Consequently, the key conditions are as follows:

Condition A:	$V = \sum_{i=1}^{12} v_i > \boldsymbol{V}^*$
Condition B:	$v_y > \boldsymbol{V}^*$
Condition C:	$v_k > \boldsymbol{V}^*$

With this in mind, the alternate hypothesis can be rewritten as follows:

Alt: The value proposition has **high efficacy** if conditions A, B, and C are concurrently satisfied:

$$V > \boldsymbol{V}^* \text{ and } \quad v_y > \boldsymbol{V}^* \text{ and } v_k > \boldsymbol{V}^*$$

Null: The value proposition has **low efficacy** if one or a combination of conditions A, B, and C are not satisfied.

$$\hat{\mathrm{v}} \leq \boldsymbol{v}^* \text{ or } v_y \leq \boldsymbol{v}^* \text{ or } v_k \leq \boldsymbol{v}^*$$

Step 5: Testing the Hypotheses

Entrepreneurs know that an experiment helps gather refined signals for hypothesis testing. The experiment provides a dataset that helps estimate the variables to be tested: V, v_y, and v_k. Their respective estimations are indicated with the symbols shown in Table 5.15.

Later, the entrepreneur uses the estimated value to test the conditions of the hypothesis.

Alt: The value proposition has **high efficacy** if the three disequalities below are concurrently satisfied:

Table 5.15 Relevant variables for the decision and their respective predicted values

	Variable to be predicted	Variable's prediction
Average of all needs satisfaction	V	$\hat{V}$
Satisfaction of the most important need	v_y	$\hat{v}_y$
Satisfaction of the second-most important need	v_k	$\hat{v}_k$

$$\hat{v} > \boldsymbol{V}^* \text{ and } \hat{v}_y > \boldsymbol{V}^* \text{ and } \hat{v}_k > \boldsymbol{V}^*$$

Null: The value proposition has **low efficacy** if at least one or a combination of the conditions above is not satisfied.

Once the experiment is complete and the data is collected, how are these estimated values calculated? The level of satisfaction expressed by physician $\boldsymbol{j}$ about medical need $\boldsymbol{i}$ is indicated as follows:

$$\boldsymbol{vj,i}$$

The average satisfaction level for need $\boldsymbol{i}$ across all the "m" interviewed physicians is indicated as follows:

$$\hat{v}_i = \sum_{j=1}^{m} \frac{v_{j,i}}{m}$$

The overall cross-needs average satisfaction level can be expressed as follows:

$$\hat{v} = \sum_{i=1}^{n} \frac{\hat{v}i}{n} = \sum_{i=1}^{n}\sum_{j=1}^{m} \frac{v_{j,i}}{m}$$

Introducing p

Let's suppose that the experimental data confirms

$$\hat{v}_y > \boldsymbol{V}^*$$

At this point, the decision-maker who follows the scientific approach seeks to calculate the probability p that, given the estimated value $\hat{v}_y$ and the satisfaction of the above inequality, v_y is higher than V* in real life. In other words, the decision-maker wants to calculate a probability expressed as follows:

$$p_y\left(v_y > \boldsymbol{V}^* \middle| \hat{v}_y\right)$$

Once the minimum acceptable threshold value of p, which we indicate with P*, is established, the decision-maker wants to verify if

$$p_y > \boldsymbol{P}^*$$

The same reasoning applies to each of the three conditions. For instance, the probability that $V > \boldsymbol{V}^*$ can be expressed as follows:

$$p = \mathrm{p}\left(V > \boldsymbol{V}^* \middle| \hat{v}\right) > \boldsymbol{P}^*$$

The decision-maker can, for instance, decide to approximate the probability p with the average of the probabilities p_i relative to all individual needs. This average and the respective inequality can be expressed as follows:

$$\hat{p} = \sum_{i=0}^{n} \frac{p_i}{\text{n}} > \boldsymbol{P}^*$$

Setting the Value of p*

Since this is not an investment decision, the entrepreneur does not dispose of the values of r and z, otherwise shown in the scenario–action table. It makes the straight calculation of p* as a function of r and z impossible.

For now, let's suppose that the decision-maker is concerned about the level of confidence that p* generates in potential investors. For this reason, the entrepreneur decides to use one of the scales called "perception-of-probability words" described in Fig. 5.6. Let's suppose that, on the basis of the observation of the elected scale, the decision-maker chooses to set p* at 70%, corresponding to the common perception of events that are "likely" to happen.

Let's now rewrite the testable hypotheses, including the conditions related to the probabilities. The final and complete hypotheses to be tested are as follows:

Alt: The value proposition has **high efficacy** if the three conditions A_p, B_p, and C_p are concurrently satisfied.

Condition A:	$\hat{v} > \boldsymbol{V}^*$ **and** $\hat{p} = \sum_{i=0}^{n} \frac{p_i}{\text{n}} > \boldsymbol{P}^*$
Condition B:	$\hat{v}_y > \boldsymbol{V}^*$ **and** $p_y = \text{p}\left(v_y > \boldsymbol{V}^* \middle| \hat{v}_y\right) > \boldsymbol{P}^*$
Condition C:	$\hat{v}_k > \boldsymbol{V}^*$ **and** $p_k = \text{p}\left(v_k > \boldsymbol{V}^* \middle| \hat{v}_k\right) > \boldsymbol{P}^*$

Null: The value proposition has **low efficacy** if at least one or a combination of the conditions A, B, and C is not satisfied.

The Experiment

The following list summarizes how the entrepreneur decides to perform the experiment:

1. Develop a draft list of needs:
   (a) Sharing and discussing the list with KOLs
   (b) Adjusting and fine-tuning the list using KOLs' feedback
2. Create a numeric scale of need satisfaction
3. Create a numeric scale for the importance of needs
4. Interview at least 10 KOLs and assess the product's ability to satisfy each need as well as the relative importance of each need.
5. Collect and analyze the experimental results

For this medical device, the entrepreneur identified 12 unmet medical needs ($n = 12$):

Need #1	Navigation toward the epidural space
Need #2	Precise identification of the epidural space
Need #3	Catheter kink avoidance
Need #4	Much longer insertion of a catheter
Need #5	Verification of the final positioning of the catheter's tip
Need #6	Avoidance of secondary catheter migration
Need #7	Compatibility with the current practice
Need #8	Sterility-friendliness
Need #9	Execution time
Need #10	Simplicity
Need #11	Training-friendliness
Need #12	Affordable device price (cost)

The numeric scale for need importance is the following:

1 = Irrelevant

2 = Not important

3 = Relevant

4 = Important

5 = Critical

The numeric scale for the product's ability to satisfy a certain need is the following:

1 = Critically detrimental

2 = Detrimental

3 = Largely misses expectations

4 = Does not meet expectations

5 = Almost meets expectations

6 = Just meets expectations

7 = Meets expectations

8 = Fully meets expectations

9 = Exceeds expectations

10 = Outstanding

Table 5.16 presents the quantitative responses collected from physicians regarding the importance of each need. Table 5.17 outlines the quantitative feedback from the same physicians regarding the level of satisfaction for each of the needs.

Step 6: Making a Decision

Figure 5.22 shows the average importance of the needs compared to the benchmark. Figure 5.23 outlines the average degree of satisfaction with respect to each need in comparison to the benchmark. Through simple statistical analysis and the use of a t-student distribution, it's easy to calculate the probability that each need's satisfaction level exceeds the benchmark.

Table 5.16 Degree of importance of all needs (0–5) as assessed by physicians

	Importance of Needs (0–5)											
	Navigation toward the epidural space	Precise identification of the epidural space	Catheter kink avoidance	Much longer insertion of a catheter	Verification of the final positioning of the catheter tip	Avoidance of secondary catheter migrations	Compatibility with the current practice	Sterility-friendliness	Execution time	Simplicity	Training-friendliness	Affordable device cost
Physician #1	3.5	5	3.6	1.9	4.3	4.3	5	2.2	4.7	3.4	3.2	3.5
Physician #2	3.5	5	3.2	2.1	4.5	4.8	4	1.7	3.4	3.3	2.7	4.5
Physician #3	2.7	5	4.8	2.3	4.6	4.2	5	2.1	4.6	3.2	3.4	4.3
Physician #4	3.4	4	3.6	1.6	4	3.4	5	2.1	4.2	3.9	2.4	3.3
Physician #5	3.5	4.3	4.3	2	4.7	4.3	4.7	2.3	3.6	3.4	2.9	3.2
Physician #6	3.2	5	3.4	1.7	3.5	3.7	5	2	4	3.9	2.8	4
Physician #7	2.9	4.4	3.8	2.2	3.4	3.8	4.6	1.9	4.6	3.6	2.4	4.7
Physician #8	3.5	4.7	3.3	2.3	3.7	3.5	5	1.9	4	3.4	3.2	4.1
Physician #9	3.5	5	4.1	1.8	4.4	4.2	5	1.6	4	3.3	2.7	4.4
Physician #10	3.3	5	3.4	1.9	4.7	4.1	5	1.9	3.5	4	3	4.1

Table 5.17 Level of needs satisfaction (1–10) assessed by physicians

	Level of Needs Satisfaction (1–10)											
	Navigation toward the epidural space	Precise identification of the epidural space	Catheter kink avoidance	Much longer insertion of a catheter	Verification of the final positioning of the catheter tip	Avoidance of secondary catheter migrations	Compatibility with the current practice	Sterility-friendliness	Execution time	Simplicity	Training-friendliness	Affordable device cost
Physician #1	8.2	8.9	8.2	9	7	7.8	7	9.2	7.7	8.3	6.8	7.8
Physician #2	8	8.7	7.9	9.3	7	7.9	7	9	7.7	8	7.1	7.8
Physician #3	8	9.2	8.1	8.8	7	8.1	7	8.7	8.1	8.1	7	8.3
Physician #4	8.2	9.2	8.3	8.8	7	8.3	7	8.6	8	7.7	6.8	7.8
Physician #5	7.9	9.2	7.8	8.6	7	7.8	6	9.2	8.3	7.7	7.1	8.2
Physician #6	8.3	9.4	7.8	9.4	7	8	7	8.6	7.7	8.3	7.1	8.3
Physician #7	8	9.3	7.9	9.2	6.1	7.9	7	9.1	8.3	8.2	7	8.2
Physician #8	8.3	8.9	8.2	9.3	7	8	7	9.2	7.8	7.7	7.2	7.7
Physician #9	8.3	8.8	8.1	9	7	8.2	6	9.2	7.7	7.9	7.1	8.3
Physician #10	7.8	9.4	8.3	8.8	7	7.8	6	9	8	8	6.8	8.1

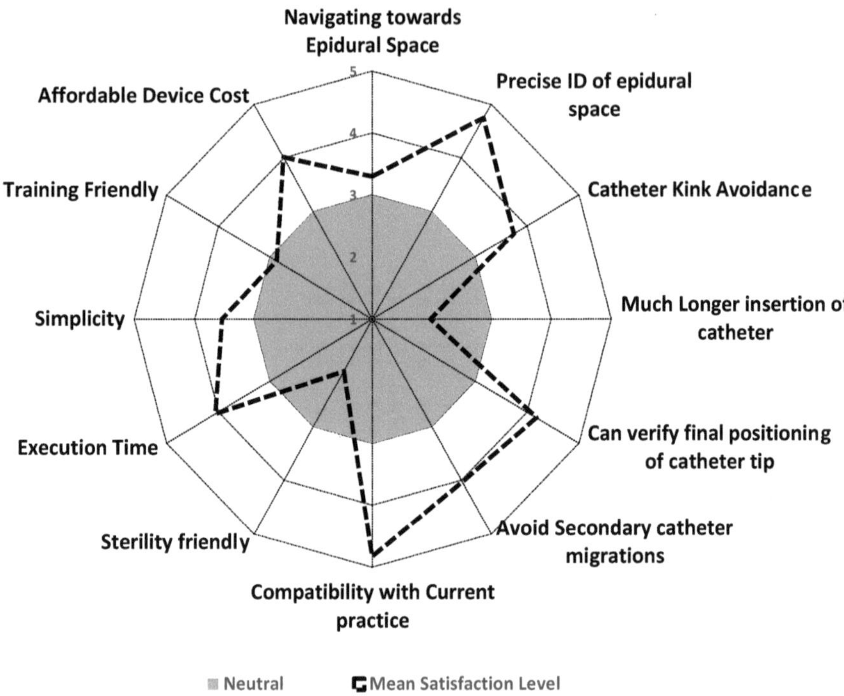

Fig. 5.22 Spider chart that shows the average importance of the needs compared to the benchmark

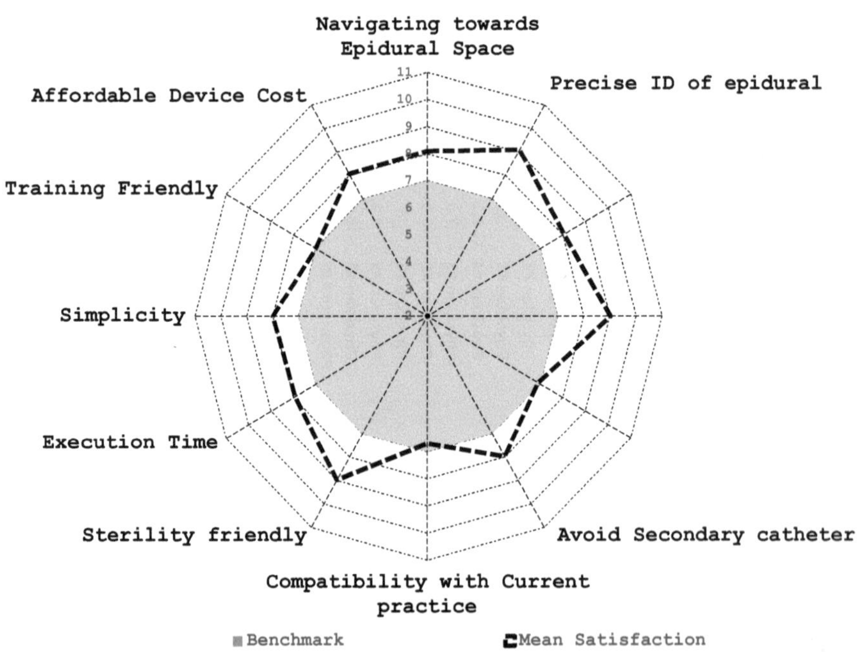

Fig. 5.23 Spider chart that compares the average degree of satisfaction in relation to each need in comparison to the benchmark

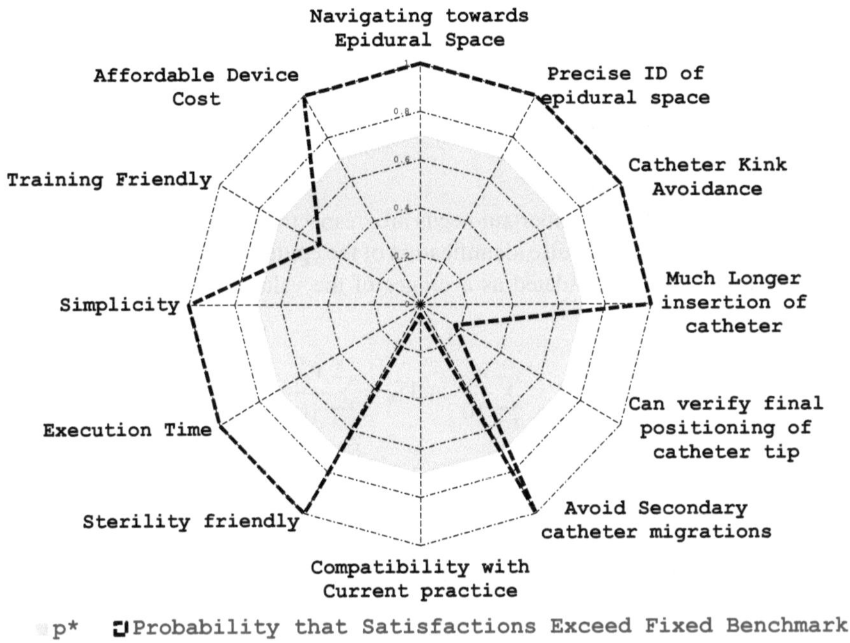

Fig. 5.24 Spider chart that shows the probability that the satisfaction of each need exceeds its respective benchmark

Figure 5.24 presents precisely the probability of each need's satisfaction exceeding the benchmark in comparison with the benchmark probability. It shows the probability that the satisfaction of each need exceeds its respective benchmark.

From Fig. 5.22, the decision-maker can visually assess the following: y = 2 and k = 7. Therefore, the alternate hypothesis to be tested must include to $\hat{v}_2$ cand to $\hat{v}_7$, and thus, the alternate hypothesis can be written as follows:

Alt: The value proposition has **high efficacy** if the three conditions *A*, *B*, and *C* are concurrently satisfied.

Condition *A*	$\hat{v} > \boldsymbol{V}^*$ and $\hat{\boldsymbol{p}} > \boldsymbol{P}^*$
Condition *B*	$\hat{v}_2 > \boldsymbol{V}^*$ and $p_2 > \boldsymbol{P}^*$
Condition *C*	$\hat{v}_7 > \boldsymbol{V}^*$ and $p_7 > \boldsymbol{P}^*$

Testing Condition A

The overall average satisfaction level across all the needs as well as all the interviewed physicians is as follows:

$$\hat{v} = \sum_{i=1}^{12} \frac{\hat{v}i}{n} = \sum_{i=1}^{12}\sum_{j=1}^{10} \frac{v_{j,i}}{m} = 8.09 > V* = 7.5$$

$$\hat{p} = \sum_{i=0}^{12} \frac{p_i}{\mathrm{n}} = 80.9\% > \boldsymbol{P}^* = 70\%$$

Therefore, condition A is satisfied.

Testing Conditions B and C

The first and second most important needs are, respectively, the compatibility with current practices and the precise identification of the epidural space (y = 2 and k = 7).

$\hat{v}_2$ and $\hat{v}_7$ can be calculated as averages of the values of the second and the seventh columns of:

$$\hat{v}_2 = \sum_{j=1}^{10} \frac{v_{j,2}}{10} \text{ and } \hat{v}_7 = \sum_{j=1}^{10} \frac{v_{j,7}}{10}$$

$$\hat{v}_2 = 9.1 \text{ and } \hat{v}_7 = 6.7$$

p_2 and p_7 can be calculated using the second and the third columns, applying a one-sided t-test with nine degrees of freedom.

$$p_2 = 4.1\% \text{ and } p_7 = 4.1\%$$

Consequently, condition B is satisfied:

$$\hat{v}_2 = 9.1 > \boldsymbol{v}^* = 7 \text{ and } p_2 = 100\% > \boldsymbol{P}^* = 70\%$$

However, condition C is **not** satisfied:

$$\hat{v}_7 = 6.7 < \boldsymbol{v}^* = 7 \text{ and } p_7 = 0.0\% < \boldsymbol{P}^* = 70\%$$

The null hypothesis is accepted since the most important need (condition C) missed both the satisfaction and the probability benchmark. Thus, the final decision is to exit.

5.5.4 Example 4: Competitiveness in Infusion Systems

The Context

In 2020, the global infusion therapy market was valued at around US$7.5 billion, showing a 15% decrease from 2019. Despite the effects of the COVID-19 pandemic, the global market size is expected to increase over the forecast period and reach US$11.2 billion by 2027. Currently, four leading companies control most of the global infusion therapy market share: Becton Dickinson, B. Braun, Baxter, and ICU Medical.

Infusion therapy devices represent an integral and essential component of many treatment plans. Most countries worldwide still use gravity drip infusions more often than pump-based infusions. However, this practice is expected to slowly shift as the sales of all varieties of infusion pumps continue to grow in many regions, including Europe. Infusion pumps offer many benefits over gravity infusions, such as increased safety and more precise control over the treatment process. The main drawbacks of these devices are higher costs and an increased necessity for clinician training.

Our observed French startup developed a small reusable medical device that allows infusion lines to be connected to syringes and other critical lines in a completely sterile manner, without needles. Despite the effort, all potential investors (mainly VCs of the largest market players) declined the offer to invest, as the connector appeared not to be better than the market average. This assessment was mainly based on the connector's usability features.

Our startup has, therefore, decided to design a new connector with the commitment of demonstrating a quantum leap of improvement compared to the previous version. Once the new connector is developed, the company must decide whether it is "better enough" to continue the project and approach potential investors or work on another product pivot. In experiments performed to assess the usability of a new, needle-less connector, the product outperformed expectations and benchmarks. With this point in mind, venture management has developed even more interesting questions.

The Decision-Related Problem

Decision-related problem: Is the new connector better than those of the competition? Is it superior enough to support the seeking of investments from strategic investors?

Or should the venture revise the product once again through a product feature pivot?

Step 1: Identifying Scenarios

The two possible scenarios identified are "H" (competitive edge is high) and "L" (competitive edge is low).

H = The product holds superior critical competitive features compared to the existing competition.

L = The product does not have any superior critical competitive features compared to the existing competition.

Step 2: Identifying Actions

These are the two possible actions selected by the decision-maker: to continue (retaining the product's current features) or to exit.

$$\text{Investment round} = \text{Actively presenting the idea to strategic investors}$$

$$\text{Exit} = \text{Exit}\left(\text{to later possibly evaluate pivoting options}\right)$$

Step 3: Building Scenario–Action Pairs

The scenario–action matrix does not contain any monetary values. Therefore, Table 5.9 can be again used to represent the qualitative scenario–action pairs.

Step 4: Developing Testable Hypotheses

In order to assess competitive superiority, decision-makers had to set the dimensions on which the products could be compared. The choice, guided by customer interviews, focused on the following nine areas of customer satisfaction:

- Cleanness
- Overall perception of controllability
- Overall perception of safety
- Overall perception of solidity
- Translucency
- Transparency

In the field of optics, transparency (also called "pellucidity" or "diaphaneity") is the physical property of allowing light to pass through the material without being scattered. Translucency (also called "translucence" or "translucidity") is a super-set of transparency: it allows light to pass through but does not necessarily (again, on the macroscopic scale) follow Snell's law; the photons can be scattered at either of the two interfaces where there is a change in index of refraction or internally. In other words, a translucent medium allows for the transport of light, while a transparent medium enables not only the transport of light but also image formation. The opposite property of translucency is opacity. Transparent materials appear clear, with the overall appearance of one color or any combination leading up to a brilliant spectrum of every color. The hypotheses defined by the venture management are as follows:

Definition: *The comparative scale ranges from 1 to 5 (corresponding to the qualitative evaluation scale: much worse, worse, same, better, and much better).*

H1 (alt): *While comparing all* ***11*** *competing products using* ***the comparative scale from 1 to 5, the average rating across competitors for each of the six identified features must be higher than 3.5****.*

H0 (null): *While comparing all* ***11*** *competing products using* ***the comparative scale from 1 to 5, the average rating across competitors for one or more of the six identified features is lower than 3.5****.*

Step 5: Testing the Hypotheses

The experiment included several large international focus groups designed to compare each of the company's products with each of the 11 competitive ones (Fig. 5.27). The experiment was called "Our Product vs. 'the Rest of the World'" (Fig. 5.25).

The experiment was designed to predict the degree of perceived superiority of the venture's new product against those of its competitors.

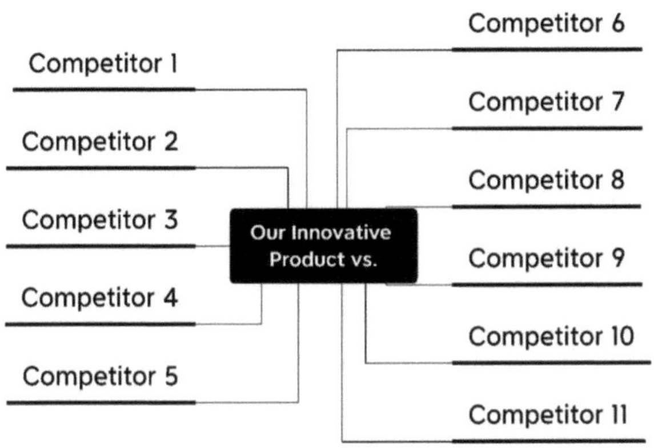

Fig. 5.25 The type of matrix that was used to collect the data during the focus groups

The variable $v_{f,\,c}$ corresponds to the degree of superiority of the venture's new product against the competitor “” concerning feature “ *f* .”

$v_{f,\,c,\,e}$ represents the evaluation performed by expert “*e*” for the product feature “*f*” compared to competitor “*c*.”

With: $f = 1,2\ldots,n = 6\ c = 1,2,\ldots,m = 11\ e = 1,2,\ldots,\ h = 20$

The overall feedback across all the analyzed product features was generally favorable to the venture's new product, but the null hypothesis must be carefully tested.

For a specific feature “*f*,” the average assessed performance of the new product against the competitor “*c*” can, therefore, be expressed as follows:

$$\hat{v}_{f,c} = \sum_{e=1}^{h} \frac{v_{f,c,e}}{h}$$

The overall average performance for feature f across all competitors can be expressed as follows:

$$\hat{V}_f = \sum_{c=1}^{m} \frac{\hat{v}_{f,c}}{m}$$

Using the t-student distribution with the probability for feature f, the performance against competitor “c” was calculated using the focus group feedback:

$$p_{f,c} = p\left(v_{f,c} > V^* \middle| \hat{v}\left(\overline{f},\overline{c}\right)\right)$$

$$\hat{p}_f = \sum_{c=0}^{n} \frac{p_{f,c}}{\text{n}}$$

The benchmark value for $\hat{V}_f$ is $\boldsymbol{V}^* = 3.5$, while for the average probability $\hat{p}_f$ the decision-maker has chosen a safe benchmark equal to 70%.

$$\boldsymbol{P}^* = 70\%$$

Table 5.18 The matrix that was used to collect the data during the focus groups

		Ours vs. *C1*	*Ours* vs. *C2*	*Ours* vs. *C3*	……	…….	*Ours* vs. *C11*
Cleanliness	***Expert 1***	$v_{1,1,1}$	…	…	…	…	$v_{1,11,1}$
	Expert …	…	…	…	…	…	…
	…	…	…	…	…	…	…
	Expert …	…	…	…	…	…	…
	Expert 20	$v_{1,1,20}$	…	…	…	…	$v_{1,11,20}$
	Expected value	$\hat{v}_{1,1}$	…	…	…	…	$\hat{v}_{1,11}$
	*Probability that V > V**	$p_{1,1}$	…	…	…	…	$p_{1,11}$
Controllability	*Expert 1*	$v_{2,1,1}$	…	…	…	…	$v_{2,11,1}$
	Expert …	…	…	…	…	…	…
	…	…	…	…	…	…	…
	Expert …	…	…	…	…	…	…
	Expert 20	$v_{2,1,20}$	…	…	…	…	$v_{2,11,20}$
	Expected value	$\hat{v}_{2,1}$	…	…	…	…	$\hat{v}_{2,11}$
	*Probability that V > V**	$p_{2,1}$	…	…	…	…	$p_{2,11}$
Safety			…	…	…	…	
			…	…	…	…	
			…	…	…	…	
			…	…	…	…	
			…	…	…	…	
Solidity			…	…	…	…	
			…	…	…	…	
			…	…	…	…	
			…	…	…	…	
			…	…	…	…	
Translucency			…	…	…	…	
			…	…	…	…	
			…	…	…	…	
			…	…	…	…	
			…	…	…	…	
Transparency	***Expert 1***	$v_{6,1,1}$	…	…	…	…	$v_{6,11,1}$
	Expert …	…	…	…	…	…	…
	…	…	…	…	…	…	…
	Expert …	…	…	…	…	…	…
	Expert 20	$v_{6,1,20}$	…	…	…	…	$v_{6,11,20}$
	Expected value	$\hat{v}_{6,1}$	…	…	…	…	$\hat{v}_{6,11}$
	*Probability that V > V**	$p_{6,1}$	…	…	…	…	$p_{6,11}$

Therefore, the experiment will reject the null hypothesis only if

$$\forall \mathrm{f}[1,6], \hat{V}_f > \boldsymbol{V}^* = 3.5 \text{ and } \hat{p}_f > \boldsymbol{P}^* = 70\%$$

Table 5.18 shows the type of matrix that was used to collect the data during the focus groups. Each column corresponds to the comparison with a specific competitor, and each group of six rows corresponds to one of the six features.

The charts shown in Figs. 5.26 and 5.27 represent a summary of the values of $\hat{v}_{f,c}$ and $p_{f,c}$, calculated using the collected assessments of the new product compared to all the other connectors.

We can observe that, for future "solidity," both inequalities ($\hat{V}_f > \boldsymbol{V}^*$ and $\hat{p}_f > \boldsymbol{P}^*$), defining the condition to reject the null hypothesis, are not satisfied.

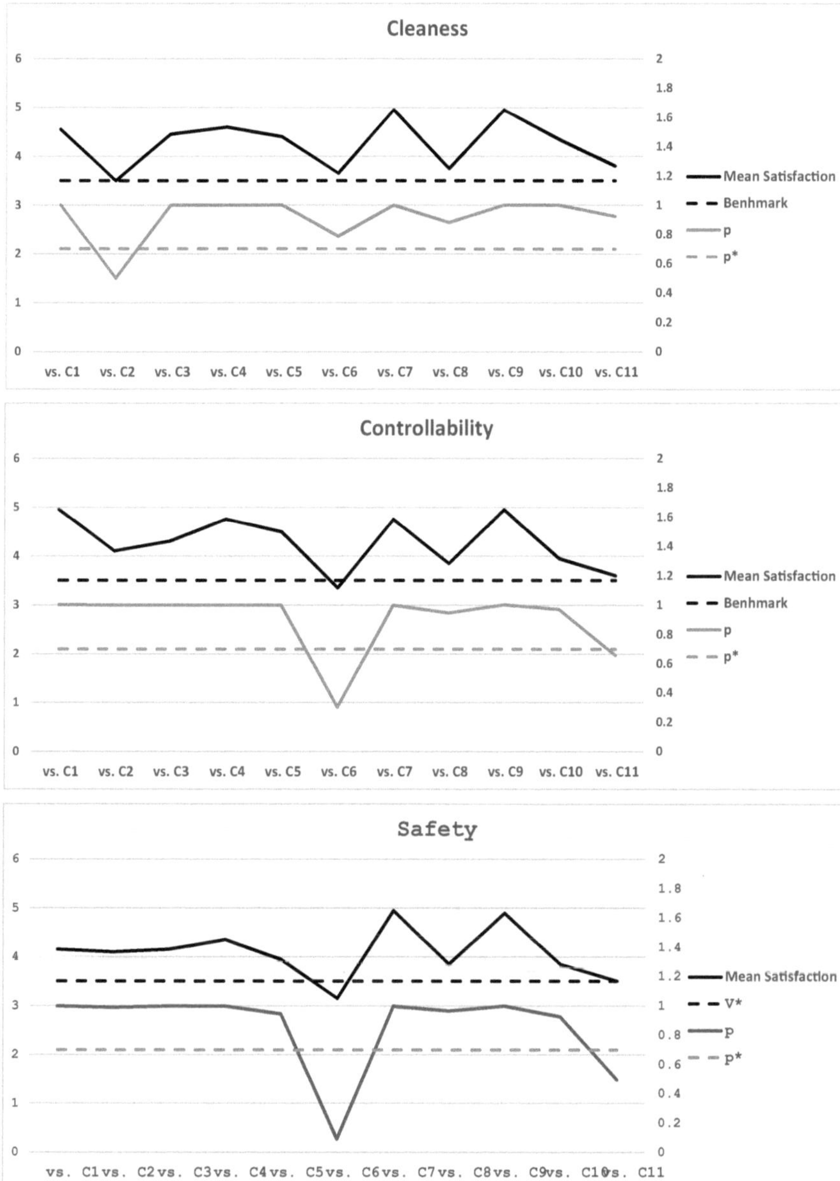

Fig. 5.26 Summary of the values of $\hat{v}_{f,c}$ and $p_{f,c}$, calculated using the collected assessments of the new product compared to all the other connectors (1/2)

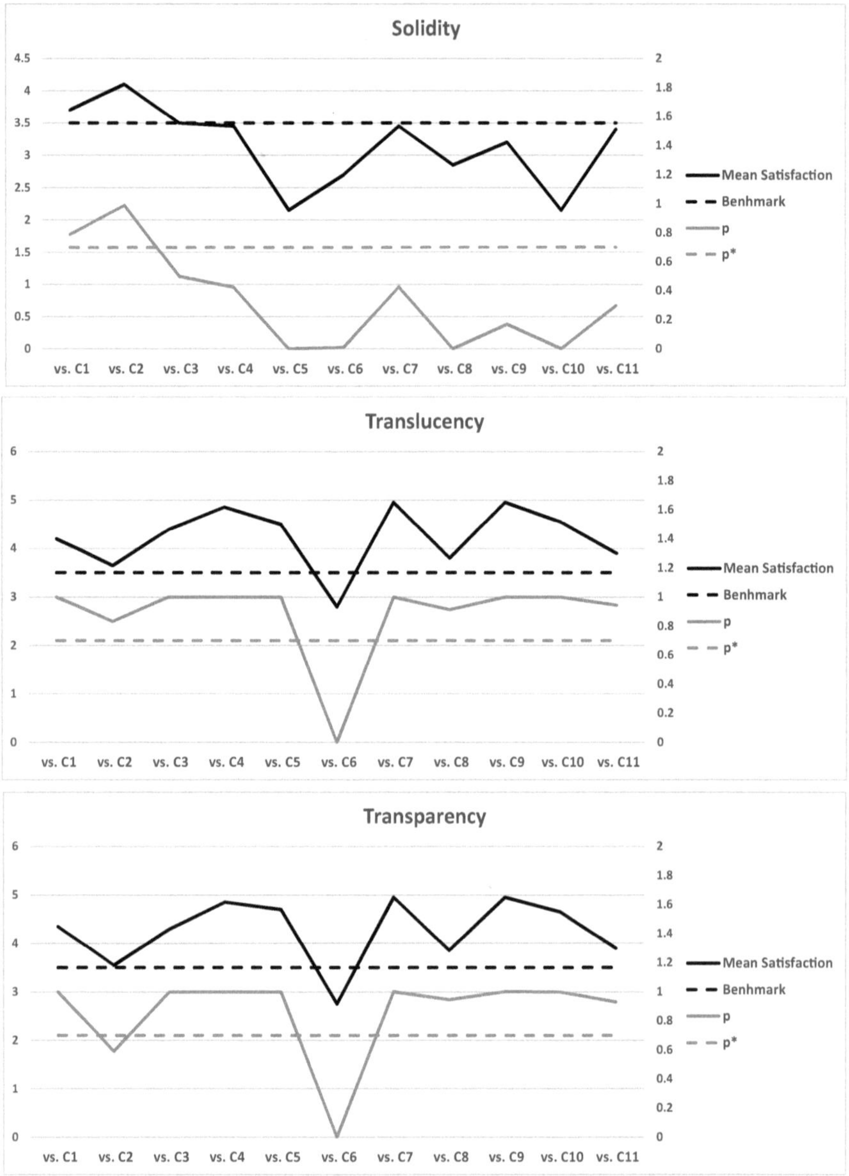

Fig. 5.27 Summary of the values of $\hat{v}_{f,c}$ and $p_{f,c}$, calculated using the collected assessments of the new product compared to all the other connectors (2/2)

Step 6: Making a Decision
The condition supporting the rejection of the null hypothesis is as follows:

$$\forall f[1,6], \hat{V}_f > \boldsymbol{V}^* \text{ and } \hat{p}_f > \boldsymbol{P}^*$$

Since, for the solidity feature, both these inequalities are dissatisfied, the experiment does not allow the decision-maker to reject the null hypothesis. The new connector is not better than those of all the existing competitors, and the decision-maker decides to exit—a decision that often leads to active product pivoting efforts focused on the underperforming features.

Bibliography

AADIl Hospital. (2025). *Epidural anesthesia.* https://perma.cc/2RFV-8MBX
Arendt, K., & Segal, S. (2008). Why epidurals do not always work. *Reviews in Obstetrics & Gynecology, 1*(2), 49–55.
Barclay, S., Brown, R., Clinton III, Peterson, C., & Phillips, L. (1977). *Handbook for decision analysis*, January, 284.
Camuffo, A., Cordova, A., Gambardella, A., & Spina, C. (2020). A scientific approach to entrepreneurial decision making: Evidence from a randomized control trial. *Management Science, 66*(2), 564–586.
Camuffo, A., Gambardella, A., Messinese, D., Novelli, E., Paolucci, E., & Spina, C. (2024). A scientific approach to entrepreneurial decision-making: Large-scale replication and extension. *Strategic Management Journal, 45*(6), 1209–1237.
Camuffo, A., Gambardella, A., & Pignataro, A. (2024). Theory-driven strategic management decisions. *Strategic Science, 9*(4), 382–396.
Davey, S. M., Brennan, M., Meenan, B. J., Mcadam, R., Girling, A., Chapman, A., & Liliford, R. L. (2011). A framework to manage the early value proposition of emerging healthcare technologies. *Irish Journal of Management, 31*(1), 59–75.
Fagen-Ulmschneider, W. (2023). *Perception of probability words | Wade Fagen-Ulmschneider.* Waf.Cs. https://perma.cc/V4QA-XW7H
Ignite Healthwise. (2024). *Learning about a leadless pacemaker.* https://perma.cc/5YWK-VPV3
NCATS, National Center for Advancing Translational Sciences. (2024). *Endpoint.* National Institute of Health. https://perma.cc/4ZT8-QXJW
Ries, E. (2011). *The lean startup.*
Sala, Pavan Kumar, Simon Philbin, and Safia Barikzai. 2022. "A Qualitative Research Study of the Tech Startup Journey through Entrepreneurial Pivoting.***"
Savage, S. (2002, November 1). *The flaw of averages.* Harvard Business Review. https://perma.cc/98NV-KCWM

6 Mapping Decision-Related Problems in Medical Innovation

Abstract

This chapter argues that disconnected facts hinder judgment and proposes a cognitively grounded, value-centric "map" of decision-related problems for medical innovation. Building on the prior chapter, this chapter explains why the scientific method—while essential for testing individual choices—is insufficient on its own: scarce resources require entrepreneurs to see all critical problems, continuously re-prioritize them, and address the most consequential first. The chapter introduces a sequential mental map (the K9 Map) that links market needs, product readiness, value proposition, competitiveness, stakeholder fit, segment attainability, commercialization viability, financials, and valuation. Aligned with Stage-Gate thinking, this structured map mitigates bias, guides scarce effort, and improves the accuracy and credibility of venture forecasts.

6.1 The Importance of Mapping Decision-Related Problems

Cognitive psychology has shown that the mind best understands facts when they are woven into a conceptual fabric, such as a narrative, mental map, or intuitive theory. Disconnected facts in the mind are like unlinked pages on the web; they might as well not exist (Steven Pinker). This chapter of the book focuses on building a rationally organized map of the key, interconnected decision-related problems in medical innovation.

In the previous chapter, we explained why the scientific approach alone is not sufficient to address the issue of decision-making biases in medical technology compellingly. The use of the scientific approach addresses only individual decision-related problems, one at a time, while there are important issues that we mentioned, such as the following:

L. Negri, *De-Risking Investments in Medical Technology Innovation*, Springer Series in Healthcare Management and Innovation,
https://doi.org/10.1007/978-3-032-15007-3_6

- In an environment of scarce resources, decision-makers cannot afford to apply the scientific approach (or any disciplined method) indifferently to all the known decision-related problems.
- Instead, decision-makers must continuously re-prioritize and identify the most relevant problems to solve at each stage of the venture's development.
- Moreover, it is important for decision-makers to be fully aware of all the critical problems of a medical technology startup without exception.

To support decision-makers in dealing with these challenges, we focus on the construction of a map that is able to lay out and link the key decision-related problems in medical technology innovation. A cognitive map of decision-related issues that needs to be approached in an orderly manner to optimize the use of scarce entrepreneurial resources. A cognitive or mental map is a mental representation that an individual uses to acquire, maintain, and retrieve information about the relative locations and attributes of phenomena in their environment (Muckli & Petro, 2017).

Figure 6.1, for instance, represents a highly simplified example of a mental map focused on predicting the number of patients potentially treated by a new therapy. The cognitive map contains an example of the key decision-related problems to be addressed in order to estimate future revenues. It also includes links between the decision-related problems that represent various types of relationships, such as dependency, causality, and simply sequence.

In the previous chapters, we explained that the category of decision-related problems contains a broad spectrum of entrepreneurial problems: from investment decisions to the scenario decisions required to elect specific business assumptions. Therefore, a complete map of decision-related problems must also include all the core business assumptions that medical entrepreneurs must consider.

Mental maps are usually formed, adapted, and memorized by individuals over the years, and depending on individual backgrounds and experiences, the accuracy of a mental map is typically skewed toward certain areas, more or less familiar, and,

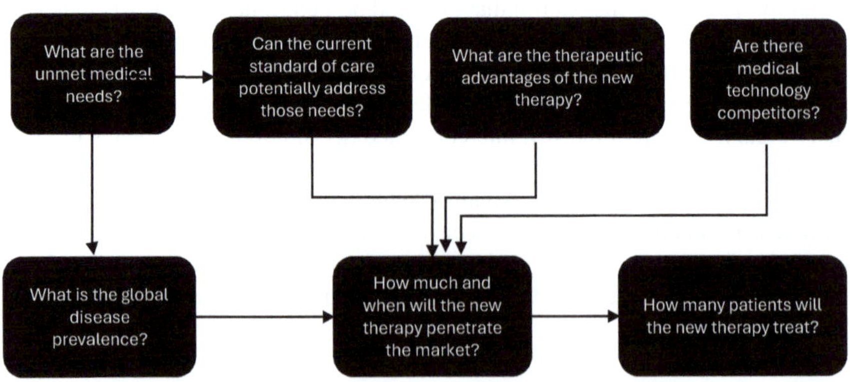

Fig. 6.1 Example of a mental map focused on predicting the number of patients potentially treated by a new therapy

in general, more or less intellectually attractive. In other words, depending on their personal experiences and backgrounds, different entrepreneurs may refer to different mental maps, which are likely to be very accurate only in certain well-known areas and often unclear or incomplete in other areas.

My colleagues and I have often observed entrepreneurs with a highly specialized academic background undermine the success of their own venture by restricting their model of reality (their mental map) to matters most known to them or by postponing for too long decision-related problems that do not belong to their area of expertise.

Typical examples include medical doctors misprioritizing or ignoring non-medical issues, and engineers often narrowing their mental model of reality to highly technical matters while overlooking the clinical and business aspects. For instance, engineers are more likely to prioritize the sophistication of certain technical solutions over the assessment of the patient benefits produced by such technical features.

Many academics fail to grasp the profound challenges and alarming failure rates linked to innovation. Securing investment is not only difficult but often daunting, yet most researchers in academia overlook the critical need to assess market potential and to devise strategies that mitigate risk. These essential steps are vital for attracting the investment necessary to bring innovative ideas to life (Behrns et al., 2021).

It is interesting to note that literature has documented "egocentric biases" in market-entry decisions, demonstrating self-focused explanations for the entry decisions made by entrepreneurs: primarily "based on evaluations of their own competence (or incompetence) and paying relatively little attention to the strength of the competition" (Moore et al., 2007).

Misprioritizing or postponing critical decision-related problems is one of the most costly entrepreneurial mistakes. For instance, many medical startups postpone marketing and pricing decisions to the very end, after new products are developed. "They embark on the long and costly journey of product development hoping they'll make money on their innovations, but not at all knowing if they will" (Ramanujam & Tacke, 2016).

Another example of misprioritization in decision-making is defining new product features before fully characterizing unmet needs, without clearly defining the specifications of "must-haves" and "desirables." This mistake is widespread, as the entrepreneur may be attracted by an academically elegant or intellectually clever solution that, on the other hand, does not actually meet the specifications defined by the broad set of unmet needs (Behrns, Gingles, and Sarr 2021).

6.2 Building a Value-Centric Map for Medical Innovation

Entrepreneurs who seek to validate their known set of decision-related problems can refer to a rich literature that, although not specifically tailored to the field of medical innovation, can be extremely helpful. There are several articles, such as

"The Questions Every Entrepreneur Must Answer" (Bhide, 1996), in which problems are organized in very high-level groups: What kind of enterprise do I need to build? Is the strategy well defined? Do I have the right resources? There are also some sources specializing in medical technology, including books such as *The Ten MedTech Commandments* (Tucker, 2022) and *Medical Innovation* (Behrns et al., 2021). These books provide invaluable guidance to medical entrepreneurs. They organize entrepreneurial decision-related problems by key functional areas, such as regulations, clinical trials, finances, licensing, and business law. In most cases, these problems are organized into broad functional areas, outlining lists of key decision-related problems supported by the author's personal experience or thorough literature research.

This book suggests a new assessment approach for medical technology innovation. Here, the use of a decision map addresses the need for a "deliberate" approach to decision-making that encompasses all the investigations and decisions faced by the medical technology entrepreneur, which remain one of the most important points of failure in medical innovation (Behrns et al., 2021).

Therefore, instead of adopting an approach structured around separate functional areas, we propose the creation of a multidisciplinary, cross-functional mental map focused on the typical decision-making problems of a medical startup, whose solution ultimately helps sharpen the prediction of the economic value of a medical venture. This map is ultimately concerned with what we identify as "the core entrepreneurial goal," which involves the provision of an unbiased valuation and the subsequent delivery of the promised value, ensuring timely monetization.

Beyond the entrepreneur's need for a rational, credible assessment of the value of a new idea, several reasons support the venture's value as the core element to which business assumptions and decisions must point. For instance, the decisions about the portfolio of new products for larger, more developed organizations are critical and largely dependent upon the predicted values of the new products.

Portfolio management for new products is deeply rooted in the work of Robert G. Cooper and colleagues. It is the systematic process by which companies decide which new product projects to invest in, prioritize, or terminate, balancing limited resources with strategic objectives.

The theory (Cooper & Edgett, 2006) is built on three main goals:

- Value maximization that focuses on selecting projects that maximize the economic value of the portfolio.
- Portfolio balance ensures a balanced mix of projects (short vs. long term, low vs. high risk, incremental vs. breakthrough).
- Strategic alignment, which ensures that the portfolio reflects and supports the company's strategic priorities.

Therefore, portfolio management for new products heavily depends on the predicted values of each new product idea. This consideration further reinforces the choice of centering the predictive efforts of the proposed methodology on valuation.

According to the elected, value-centric approach, all individual decision-related problems can be considered as value drivers contributing to the realization of the venture's value.

6.2.1 Building a Value-Centric MAP

While multiple well-known frameworks already exist to assess the value of a life sciences product, many of these were specifically developed to assess BioPharma drugs. In the context of MedTech, however, the simple implementation of these frameworks cannot lead to consistently appropriate business assumptions and, therefore, decisions pertaining to the development and adoption of high-value medical technologies that improve patient lives. Value assessment in MedTech must encompass the medical technology's comprehensive set of potential value drivers, including those captured by patients and other stakeholders, such as users and payers (Deloitte, 2021).

The search for an organic approach built on the contribution of value drivers, such as individual decision-related problems, requires a valuation process based on future revenues. For this reason, in this book, venture valuation is based on the challenging but fundamental prediction of cash flow rather than on other methods such as scorecards.

Methods such as valuation scorecards cannot support this working philosophy, as they are not based on the venture's financials but on comparisons with other similar companies in the same sector, geographical area, or stage of development (Puca, 2020).

In complete harmony with the Stage and Gate approach (Cooper, 2001) described in section 4.4, in 2010, Alan Girling, from the Institute of Applied Health Research at the University of Birmingham (UK), proposed a decision-making approach for the medical device industry that involves linking the issue of valuation with the key management decisions that lead to the commercial launch (Girling et al., 2010).

This valuation model refers to a sequential map of entrepreneurial decisions and links each decision-related problem to the venture's valuation. In particular, Girling proposes a simplified, map-based framework for the valuation of early-stage innovative medical devices: It is based on a stage-gated model of the development cycle for new products, as presented in Fig. 6.2.

The sequential map consists of three key phases that are separated by two decision gates. In the first phase (P1), the entrepreneur originates an initial concept for a medical device. At G1, the entrepreneur decides whether to enter the development phase (P2). Once the product has been developed, a final decision (G2) is required to take the product to the market and to enter the post-market phase (P3).

This type of model links the valuations of the venture to the decisions made over time, as it links the sequence of decisions to the achievement of a future cash flow. For instance, in G1, the entrepreneur decides to invest in product development only if the investment is "rewarded" by the post-market cash flow (PostMCF), which, with a significant degree of simplification, can be expressed as follows:

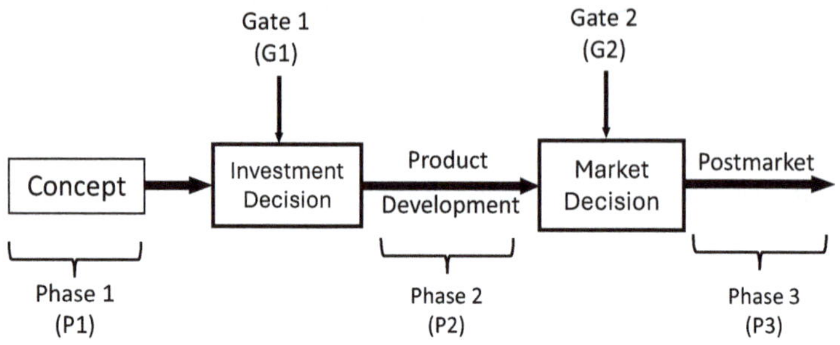

Fig. 6.2 The Stage-Gate Model from (Girling et al., 2010)

$$\text{PostMCF} = \text{Sales Revenue} - \text{Costs}$$

At G1, a decision must be made despite the high degree of uncertainty about future revenues, product costs, and commercial costs. Decision G1 is driven by the expectation that the PostMCF generated in the post-market phase will be sufficient to justify the investment required for the product's development.

If decision G1 brings the venture to product development, the entrepreneur will be in a position to gather much more information about the product cost and mitigate this uncertainty component. For instance, at G2, the decision between going to market and exiting will enjoy a diminished degree of uncertainty in relation to the cost component of the PostMCF.

6.2.2 Mapping Key Decision-Related Problems Relevant to the Predicted Cash Flow

We have chosen to embrace Girling's approach to valuation philosophy, which links decisions to the realization of value. This philosophy is based on the Stage and Gate System (Cooper, 2001) and relies on the availability of a complete map of sequentially interconnected decision-related problems that are somehow linked to the venture's value.

To ensure the map's completeness, we use a reverse approach. Instead of establishing the key problems ex ante (Garling's decision gates), we identify them by reverse engineering the determination of the venture value.

Therefore, we start from the venture value, which is based on the future cash flow. As visualized in Fig. 6.3, predicting a venture's cash flow involves estimating two key components: the pre-market cash flow (PreMCF), characterized by more predictable operating expenses and investments, and the PostMCF, driven by revenues, product cost, and operating expenses. The sum of the two is equal to the venture's cash flow.

$$\text{Cash flow} = \text{PreMCF} + \text{PostMCF}$$

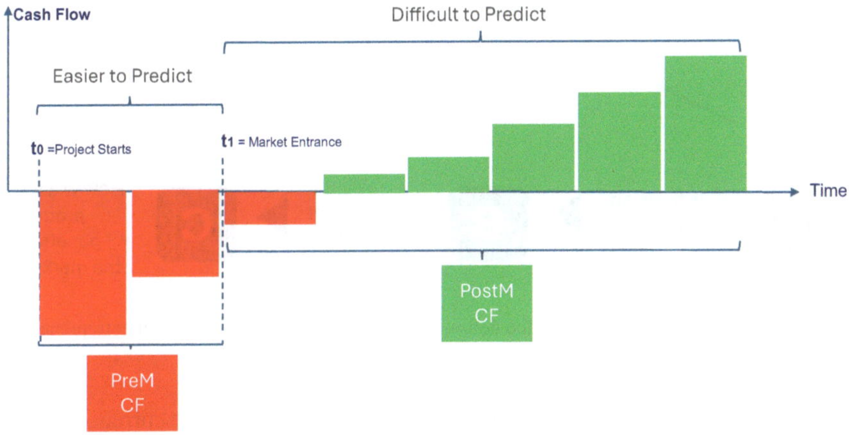

Fig. 6.3 A high-level summary of the typical startup cash flow overtime

It is important to emphasize that predicting the PreMCF is much easier than predicting the PostMCF. For this reason, the core challenge of valuation accuracy lies in predicting future profits. For this reason, the construction of the decision map is mainly based on the value drivers of the PostMCF.

6.2.3 Predicting Post-Market Cash Flow for Medical Innovation

The Post-Market Cash Flow, driven by product sales and profitability after launch, during a defined time window, can be expressed as follows (Girling et al., 2010):

$$\text{PostMCF} = \text{MS}^{*}\text{MP\%} \times (\text{P} - \text{L}) - \text{F}$$

Here, P is the unit price, F is the fixed (production/marketing) cost, L is the variable (production/marketing) unit cost, MS is the market size, which denotes the total number of units in the entire market, and MP% is the percentage of market penetration of the venture's product.

Closely considering the expression of PostMCF, what appear to be the main top-level decision-related problems that entrepreneurs must solve in order to predict the value of each variable in these equations? It is key to look at each component separately.

Figure 6.4 provides a visual description of the three components.

The first component, which is the market size (MS), refers to the size and attractiveness of the business opportunity. Several elements and measures, such as the existence and importance of unmet medical needs, the patient population relevant to these needs, and the socioeconomic burden, determine it.

The second component, which includes MP% and P, refers to the venture's ability to penetrate the market at a certain price P. It depends on the ability of the venture to satisfy the existing medical needs in a competitive manner.

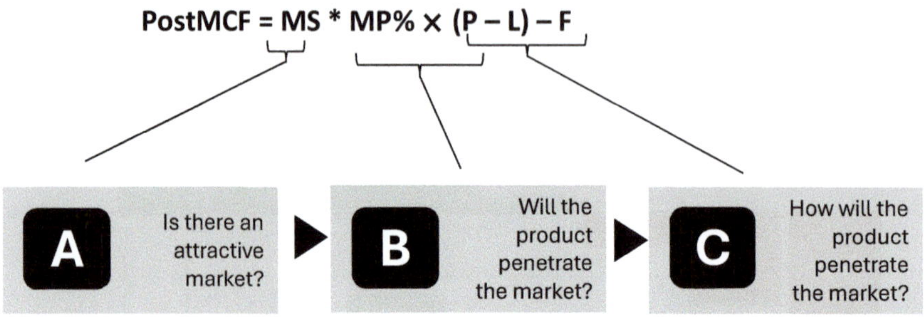

Fig. 6.4 The relationships between the definition of PostMCF and the three main entrepreneurial questions

Indeed, this degree of satisfaction must be superior to that of the current standard of care, and of the direct competitors, at a price that fits the health economic fundamentals. Predicting penetration also means predicting who will pay for the new technology—in other words: to what extent the innovative technology fits the priorities of the stakeholders critical to adoption decisions.

The third component, in partial overlap with the second, is represented by the expression "(P–L)–F," which includes both the fixed and variable resources that must be employed to penetrate the global market, and the average selling price P that must be sufficient to cover the costs of those resources abundantly.

Therefore, the prediction of the PostMCF is delegated to the solution of these three high-level problems:

A. Is there an attractive market?
B. Can the product penetrate that market?
C. How will the product penetrate the market?

6.2.4 Exploring the Three Top-Level Problem Clusters

In order to answer each of the high-level questions (A, B, and C) mentioned above, entrepreneurs need to answer other, lower-level questions such as the following: Which market (and geographical) segments are the most attractive? What business model can maximize penetration? What is the pricing for each market segment? What commercial organization is required? What organizational competencies are needed? A literature review and our field experience suggests that questions A, B, and C can be broken down into lower-level problems:

A. **Is there an attractive market?**

   The issue of determining and assessing target market attractiveness encompasses the following:

   - Understanding all the unmet medical needs that the product can address. Indeed, the issue of unmet medical needs leads to any consideration of market (Vreman et al., 2019).
   - Assessing the market value and size based on the unmet medical needs.

B. **Can the product penetrate the market substantially?**

The answer to this question involves the following problems:

(i) It is essential to define a detailed target product profile (TPP), including indications and clinical outcomes, and to establish the required usability and clinical trials to prove the value proposition claims and successfully identify the pathway to accomplishing global regulatory duties (Gupta, 2016).
(ii) It is necessary to explicitly formulate a value proposition that can demonstrate how the new product successfully addresses the identified unmet needs (Frow & Payne, 2011), at what price, and why it deserves reimbursement.
(iii) It is important to identify and assess the role of existing and future competition focused on the same unmet medical needs.
(iv) It is crucial to assess how the competing priorities of market stakeholders (referred to as the "status quo") will facilitate, hinder, or impede new product penetration.

C. **How will the product penetrate the market?**

The answer to this question involves the following third-level problems:

(i) The issue of segmentation and positioning.
  (a) It is important to select the appropriate geographies and (within geographies) identify and target the most promising segments (Dolnicar et al., 2018).
  (b) It is necessary to correctly position the product within the identified target segments (Ostasevičiute & Šliburyte, 2008) fine-tuning the pricing considerations of point A.

(ii) The issue of determining the optimal monetization and commercial models and their implementation costs.

Figure 6.5 visually summarizes the main characteristics of each of the above decision-related problem clusters and their relationships.

It is worth noting how this first basic version of the map already suggests a possible order of approach to each problem cluster. At a high level, the entrepreneur must first understand whether there is an attractive market for the proposed innovation (cluster A).

Then, the entrepreneur must concurrently address the questions about the product's ability to satisfy unmet medical needs, its ability to penetrate the market, and the question regarding the identification of a price consistent with the desired market penetration (cluster B).

Finally, the entrepreneur must answer questions about exactly where, how, and at what cost the market penetration will take place while ensuring an attractive profitability.

Addressing question B before answering question A makes little sense, and the same is true for addressing question C before questions A and B. This observation

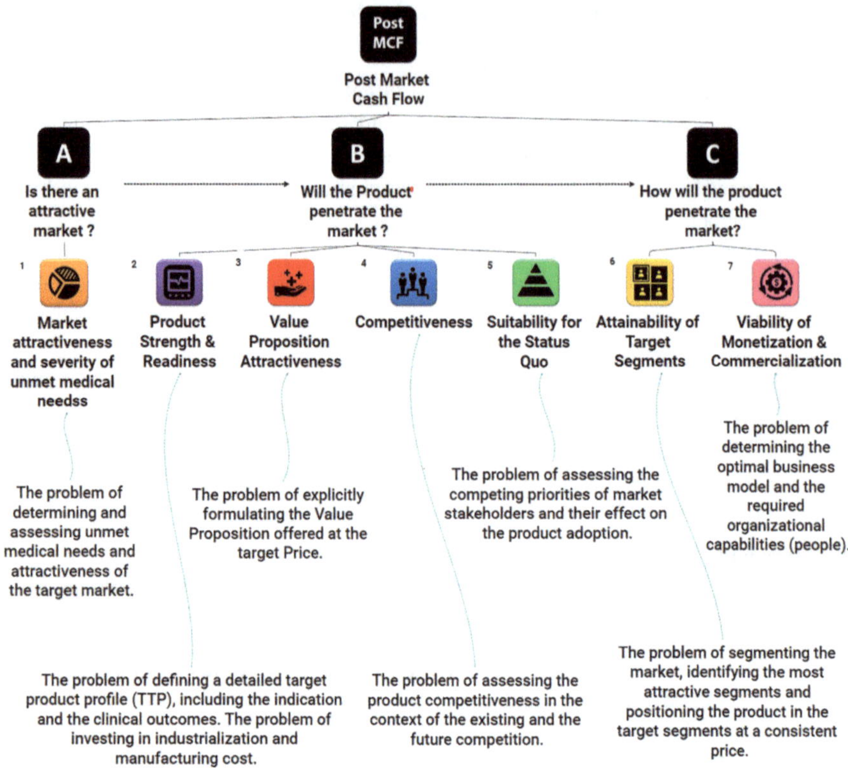

Fig. 6.5 The hierarchical relationships between the Post-Market Cash Flow and six of its main drivers

suggests an intrinsic sequence useful when approaching decision-related problems: a preliminary, rough prioritization criterion of the mapped problems. The order determined by sequences A, B, and C affects the order of the six lower-level clusters outlined in Fig. 6.5.

Indeed, it is not truly feasible to focus on product features and the related value proposition before there is some clarity about the specific unmet medical needs being addressed and the degree of attractiveness of the market opportunity. Therefore, clusters 2 and 3, respectively called "product strength and readiness" and "value proposition attractiveness," can be approached only after having resolved most of the decision-related problems of cluster A: "Market attractiveness and severity of unmet medical needs."

The fourth cluster concerns competitiveness, and it can be approached only when the new product features and the new value proposition have been decided: Only at this point can an entrepreneur start a thorough comparison of the new device with both the current standard of care and other competing devices.

The fifth cluster, "Suitability to status quo," refers to the ability of the new device to blend into the complex stakeholders' priorities. Approaching this cluster requires all the information generated in the first four clusters. For instance, it is difficult to predict the likelihood of any public or private reimbursement without a clear understanding of the current reimbursement status for other competing solutions or for the current standard of care.

The sixth cluster, "Attainability of the target segments," groups the decision-related problems concerned with segmentation, segment prioritization, and penetration of the new device at segment-specific prices. It will be very difficult to approach all these decision areas without gaining clarity on the issues addressed in the previous clusters, such as the geographical presence of the main competitors, the countries with a higher likelihood of reimbursement, the regional clinical practices, and the therapeutic indications characterized by the highest unmet medical needs. Once the entrepreneur has decided which geographies and market segments to prioritize and what penetration targets to set at specific prices, it is time to look at both the business model and the organization capable of supporting the commercial aspirations.

The seventh cluster, "Viability of monetization and commercialization," includes all the decision-related problems related to the choice of the business model supporting the new product's monetization, acquisition, and retention of the talents required to pursue the commercial and operational challenge. This seventh cluster is crucial because it also deals with the investments and expenses that the venture must support to secure the desired revenues.

6.2.5 Predicting Pre-Market Cash Flow (preMCF)

Most of the components of PreMCF are related to investments and expenses belonging to two of the clusters described above: product strength and readiness and business model.

All decisions about the investments and the expenses required for proving product efficacy and effectiveness (such as clinical and usability trials), realizing a proof of concept, industrializing the product, protecting the intellectual property (IP) (i.e., filing of patents and payment of their maintenance fees), and filing the international regulatory applications are grouped under product strength and readiness. Under the cluster concerned with monetization and commercialization, we find all the decision-related problems relating to the expenses required to attract, retain, and compensate human capital capable of bringing in the necessary set of non-technical and non-medical competencies. Many of these talent acquisitions must be finalized before market entry, not so much to prepare the commercial organization but to acquire the competences necessary to understand the market dynamic and the intricate relationships among key stakeholders (Fig. 6.6).

Fig. 6.6 The decision problem clusters that most affect the Pre-Market Cash Flow

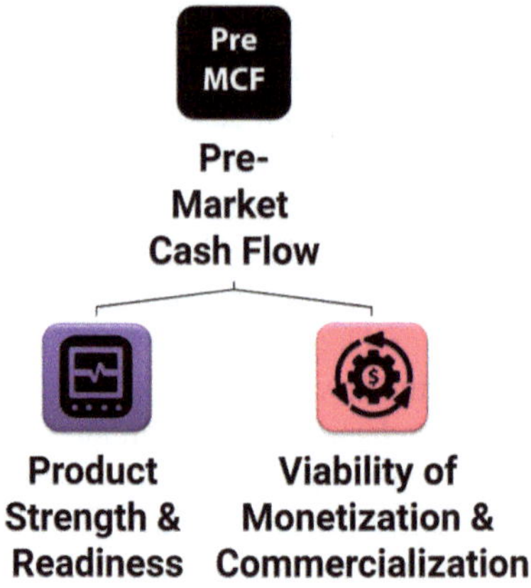

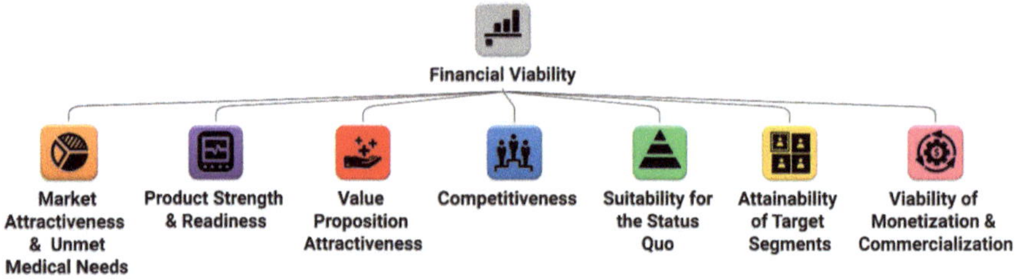

Fig. 6.7 A simplified framework illustrating how the information generated in the seven identified decision clusters is utilized to predict the venture's cash flow and facilitate the evaluation of its financial viability

6.2.6 The Map and Prediction of the Venture's Cash Flow

The total cash flow of the venture can be expressed as follows:

$$\mathrm{CF} = \mathrm{Pre} - \mathrm{MCF} + \mathrm{Post} - \mathrm{MCF}$$

Therefore, the cash flow can be determined using the information generated by all the decisions taken in the six previous clusters, as described in Fig. 6.7. All the efforts that the entrepreneur makes in prioritizing and resolving the problems encompassed in each of the first seven decision clusters generate highly valuable information for the determination of future revenues and profits. The eighth cluster, called "financial viability," outlined in Fig. 6.7, contains a simplified framework illustrating how the information generated in the seven identified decision clusters is utilized to predict the venture's cash flow and facilitate the evaluation of its financial viability.

In other words, the proposed map-driven process supports a type of bottom-up exercise, where individual value drivers contribute to the estimation of both short-term and long-term business outcomes, with a specific focus on post-market revenues and profits. Since matching the venture's valuation with actual financial returns

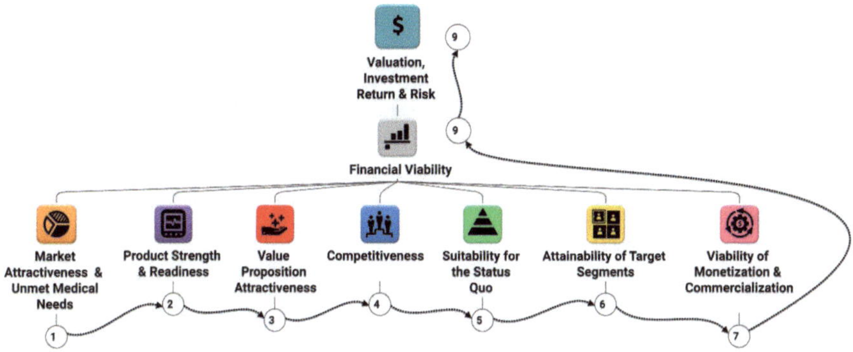

Fig. 6.8 How the identified clusters, placed in sequential order, contribute to predicting the venture's value

is the final goal of this exercise, a final cluster including the valuation exercise must be added to complete the map. This last cluster of the map, called "Valuation, investment return, and risk," pertains to all the decision-related problems involving valuation, return on the investment, and the framing of exit options at different stages of the venture's development. This last cluster represents not only the "end" of the sequential matrix, but also the end goals of all the decisional efforts invested in the previous problems: the maximization, the prediction, and the actual monetization of the venture's value. Figure 6.8 represents how the identified clusters, placed in sequential order, contribute to predicting the venture's value and support the entrepreneur in realizing that monetary value.

6.3 The Key Nine Clusters: The "K9 Map"

Table 6.1 visually summarizes the nine clusters with their main descriptions and characteristics.

A more compact and sleeker version of the nine clusters is provided by the K9 map, which is shown in Fig. 6.9. It is important to memorize or become acquainted with these nine icons and their meaning, as they will frequently appear in the following sections.

6.3.1 The Role of the Clusters' Sequence in the K9 Map

As mentioned previously, the sequential order of the nine clusters map is not merely a result of field practice; it also owes to an extensive literature review, with particular emphasis on the Stage and Gate system (Cooper, 2001) and on Girling's approach (Girling et al., 2010). This sequence offers basic support to the clusters' prioritization efforts. For instance, it considers the decision-maker's design of innovative infusion pumps to avoid spending significant resources in making firm penetration

Table 6.1 Visually summarizes the nine clusters with their main descriptions and characteristics

The Nine Sequential Key Decision Clusters in Medical Innovation		
	Clusters	**Involved Decision-related Problems**
1	Market attractiveness and severity of unmet medical needs	Is there an actual market? What are the needs that the innovation must address? How important are they? What is the size of the market of interest?
2	Product strength and readiness	Is the product defined? What is the exact definition of the product? What are the indications? What are the accountabilities in terms of efficacy and effectiveness? What is the level of readiness?
3	Value proposition attractiveness	Does the proposed value fit the highest-priority needs? Is the value added truly innovative? At what price is the net added value still attractively positive? Are there collateral downsides attached to the innovative features?
4	Competitiveness	Is the proposed solution competitive compared to the current standard of care and the existing and emerging industrial competitors?
5	Suitability for the status quo	Does the product/solution blend into the complex, conflicting stakeholders' priorities? Are the stakeholders ready to accept this product/solution? Are there significant barriers?
6	Attainability of the target segments	Which segments of the market are the most attractive? What are the price and positioning of the new device compared to the competition in these segments? What are the segment-specific penetration targets?
7	Viability of monetization and commercialization	What is the cash-generating concept? Who pays? What type of sale, lease, or subscription will generate revenues? Is licensing the revenue generator? What type of commercial organization will support revenue growth?
8	Financial viability	What is the expected stream of P&L performance? What are the NPV and the IRR of the cash flow? What is the sensitivity of these values to the key assumptions made in nodes 1–7? What is their distribution?
9	Valuation, investment return, and risk	When and how will the investor be in a position to collect their returns and make an exit? What types of exit are possible for investors? What are the developmental milestones critical to the achievement of the exit options?

assumptions and pricing decisions (cluster 6) before gaining clarity about the competitive issues in the bedside arena (cluster 3) and before field-testing (cluster 2) the device's ability to support the value proposition promises.

On the other hand, it is essential to underline that this suggested sequence represents only one helpful form of support for prioritizing clusters and problems. It is neither the only tool available to the decision-maker, nor a license to completely ignore or indefinitely postpone clusters positioned toward the end of the map. Nor

Fig. 6.9 A compact and sequential version of the nine clusters is provided by the K9 map

is the sequential order a rule that forbids decision-makers from considering clusters positioned well ahead of the current decision-related problems.

Moreover, in this regard, it is essential to recall that investigations most frequently precede all decision-making processes. Therefore, during such investigative phases, it is normal and even beneficial for the entrepreneur to constantly browse the entire map and perform investigative activities virtually on all clusters. The only thing that the K9 map truly forbids is consistently employing a straight reverse prioritization of decision-related problems: from latter clusters backward.

Typically, entrepreneurs using the K9 map pass over the same cluster several times, first performing investigations, then making business assumptions (scenario decisions), and finally making an investment decision.

Important Note

A thoughtful reader may posit that, following the most common sequence used to commercialize innovative solutions, the product cluster (product strength and readiness) should follow the value proposition analysis (value proposition attractiveness). In this book, the value proposition cluster follows the product cluster instead. Indeed, this book is concerned with the real world of early-stage startups, specifically those in the stages that follow the creative phases. Most startups at this stage have already, somehow, defined a new target device idea or a new target service idea, and they hold relatively strong opinions about the ability of their product or service to satisfy unmet medical needs compellingly. For this reason, this book focuses first on the product-related cluster to clarify the indications, use cases, and definitions of therapeutic or diagnostic features. Only later does the book address the value proposition to verify the product's ability to deliver the promised value and that the value proposition truly satisfies the unmet medical needs.

6.3.2 Relevance of the K9 Map in Other Valuation Methods

Deterministic discounted cash flow analysis is a well-accepted technique in the appraisals of developed companies. The common practice incorporates all uncertainty influences within a single variable—namely, the discount rate—which also represents the time value of money (Carmichael, 2017). The ultimate goal of the scientific approach and the K9 map is to support entrepreneurs in formulating more rational and credible revenue and margin expectations, thereby enabling the rational prediction of evaluations based on discounted cash flow. However, in the pre-seed phase, non-financial methods are often used and they are not based on discounting the future cash flow but on comparative scorecards. Does the K9 map help entrepreneurs assess their startup's values while using non-financial methods such as scorecards? Or is the K9 map useful only for cash-flow-based valuations? This section will prove that the K9 map is an essential tool also for non-financial valuations, as it addresses all the critical dimensions of scorecard methods.

Scorecard methodologies do not rely on the company's financial information but rather benchmark the company relative to other companies in its sector, geographical area, and stage of development. Scorecard methods assign a relative score to reflect the company's performance in several areas that are deemed strategic (Festel et al., 2013). Without discussing the details of these methods, let's focus on the two most popular methods and analyze their respective valuation drivers.

The Payne Method

The Payne method is a scorecard-based approach that determines the pre-money valuation of pre-revenue ventures by comparing the company to similar companies, from which venture's valuators find out the actual market value. This comparison is performed considering the following factors (Payne, 2011):

Factor	Weight
Strength of the management team	0–30%
Size of the opportunity	0–25%
Product or technology	0–15%
Competitive environment	0–10%
Marketing, sales channels, or partnerships	0–10%
Need for additional investment	0–5%
Others	0–5%

These seven qualitative factors are strategic for the company's ability to deliver financial returns to investors. Each factor is weighed based on its relative importance in the company's valuation.

The Risk Factor Summation Method

The risk factor summation (RFS) method is a different type of scorecard method used for the most difficult valuations: the pre-money valuation of pre-revenue seed companies. It is based on the assessment of risk and the relationship between risk and company value. The RFS method compares the target company against a peer group using a list of 12 risk factors. Each risk factor is evaluated on a scale, where a low score indicates that the company provides effective mitigation of the respective risk factor compared to the peer group, while a high score indicates that the target has poor risk mitigation (Table 6.2).

Figure 6.10 shows how the list of nine clusters outlined by the K9 map performs against the list of valuation drivers of the Payne method and the RFS method. We observe that the K9 map's list of nine clusters performs rather well in offering at least a corresponding cluster for each factor of both scorecard methods. This short analysis provides an additional degree of confidence when using the K9 map to build credible funding knowledge, leading to more rational and transparent valuations for the entrepreneur.

The Expected Commercial Value Method

The expected commercial value (ECV) is not a typical evaluation method; it is more an option pricing approach that captures the decision situation in a new product (NP) project. It is more commonly used to support decisions related to the portfolio of new products. This method uses a decision tree to map out the Go/No Go decisions in a new product project, their outcomes, and the economic consequences of each outcome, with probabilities of outcomes occurring built in. It is based on a binary model that determines the expected value (EV) of a decision with two possible outcomes (e.g., "success" or "failure"); a simple decision tree can represent

Table 6.2 An example of application of the risk factor summation method

Risk Factor	Risk Assessment				
	-2	-1	0	1	2
Management risk					✓
Stage of the business				✓	
Legislation or political risk			✓		
Manufacturing or supply chain risk				✓	
Sales and marketing risk		✓			
Funding- or capital-raising risk	✓				
Competition risk	✓				
Technology risk		✓			
Litigation risk			✓		
International risk				✓	
Reputation risk					✓
Exit value risk				✓	

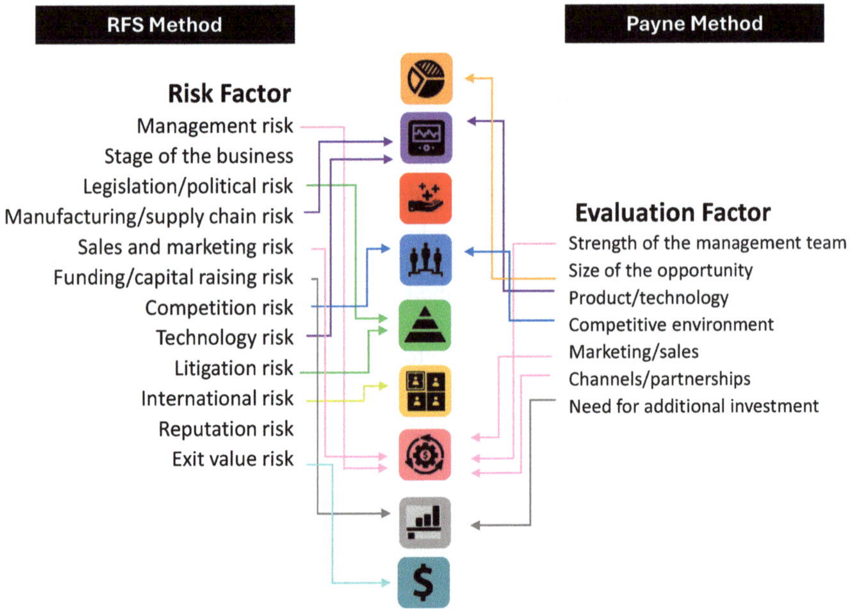

Fig. 6.10 A representation of how the K9 map's nine clusters perform well in offering at least a corresponding cluster for each factor of both scorecard methods

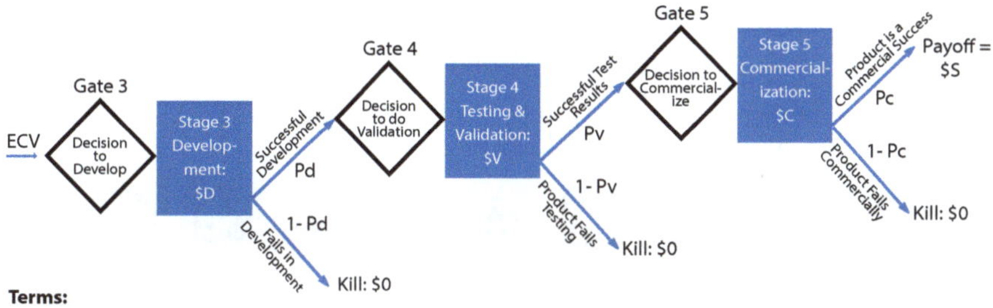

Terms:

D = Development cost
V = Testing & Validation cost
C = Commercialization cost
Pd= Probability of Development success
Pv = Probability of Validation success (product passes tests & trials}
Pc= Probability of commercial success, given product is a technical success
Pt= Probability of technical success = Pd * P v
Ps = Overall probability of product success = Pc * Pt
S = the Success Payoff, the Present Value of future income stream, discounted at risk free rate

Resulting Equations are:

ECV = S * Ps - C * Pt - V * Pd - D

Value of Stage-wise Risk Mitigation = V * (1- Pd) + C * (1- Pt)

Fig. 6.11 The framework of the expected commercial value method (Cooper, 2023)

the binary model and then can be extended to capture the multiple-gate decision process that occurs in new product development (NPD) (Cooper, 2023).

The entire decision model shown in Fig. 6.11 is based on an elaboration of Cooper (2023), and it is applied to the following sequence of decisions: product development, product validation, and product commercialization. Following this sequential decision model, the expected commercial value can therefore be expressed as:

$$ECV = NPV^{*} Pc^{*} Pd^{*} Pv - C^{*} Pd^{*} Pv - V^{*} Pd - D$$

With simple formula manipulations, the ECV can also be expressed as follows:

$$ECV = \left[\left(NPV^{*} Pc - C \right)^{*} Pv - V \right]^{*} Pd - D$$

This expression of ECV facilitates the research of its links with the K9 map's clusters and, therefore, with the information that a rational use of the map can provide to an estimation of ECV. For instance, the parameter NPV is the same as that obtained in the last two clusters of the K9 map. The probability of commercial success and the cost of commercialization can be determined from the business assumptions and the decisions made in clusters such as value proposition attractiveness, competitiveness, suitability to the status quo, attainability of the target segments,

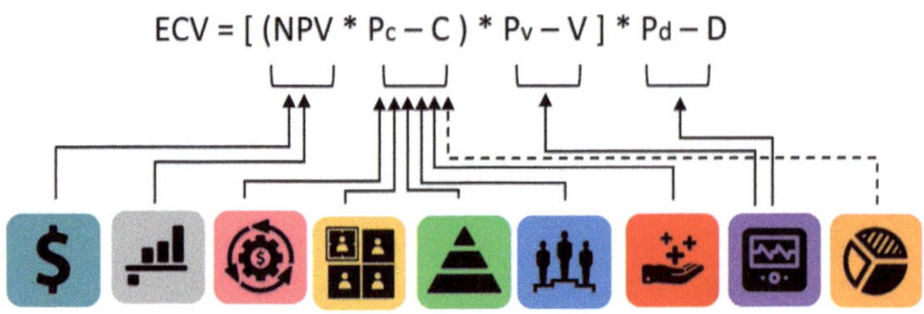

Fig. 6.12 The explicit relationships between each map's cluster and elements of the ECV expression

and viability of monetization and commercialization. The probability of successful product development can be derived from the assumptions and decisions made within the cluster regarding product strength and readiness. Figure 6.12 shows the explicit relationships between each map's cluster and elements of the ECV expression, supporting the hypothesis that a disciplined approach to the K9 map can also support ECV estimations.

6.3.3 What Now? "Exploding" the Nine Decision Clusters into Sub-Clusters

The backbone of problem-mapping in medical technology innovation has been summarized by nine decision-making clusters that, when connected in sequence, build the decision-making roadmap (the K9 Map). Fig. 6.9 presents this sequential map. Each of the nine main problem clusters encompasses several lower-level clusters or sub-clusters. Therefore, in the following chapter, we explore these lower-level clusters and explain how each of the top-nine problem clusters generates between 10 and 15 sub-clusters.

In other words, each of the nine top clusters leads a hierarchically structured map of sub-clusters composed of sub-groups of problems as well as individual decision-related problems. This approach leads to a mapping of decision-related problems that are composed of more than 100 clusters, between sub-clusters and individual problems. Moreover, in line with the problem prioritization needs discussed earlier, the book dedicates the entire 8 to the prioritization of clusters.

Bibliography

Behrns, K., Gingles, B., & Sarr, M. (2021). *Medical innovation*.

Bhide, A. (1996). The questions every entrepreneur must answer. *Harvard Business Review, 74*(6), 120.

Carmichael, D. G. (2017). Adjustments within discount rates to cater for uncertainty—Guidelines. *The Engineering Economist, 62*(4), 322–335.

Cooper, R. G. (2001). *Stage-gate systems: A new tool for managing new products*.

Cooper, R. G. (2023). Expected commercial value for new-product project valuation when high uncertainty exists. *IEEE Engineering Management Review, 51*(2), 75–87.

Cooper, R. G., & Edgett, S. J. (2006). Portfolio management for new products product innovation best practices series picking the winners. *PDMA Visions Magazine*. https://perma.cc/4R4T-QCFX.

Deloitte. (2021). *Value-based care and medical technology: Assessing the value of MedTech products and offerings*. https://perma.cc/2PUG-E83Z

Dolnicar, S., Grün, B., & Leisch, F. (2018). *Market segmentation analysis*.

Festel, G., Wuermseher, M., & Cattaneo, G. (2013). Valuation of early stage high-tech start-up companies. *International Journal of Business, 18*(3), 216–231.

Frow, P., & Payne, A. (2011). A stakeholder perspective of the value proposition concept. *European Journal of Marketing, 45*(1), 223–240.

Girling, A., Young, T., Brown, C., & Lilford, R. (2010). Early-stage valuation of medical devices: The role of developmental uncertainty. *Value in Health, 13*(5), 585–591.

Moore, D. A., Oesch, J. M., & Zietsma, C. (2007). What competition? Myopic self-focus in market-entry decisions. *Organization Science, 18*(3), 440–454. https://doi.org/10.1287/Orsc.1060.0243

Muckli, L., & Petro, L. S. (2017). The role of mental maps in-making. *Trends in Neurosciences*. Elsevier Ltd.

Ostasevičiute, R., & Šliburyte, L. (2008). Theoretical aspects of product positioning in the market. *Engineering Economics, 56*(1), 97–103.

Payne, B. (2011). *Scorecard valuation methodology establishing the valuation of pre-revenue, start-up companies*.

Puca, A. (2020). *Early stage valuation: A fair value perspective*. Wiley.

Ramanujam, M., & Tacke, G. (2016). *Monetizing innovation: How smart companies design the product around the price*.

Tucker, Matthews. 2022. *The Ten Medtech Commandments: Ten Imperatives to Achieve Commercial Success with Products That Improve Care so More Patients Get the Treatments They Deserve*. Breakthrough Holdings LLC.

Vreman, R. A., Inkatuuli Heikkinen, A., Schuurman, C. S., Garcia, J. L., Hedberg, N., Athanasiou, D., Grueger, J., Leufkens, H. G. M., & Goettsch, W. G. (2019). Unmet Medical Need: An Introduction to Definitions and Stakeholder Perceptions. *Value in Health, 22*(11), 1275–1282.

Deep-Dive into the Mapping of Decision-Related Problems

7

Abstract

Problem visualization helps decision-makers navigate a multitude of concurrent decision problems. In the case of medical ventures, the ultimate problem lies in delivering the promised return to investors. The K9 map was designed by reverse-engineering this objective; from this standpoint, it can be considered fit to solve that ultimate problem. This chapter explores more profound layers of the K9 map, breaking each of the nine clusters into sub-clusters and graphically organizing their hierarchical relationships. Doing so supports and orients decision-makers in appraising their venture's most critical decision-related problems and predictive assumptions. This exercise originates a significantly more detailed decision map, constructed with the guiding support of extensive literature reviews and validated by field experience. This larger, "exploded map," places particular emphasis on problems that frequently appear in medical innovation, including those that entrepreneurs often overlook. Regardless of whether this more extensive and detailed map directs the entrepreneur's attention to necessary investigations, scenarios, or investment decisions, it should be routinely browsed to prioritize the decision-related problems requiring higher-quality attention and resources.

7.1 Market Attractiveness and Unmet Medical Needs

Life sciences startups must minimize their innovation-related risk by creating products that deliver unique value to large, growing, and profitable market segments characterized by high unmet medical needs. For this reason, an unbiased estimation of the market attractiveness is essential, constituting an invaluable foundation for success.

L. Negri, *De-Risking Investments in Medical Technology Innovation*, Springer Series in Healthcare Management and Innovation,
https://doi.org/10.1007/978-3-032-15007-3_7

This process requires the making and testing of numerous assumptions related to market size and unmet medical needs. As shown in Fig. 7.1, the four proposed founding pillars of market attractiveness are market size, degree of competitiveness (determining both barriers to entry and profitability), the patient's unmet medical needs, and the unmet needs of the health care system together with society. There are other known elements that companies should consider, for instance, potential regulatory, reimbursement, or affordability barriers. These are essential but highly complex elements that are better placed in dedicated, subsequent clusters; this chapter focuses only on three basic drivers: need, volume, and degree of competitiveness (or degree of rivalry).

A fact-based estimation of the market size helps the venture determine the viability of any business plan. If the target market is excessively small, the effort may not be rewarded with attractive returns. Market size also affects the probability of raising capital: A large market is more likely to attract investors than a smaller one. On the other hand, many small markets can support a healthy private company, which can be an attractive acquisition target (MDDI, 1997). In general, incorrectly

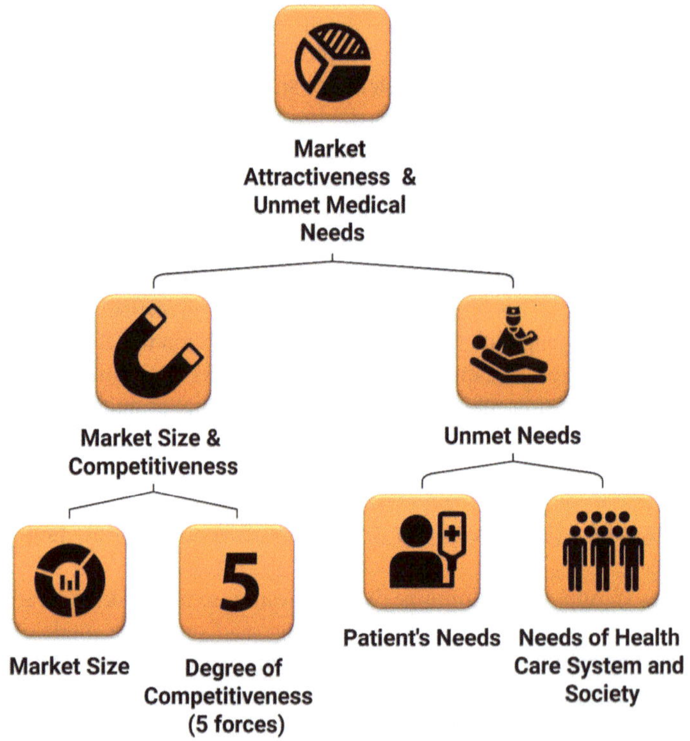

Fig. 7.1 The four proposed founding pillars of market attractiveness: market size, degree of competitiveness, patient's unmet medical needs, and the unmet needs of the health care system and of society

estimating the market size entails compromising the validity of all subsequent commercial and financial predictions in the K9 map up to the final valuation cluster.

7.1.1 Market Attractiveness

Medical technology entrepreneurs can therefore assess market attractiveness—similar to entrepreneurs in other industries—by weighing the size of the market (in units and in economic value) and the degree of its competitiveness.

7.1.1.1 Total Market Size and Value

To assess the market size, several standard methods are used, often in combination, depending on data availability, the maturity of the sector, and the product's characteristics. Broadly, these approaches fall into two categories: top-down and bottom-up.

The Top-Down Market Analysis

This approach often begins with high-level datasets, such as epidemiological indices, as the primary drivers of the respective overall healthcare spending and market size. Reports from reputable organizations, such as the World Health Organization (WHO) or established market research firms, often serve as foundational sources for this analysis.

The determination of the market size is often linked to the analysis of disease prevalence. Prevalence refers to the proportion of individuals in a population who have a particular disease or attribute at a specific time or over a specified period.

It is essential to distinguish whether quantitative market estimation focuses on an existing market or a new one. Existing markets are those whose development has been, to a large extent, achieved. For instance, the demand for endotracheal tubes is currently well established, but it has not always been the case. The endotracheal tube is a result of the ideas and innovations of physicians and entrepreneurs across multiple centuries, which went through numerous iterations before arriving at its current form (Constantino Jr. & Waisel, 2018). During those years, a growing existing market took over a potential market that was increasingly being exploited.

Prevalent and known transactions characterize existing markets. For their estimation, entrepreneurs can utilize published market surveys that exist for various diseases and their related therapeutic and diagnostic devices. For instance, US market data on medical products are available from IMS Global Services and consulting firms such as Frost & Sullivan. Other sources include research reports published by investment banking firms, information from trade associations, such as the Health Industry Manufacturers Association, and even annual reports from companies in the industry (MDDI, 1997).

On the contrary, estimating potential market size entails predicting future demand, which is often based on disease prevalence and incidence in the field of medical innovation. In many cases, a potential market can be estimated as a portion or an extension of an existing market. For instance, in the very early stages of automated intraoperative blood recovery equipment, the market was mainly estimated to be a portion of blood-intensive surgeries, such as open-heart surgeries, liver transplants, and hip replacement surgeries.

Another example comes from the home-based robotic rehabilitation market, which is currently hardly an existing market, but rather a portion of the existing market for robotic rehabilitation devices used in rehab clinics. In summary, determining a potential market size is considerably more complex than determining the size of a current market.

Early Segmentation Helps Avoid Market Overestimation

When applying top-down approaches, we have often observed medical technology entrepreneurs overestimating the patient population (particularly in the area of rare diseases). One reason is that, in daily practice, the number of patients requiring treatment rarely corresponds to the one mathematically extrapolated from disease prevalence (Bratic et al., 2014). For instance, overestimating disease prevalence is common for many rare diseases, probably because global prevalence estimations are obtained by extending the prevalence rates observed globally in regions where prevalence is known, which often bear higher prevalence rates. Overestimation of the patient population can significantly affect the accuracy of revenue forecasting and, consequently, that of the startup's value.

Another frequent cause of overestimating the total market size is confusing actual and potential markets. For instance, startups commercializing devices to support the correct positioning of prosthetic knee replacement during knee surgery may be tempted to hold the total market as equal to the projected global forecast for surgical knee replacements. However, in most cases, these medical products have specific fields of application within the named surgery. While global surgical knee replacements may represent the upper ceiling, products like navigation systems, positioning guides, or sensor-based alignment devices rarely penetrate all cases.

Other, often overlooked, market-restricting factors come from the number of insured and non-insured patients who can afford an out-of-pocket payment.

Ignoring market-restricting factors is a frequent mistake that provides artificially large and misleading market estimations. Entrepreneurs falling into the trap of using unrealistically large market estimations can predict attractive revenues even while applying very conservative penetration rates. This predictive bias hides additional collateral damages. Indeed, conservative penetration rates often give the entrepreneur a false sense of "safety" and "honesty" regarding the accuracy of optimistic revenue forecasts. Showing attractive revenues by applying merely a few percentage points of penetration is indeed perceived as a safe predictive approach that does not require further validation. When these generous, global revenue predictions are confronted with more detailed, country-based, region-based, or hospital-based bottom-up analyses, they immediately exhibit their weaknesses.

Suppose we apply these considerations to the medical use of augmented reality (AR)—a relatively new technology with several medical applications, including the execution of distance-proctored surgeries. As one of its possible applications, AR allows surgeons to incorporate data visualization into diagnostic and treatment procedures to improve work efficiency and safety and reduce cost, enhancing surgical guidance and training (Grimaldi et al., 2022). For a startup operating in this field, estimating the total available market using the figures of the entire surgical market constitutes a critical error in prediction. The global market size of surgical procedures was roughly 300 million units in 2022, and it is predicted to reach almost 400 million units by 2030 (Globe Newswire, 2022). From a revenue prediction standpoint, these are very attractive figures, but in reality, the actual necessity for remote surgical guidance is likely to be rather limited to geographies characterized by highly concentrated competencies in a few metropolitan areas surrounded by large rural areas with an acute unmet demand for highly specialized surgeons. In this case, a more accurate total market prediction for the proctoring and training market in the USA should consider only the number of specialty surgical procedures currently performed in rural areas lacking highly skilled surgeons.

Bottom-Up Approaches

In contrast to the top-down approach, bottom-up market estimation begins at the individual product or customer segment level to create an overall market size. These methods are more focused on elements like the patient journey or on the specific needs of individual healthcare organizations.

These methods utilize techniques such as direct surveys, focus groups, and interviews with stakeholders, which provide insights into actual usage and potential demand for specific products. These techniques can also be viewed as refining and filtering processes that enhance the top-down approaches. Bottom-up approaches should therefore be considered as a necessary complement to top-down approaches. In other words, after top-down market estimations, forecasters should always utilize an explicit refining process.

For instance, the map shown in Fig. 7.2 outlines the typical assumptions that must be made (and tested) to refine the number of patients requiring vascular stenting or hip replacement in a specific geographical area. Epidemiological parameters such as prevalence (NIMH, 2022) are used to calculate the number of individuals fitting the generic indications for the procedure. Other epidemiologic and demographic data can further support the forecaster, indicating the growth rate of the segment. These initial assumptions provide only a rough estimation of the market size. Indeed, not all these patients will undergo a surgical procedure. Only patients suffering from specific symptoms or those requiring regular medical checkups may undergo a surgical procedure. It is a smaller group of patients, indicated in Fig. 7.2 with the "potential patients" icon. Of this group, only a subsegment (indicated with the icon "treated patients") actually undergoes a stenting procedure: patients with actual access to the therapy. This additional filter is primarily defined by the presence of a public reimbursement (in countries with more "socialized" healthcare systems) or by the percentage of individuals who are privately insured or can afford to pay out-of-pocket for the procedure. Moreover, most early-stage startups do not enjoy commercial access to a global market on the very day of the product launch. Therefore, an additional filter must be used to reduce the available market to regions offering actual commercial reach.

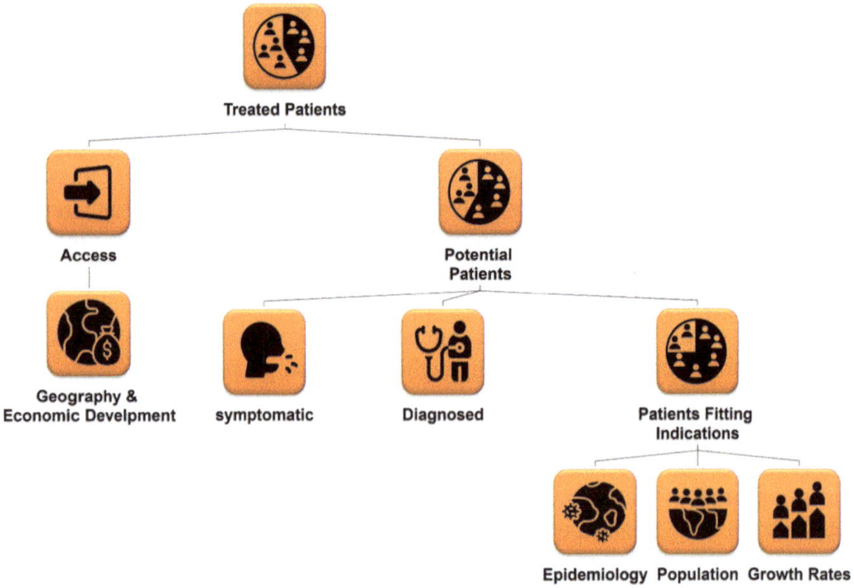

Fig. 7.2 A systematic process to quantify the number of patients potentially treated with a new technology

TAM, SAM, & SOM

In the context of market refining size analysis, the concepts of total addressable market (TAM), serviceable available market (SAM), and serviceable obtainable market (SOM) are often used to distinguish the potential market size by the market portion that a company, with its products, can actually serve and by the realistic share that can be captured. TAM represents the overall revenue opportunity available if a product or service achieves 100% market share. SAM restricts it to the segment of the TAM targeted by a company's products and services, while SOM further refines it to the subset of the SAM that a company can realistically capture in the short term.

In general, employing a multiple-filter approach ensures a finer segmentation of the total market and a sharper identification of the market relevant to a specific medical innovation.

In summary, while estimating market sizes, the medical entrepreneur must adopt a rational position in answering the following questions:

- What is the top-down basis for the market size calculation—prevalence, incidence, or the number of currently performed diagnostic or therapeutic procedures?
- What bottom-up considerations can refine this estimation? For instance:
  - Which additional filters best extract the number of patients, physicians, or hospitals that are potentially in need of or attracted by my medical innovation?
  - Are there multiple layers of filtration?

- How does patient access (to therapy or diagnosis) influence the size of the available market?
- In which way does the venture's commercial reach influence the definition of the reachable market?

7.1.1.2 Degree of Competitiveness (or Rivalry)

It is essential to note that this section focuses solely on the general degree of market rivalry or competitive intensity, taking into account contextual elements such as price pressure, the number of competitors, and the degree of commoditization of the product or service type. It does not account for head-to-head competitiveness, which is intended to represent the competitive edge of a new medical product or service idea compared to each competing offer.

Section 7.4 is entirely dedicated to the consideration held by the fourth cluster of the K9 map, which groups all the issues related to the competitiveness of the new product or service compared with each individual rival.

A high degree of competitive intensity can be defined as "a situation where competition is fierce due to the number of competitors in the market and the lack of potential opportunities for further growth" (Auh & Menguc, 2005).

Early-stage medical startups often overlook both the degree of current competitiveness in the target market and the barriers that can be faced by a young startup entering this market. For a medical entrepreneur, it is essential to characterize the target market segment and make correct assumptions about its degree of rivalry.

For instance, many startups are currently in the process of entering the global segment of wound management. This market segment is worth around US$30 billion and is growing at a low, single-digit level in a highly competitive context due to the many established companies offering a variety of solutions. This competition is characterized by a wide spectrum of participants, ranging from well-established multinational corporations to innovative startups. All these participants are driven by the constant pursuit of innovation for the purpose of better meeting current needs and the changing requirements of physicians and patients.

As a consequence, for many new medical ventures, entering this market could therefore be considered unrealistic or likely to require quantum-leap innovations that may credibly disrupt the current medical practices.

Another example comes from a few startups that are still playing in the Foley catheter segment, which are thin, flexible tubes that drain urine from the bladder into a collection bag. These are also known as indwelling catheters, as they can remain in place for extended periods. This product category has been considered a commodity since the early 1990s; unless an entrepreneur is able to revolutionize the core concepts of the products currently available, entering a market where the leading technologies have already reached the end of their lifecycles is inadvisable.

These markets are characterized by declining prices and a desperate search for profitability, often sought through high volumes or manufacturing cost advantages. In many cases in these markets, there are reimbursement prices that, over time, have slowly adapted to the low-price and high-commoditization trend. These prices became references that are hard to change and intrinsically "hostile" to the higher prices of more innovative products. When evaluating their entry into these highly competitive contexts, entrepreneurs must ensure that they intend to launch a quantum-leap innovation that concurrently addresses critical needs and impacts a vast population.

Reality has taught us that, in markets with high competitive rivalry, a new technology characterized by marginal benefits for the patient or for the overall healthcare delivery cost is likely to miss the opportunity to participate in the global market completely. This consideration becomes even more relevant when a small, young venture commercializes the technology.

In summary, while evaluating the degree of competitiveness, the medical entrepreneur must adopt a rational position in answering the following questions:

- How many companies are already commercializing solutions to the same set of unmet medical needs?
- What is the stage of the space's technology life cycle?
- To what extent is price the current driver for adoption decisions?
- What are the sources of price pressure on the target space?
- Are past reimbursement prices setting low product or service costs?
- What is the organizational development stage and the commercial reach of the main current players?
- Does the new medical idea represent a quantum-leap innovation or only a marginal one?

And, as already discussed, if the answers to the above questions paint a picture of a high degree of market rivalry, then the next key questions must focus on the ability of the venture's technology to be disruptive.

7.1.2 Unmet Medical Needs

Qualifying and quantifying the unmet medical needs is a critical phase for a correct assessment of the market attractiveness.

The willingness to adopt and pay for new and often more expensive medical technologies is influenced not only by their improved therapeutic effectiveness or diagnostic accuracy, but also by how patients, key stakeholders, and society assess the urgency, impact, and relevance of an unmet medical need. Key elements, such as urgency, impact on quality of life, health-economic considerations, and the overall significance of unmet needs, play a crucial role in shaping the willingness to adopt and pay for these technologies (Stein et al., 2016).

When a medical condition is perceived as urgent or life-threatening, individuals and stakeholders are more inclined to accept and invest in new treatments and technologies. This sense of urgency creates an immediate motivation to improve health outcomes, as evidenced by patients' willingness to seek innovative therapies that promise rapid relief or enhancement of their quality of life. Researchers have noted that urgency influences not only patient decisions but also how healthcare providers prioritize certain treatments (Toloo et al., 2016; Coster et al., 2017).

The potential impact of a medical technology on a patient's quality of life significantly affects their willingness to pay. For instance, patients often show a readiness to adopt technologies that manage chronic conditions, such as diabetes management tools, due to their direct effects on daily living. Healthcare stakeholders—including policymakers, providers, and patients—are more likely to invest in and adopt technologies that address known unmet medical needs where there is a clear gap in existing treatment options. When the medical community rallies around a particular need, as seen in the push for more effective treatments for conditions like Alzheimer's disease, the incentives for adoption and funding become significantly stronger.

Economic factors, including current societal spending and existing reimbursement policies for specific diseases, also influence the willingness to adopt medical technologies. When society allocates significant funds for the prevention and treatment of certain diseases, there is a compelling unmet need for solutions that offer robust health-economic justification in terms of long-term cost savings or improved patient outcomes. As a result, the willingness to invest tends to rise.

One important and unfortunately recognized proof of the role of patients' and stakeholders' perceptions regarding the relevance of needs is the underinvestment in preventive medicine.

Traditionally, healthcare financing models prioritize curative treatments over preventive measures. This approach is largely due to the immediate benefits that curative therapies provide, leading healthcare providers and institutions to invest more in solutions that demonstrate a clear and prompt impact on health outcomes. For example, treatments for diseases like diabetes and cancer often receive priority because they require urgent interventions, while preventive measures may be viewed as less pressing. In low-income settings, households frequently neglect preventive care, as the urgent need for curative solutions overshadows options for prevention (Baulia, 2020).

Preventive healthcare is often underestimated in terms of its benefits. Both individuals and policymakers tend to see the effectiveness of preventive measures, such as vaccinations, as reliant on the absence of disease, which makes their advantages less apparent. Consequently, society's willingness to fund or adopt preventive interventions tends to be lower. While preventive health measures can generate substantial health benefits and cost savings over time, societal focus remains significantly oriented toward treatments, especially when immediate effects are observable (Masters et al., 2017).

Medical entrepreneurs must investigate, analyze, and validate every aspect of the unmet medical needs. Characterizing unmet medical needs is difficult; it is a nonlinear, iterative process that must be regularly revisited and re-validated. In software innovation, pivots related to reassessments of customer needs are the most common among all pivot types (Bajwa et al., 2017) and are considered very healthy processes. In medical technology innovation, pivots linked to changes in the understanding of patients' and stakeholders' needs are much less common.

In the medical technology industry, an unmet need has multiple dimensions, from clinical to personal and social to economic: It is not defined merely by patients but also by the combined individual requirements of many stakeholders with various roles and often competing priorities.

A profound understanding of unmet needs requires the input and observations of physicians, patients, nurses, and other healthcare professionals. It requires knowledge of the anatomy, physiology, pathophysiology, and etiology of the disease. The prevalence and incidence of the disease, as well as the existing therapeutic market, must be clear. Segments where market entrance may be easier must be precisely identified and quantified. The context of available treatment options and the patient's journey must be rationally framed (Behrns et al., 2021).

As already and tirelessly reinforced, the product development process in life sciences is a very high-risk exercise due to the fast-changing industry and market landscape. Thus, both startup and established companies must minimize their innovation-related risk by creating products that precisely target market segments characterized by a high degree of unmet needs and deliver unique value to these markets. Doing so is critical for startup ventures since they deal with highly uncertain and constrained financial resources (Jambulingam, 2018).

There is no universally accepted definition of "unmet need," and different commentators use the term in different ways (S. Smith & Connolly, 2020). In a popular and general definition, unmet medical needs refer to the state in which a patient's medical care or service is insufficient, inadequate, or lacking (Yoon et al., 2019). In other words, "unmet medical needs means a condition for which there exists no satisfactory method of diagnosis, prevention or treatment (…) or, even if such method exists, concerning which the medicinal product concerned will not be of a major therapeutic advantage to those affected" (European Medicines Agency, 2019).

Other sources suggest that patients are not the only relevant stakeholders influencing and determining unmet medical needs; researchers have identified a larger group of stakeholders, including policymakers and the broader society (S. Smith & Connolly, 2020). The satisfaction of the criterion of unmet medical needs for one stakeholder does not necessarily mean the satisfaction of the criterion for another; therefore, context is essential while applying the concept and definition of unmet needs.

The challenge for life sciences entrepreneurs attempting to assess unmet global needs relevant to their innovative idea is that unmet needs constitute a complex, multidimensional concept that can be neither captured nor measured by a single indicator or metric (Vreman et al., 2019) (Fig. 7.3). To offer guidance on this complex issue, some authors suggest different tools that facilitate data collection in an attempt to measure unmet medical needs. For instance, tools to measure important indicators of patients' unmet needs are individual, self-reported questionnaires, disease-specific scales, and the disability-adjusted life in years (DALYs). This book embraces the orientation of the authors (Vreman et al., 2019) who identified three key recurring elements as the defining pillars of unmet needs (Fig. 7.3):

1. Severity of the disease
2. Existence of alternative treatments
3. Patient population

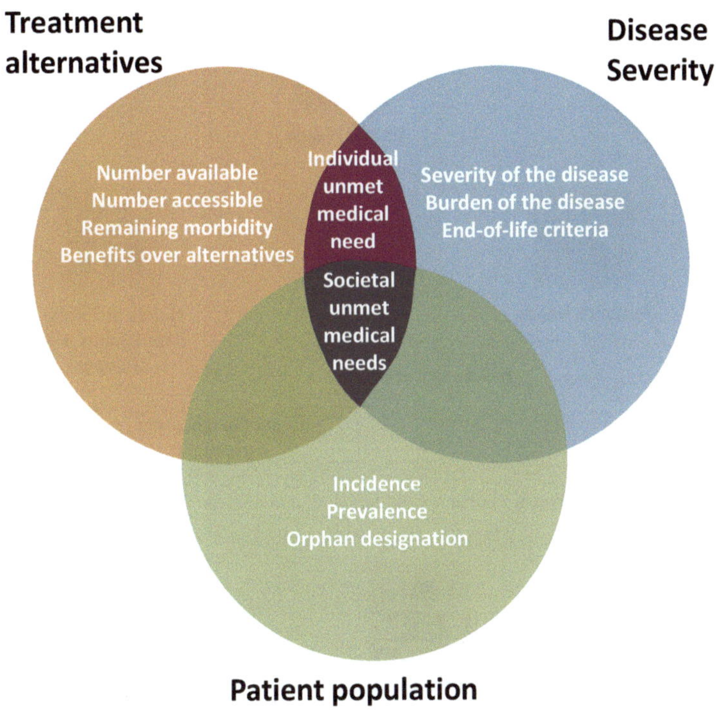

Fig. 7.3 The defining pillars of unmet needs (Vreman et al., 2019)

This framework suggests that, in the context of a specific patient population, the interaction of disease severity and availability of treatments defines individual unmet needs (Vreman et al., 2019).

Furthermore, we may refine the definition of patients' unmet needs by identifying sub-types of need, such as unperceived unmet need, which exists when individuals are unaware that they have that particular need. Subjective unmet need is typical of an individual who perceives a need but does not seek medical treatment. Subjective unmet expectations emerge when an individual perceives a need for treatment and subsequently seeks and obtains treatment but does not see it as adequate (S. Smith & Connolly, 2020).

The book further distinguishes between the unmet medical needs and their drivers (Fig. 7.4): drivers that generate, reinforce, or mitigate the unmet needs, and unmet needs which are a specific reflection of these drivers on the patients and each stakeholder.

7.1.2.1 Patients' Needs

In alignment with the mentioned approach, patients' needs are characterized by the overall need drivers and detailed at a personal level by specific individual needs, which are further categorized as clinical and non-clinical (Fig. 7.5).

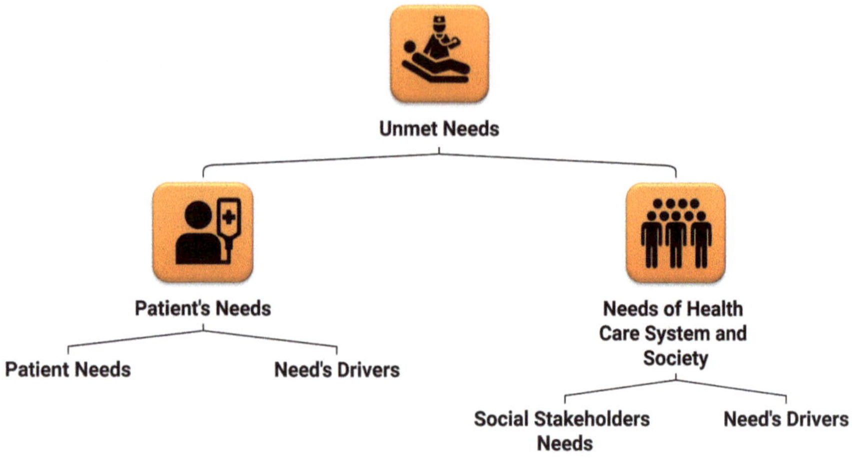

Fig. 7.4 The basic framework to observe, understand, and analyze unmet medical needs

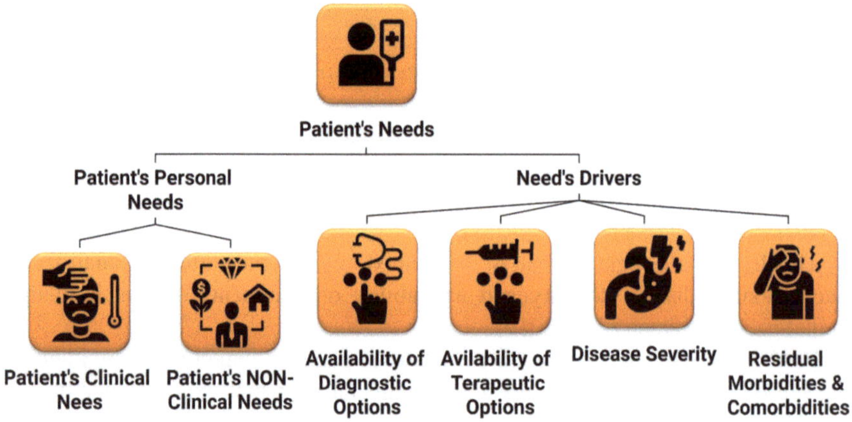

Fig. 7.5 A framework to observe, understand, and analyze patient's unmet medical needs

Patients' Personal Needs

Patients' personal unmet needs can be categorized as clinical needs and non-clinical needs (Fig. 7.6). Unmet clinical needs are gaps resulting from unsatisfactory clinical outcomes of the available diagnostic and therapeutic techniques. Diagnostic unmet needs concern the safety, sensitivity, and specificity of the available diagnostic platforms. Therapeutic unmet needs pertain to the safety, efficacy, and effectiveness of the available therapies.

On the other hand, patients' non-clinical needs are individual, social, and economic needs, such as housing, food, transportation, and protection from violence. Social needs are major contributors to poor health outcomes and are estimated to impact up to 50–60% of health outcomes (Joseph et al., 2023). In an effort to

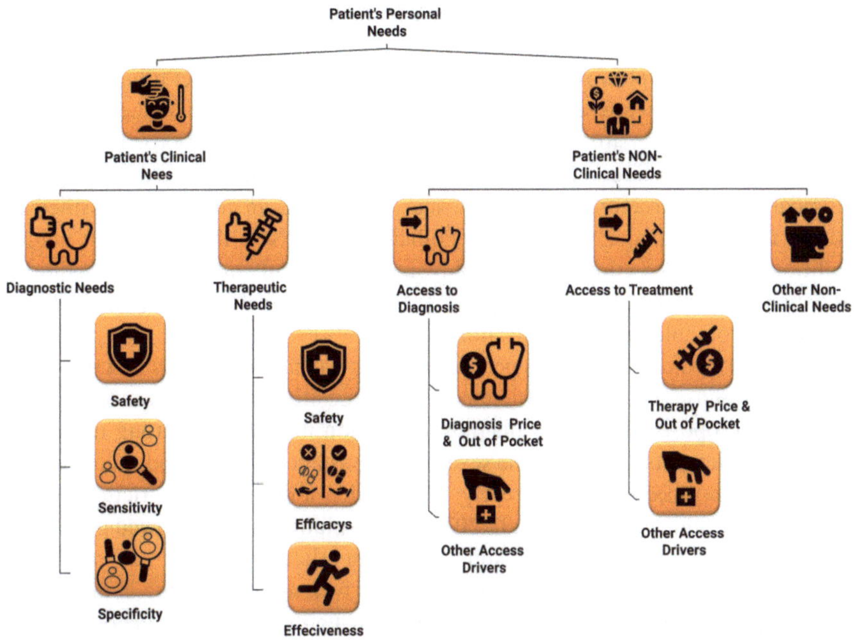

Fig. 7.6 The framework of the patient's personal unmet medical needs

simplify the matter, we include the issues of access to existing therapy and diagnosis in the category of non-clinical needs. Indeed, the drivers of accessibility are led by a multitude of non-clinical factors, such as insurance decisions, patients' economic ability to pay for certain unreimbursed procedures, and even patients' residence in rural areas that do not have access to certain specialized centers (Fig. 7.6).

Drivers of Patients' Needs

Under European Community regulations (Article 4 paragraph 2 of EC No. 507/2006): patient's unmet medical needs means a condition for which there exists "no satisfactory method of diagnosis, prevention or treatment in the Union or, even if such a method exists, in relation to which the medicinal product concerned will be not of major therapeutic advantage to those affected."

Therefore, the first driver of unmet patient needs is the availability of diagnostic and therapeutic options. These drivers depend on the number of alternative diagnostic or therapeutic solutions available, the number of diagnosis options or treatments that are realistically accessible, the residual morbidity after these treatments, and the actual medical benefits of these treatments.

An essential example of the unsatisfying availability of therapeutic alternatives for a patient may be seen in rheumatoid arthritis (RA). Despite the wide range of treatments available, RA continues to pose a substantial human and economic burden on patients, and unmet needs, such as pain, physical function, mental function, and fatigue, continue to persist across key domains. These findings suggest a need for further advances in treatment for RA to address these contemporary unmet needs (Taylor et al., 2016).

This example emphasizes the importance of another vital driver of unmet needs: disease severity. Disease severity is the total effect of the disease on the individual, including both irreversible and reversible components (Liang, 2006). The severity of the disease, therefore, refers to the extent of organ system derangement or physiological decompensation of a patient, and in general, an illness is classified as minor, moderate, major, or extreme (Pollack, 2016). New therapies with unique abilities to diagnose, cure, or measurably and significantly alleviate the conditions of patients affected by high-severity illnesses are more likely to experience a faster and more profitable market penetration.

From the medical innovator's perspective, it is crucial to highlight that the severity of the disease alone is insufficient to create positive market entrance expectations. The proposed therapy or diagnostic solution must have a strong impact on most of the elements that define the severity. For instance, the rapid adoption of total hip arthroplasty occurred in the early 1960s when a durable, low-friction arthroplasty was developed (Knight et al., 2011). This solution finally allowed patients to walk again, and for decades, with significant pain relief.

However, the clinical impact on patients is driven not only by the ability of the proposed technology to diagnose, avoid, or mitigate the severity of the disease but also by its capacity to reduce residual morbidities or comorbidities. The non-clinical benefits for the patient (or the caregiver) include improved accessibility to diagnosis and/or therapy, patients' therapeutic experience, and the impact on patients' social life, family life, professional life, and personal financial situation. Figure 7.6 offers a visual description of the components that are of critical value to the patient. Access to both therapies and diagnosis is affected by several factors, but one of the most socially difficult to accept is their affordability; indeed, in many cases, these procedures must be paid for partially or in full out-of-pocket by the patients themselves.

Identifying Unmet Needs Through the Patient's Journey

The patient's journey can be defined as the "spatio-temporal distribution of patients' interactions with multiple care settings over time" (Carayon & Wooldridge, 2020). It describes the series of steps and personal experiences that a patient encounters while interacting with different healthcare providers and services, from diagnosis to treatment, and post-treatment processes. Mapping the patient's journey is a useful way to visualize key sequential experiences of a patient seeking care and relief. Patient journey mapping "describes the patient experience, including tasks within encounters, the emotional journey, the physical journey, and the various touch points"(McCarthy et al., 2016). It includes not only clinical aspects but also patients' emotional and psychological experiences, which are critical to understand both unmet needs and preferences (Oben, 2020). McCarthy suggests a handy way to frame the components of the patient's journey mentally (McCarthy et al., 2016) that helps the medical entrepreneur build the journey's map. The framework consists of seven main blocks: encounters, task, goals, constraints, actors, emotional journey, physical journey, and device touch points. Encounters are the sequential touch points patients have with different healthcare delivery points. Tasks are the specific assignments to be delivered by each actor in the healthcare service in an attempt to obtain certain clinical or non-clinical goals without violating constraints such as medical guidelines, safety, and managerial governance. Actors are those individuals directly involved in the delivery of clinical and administrative healthcare services. The emotional journey tracks a patient's emotional state throughout the journey's timeline, marked by events of specific physical interactions between the patient and healthcare professionals or organizations. Interestingly, for the medical entrepreneur, the last component of the patient's journey concerns device touch points, and it focuses on all the technologies employed by the different healthcare professionals, caregivers, and patients at each touch point.

Complete Map for Patient's Unmet Medical Needs

Figure 7.7 summarizes the key elements relevant to the determination and characterization of patients' unmet medical needs.

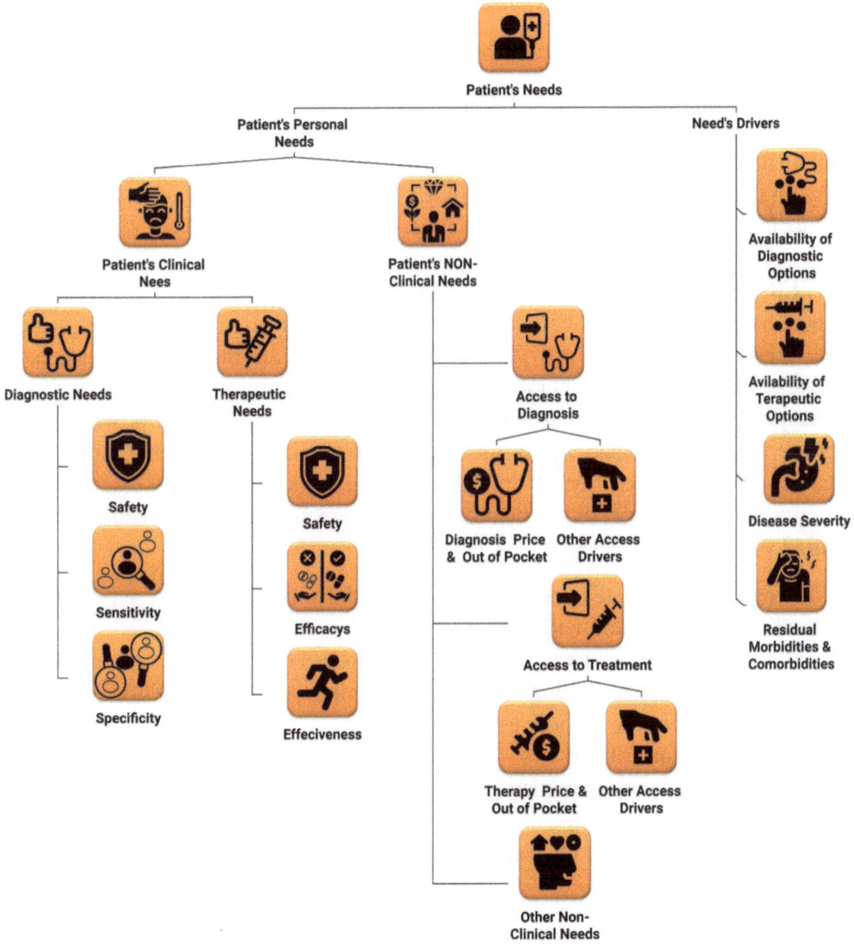

Fig. 7.7 The complete framework of Patient's unmet medical needs

Fig. 7.8 The unmet needs of social stakeholders and society

7.1.2.2 Needs of HealthCare Systems and Society at Large

Following the approach used for patient needs, we characterize the unmet needs of social stakeholders and society, distinguishing their drivers by the specific needs of individual stakeholders (Fig. 7.8).

Unmet Needs of Social Stakeholders

The study and characterization of the stakeholders contributing to decisions relevant to patients and therapies are presented in Section 7.5. This chapter, instead, focuses only on the three most critical stakeholders for the identification of unmet medical needs: healthcare providers, intermediaries, and payers.

7.1.2.3 Unmet Needs of Healthcare Providers

A healthcare provider is a licensed individual or organization that provides healthcare services. Healthcare providers encompass a diverse range of professionals, including doctors, nurses, therapists, pharmacists, laboratory technicians, and staff at hospitals, clinics, and other healthcare centers. The goal of these providers is to deliver healthcare to patients in an effective, financially sustainable, and timely manner. With a high degree of simplification, the priorities (and, by extension, the determination of the needs) of healthcare providers can be summarized as follows (Fig. 7.9):

- To ensure an adequate operating capacity for delivering healthcare services to all the patients associated with the provider
- To constantly deliver the highest-possible standard of care
- To maintain the costs within the available reimbursement framework
- To advocate for new or more generous reimbursement when a new technology is evaluated and prioritized
- To negotiate economic risk-sharing with insurance companies and other intermediaries
- To project an image of technological advancement with the highest standard of care

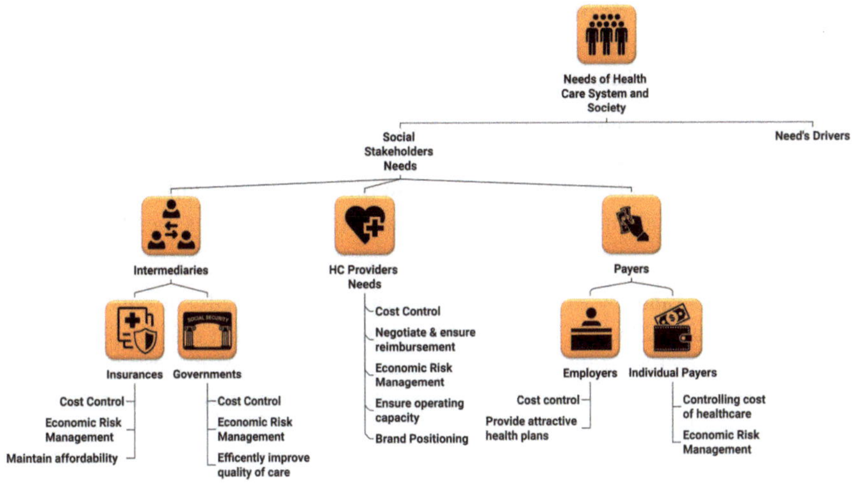

Fig. 7.9 A visual framework of the unmet medical needs of healthcare systems and social stakeholders

7.1.2.4 Unmet Needs of Intermediaries

Intermediaries are entities that collect funds from a group of people, pool the funds, and use them to pay for healthcare on behalf of individuals who are covered. They are also referred to as private insurance companies, insurers, and health plan providers (NCI 2024b). There are two types of intermediaries: insurance companies and government payers.

Insurance companies are private organizations selling insurance policies: contracts that promise the payment of the holder's medical bills in exchange for the payment of insurance premiums. These premiums are known expenses that individuals are willing to bear in exchange for mitigating the expenditures caused by possibly expensive and unforeseen medical bills. Governments function as a public intermediary: they collect funds from the population and, in exchange, cover healthcare costs for some groups of people; the funds can be collected through the tax system.

Due to the role of the intermediaries, most funds flow through them from the patients to the healthcare providers. The patients pay money for private insurance in the form of premiums or to a government intermediary through taxes or other fees. The intermediaries pay healthcare providers to provide care to patients. There are also cases in which patients directly pay the healthcare delivery institution; they are called co-payments or out-of-pocket payments.

The high-level priorities (and, by extension, the determination of the needs) of intermediaries can be summarized as follows:

- Constant improvement, management, and negotiation of risk-sharing with healthcare delivery organizations and patients
- Mitigation of the effects of growing healthcare costs due to new technologies
- Promotion of the expansion of technologies that improve the efficiency of affordable healthcare delivery by adopting value-based care

7.1.2.5 Unmet Needs of Payers

This book uses the term "payers" to refer to all the individuals and organizations that actually finance the entire healthcare system:

- Individual payers make payments to governments (through taxation), to private insurance companies, and, sometimes, directly to healthcare providers in the form of co-payments or direct payments in the absence of insurance.
- Employers (mainly in the USA) may sponsor their employees' health insurance policies.

The interests (and, by extension, the determination of needs) of payers involve the search for and selection of affordable insurance plans that effectively mitigate financial risks to health while providing satisfactory quality healthcare services. Figure 7.9 provides a visual summary of the unmet medical needs of healthcare and social stakeholders.

Drivers of Stakeholders' Needs

Figure 7.10 provides a snapshot of the drivers of social stakeholders' unmet needs. Most stakeholders agree that, from a societal perspective, unmet medical needs increase with the growing size of the patient population (Vreman et al., 2019). The size of the patient population remains, indeed, irrelevant to the direct determination of individual patients' unmet medical needs. Nevertheless, a patient may be indirectly affected if an unexpectedly fast-growing population size puts the healthcare system under pressure for service supply. Another important case occurs when an individual patient is part of a very small population. In such a situation, the patient may be confronted with the limitation or unavailability of treatment options and adequate healthcare organization competencies.

It is necessary to remember that this situation changes significantly when a rare disease is classified as an orphan disease, thereby receiving special attention from healthcare authorities. According to the FDA, an orphan disease is defined as a condition that affects fewer than 200,000 people nationwide in the USA (The Orphan Disease Collective, 2025). They include diseases as familiar as cystic fibrosis, Lou Gehrig's disease, and Tourette syndrome and as unfamiliar as Hamburger disease, Job's syndrome, and acromegaly or "gigantism." Some diseases have patient populations of fewer than a hundred. The degree of access to diagnosis is also an important variable for institutional stakeholders. Indeed, access to diagnosis defines the ability of stakeholders to accurately estimate the size of the patient population and the required capacity for delivering healthcare services and pharmaceutical products. Obviously, and independently by the degree of accessibility to certain diagnostic procedures, the existence of valid diagnostic options is an even stronger determinant of the mentioned stakeholder needs. We may recall the public health issues encountered by both public and private healthcare institutions when reliable tests for COVID-19 were not available and how long it took between their availability and their more extensive accessibility.

The availability of appropriate therapeutic options and the degree of access to them are also important drivers for stakeholders and their ability to finance and

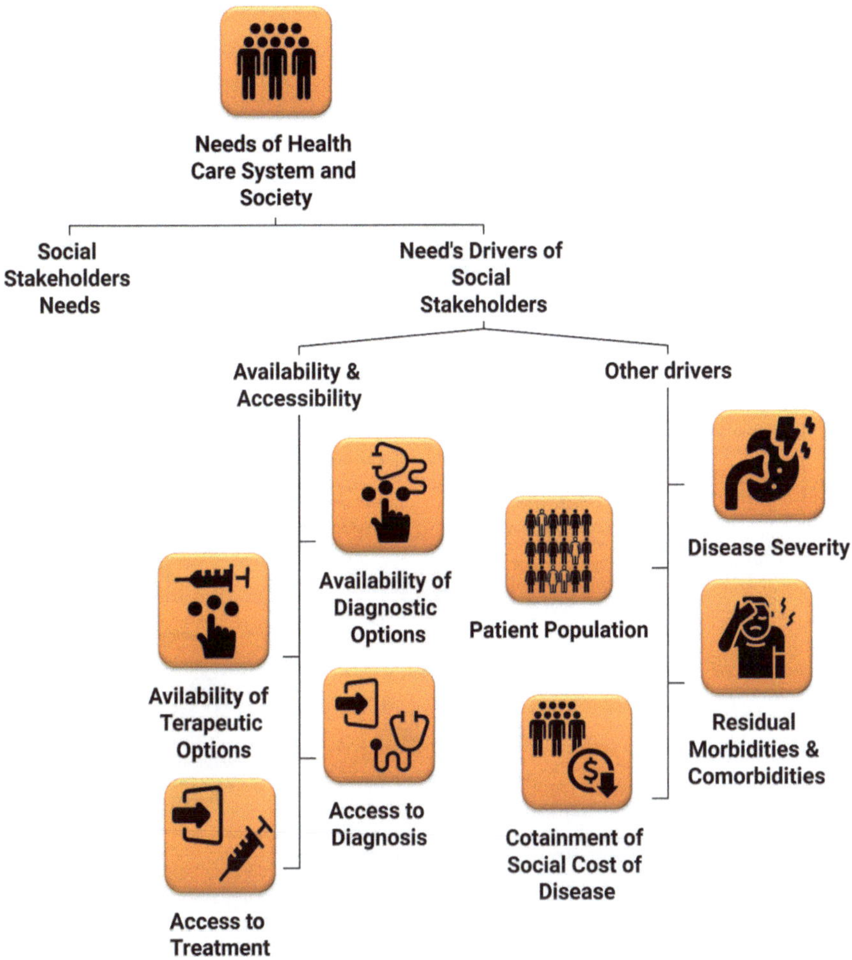

Fig. 7.10 A snapshot of the drivers of social stakeholders' unmet needs

provide adequate care. Additional reviews of the definitions of healthcare and social stakeholders' unmet needs confirm the shared dependence on the following elements: the adequacy of available treatments, the severity or burden of the disease, and the size of the patient population. It is also important to note that the quantification of unmet needs is highly dependent on the scope and value framework in which it is applied, based on different stakeholder preferences and responsibilities (Vreman et al., 2019).

Complete Map of Stakeholders' Unmet Medical Needs

Figure 7.11 summarizes the key elements relevant to the determination and characterization of stakeholders' unmet medical needs.

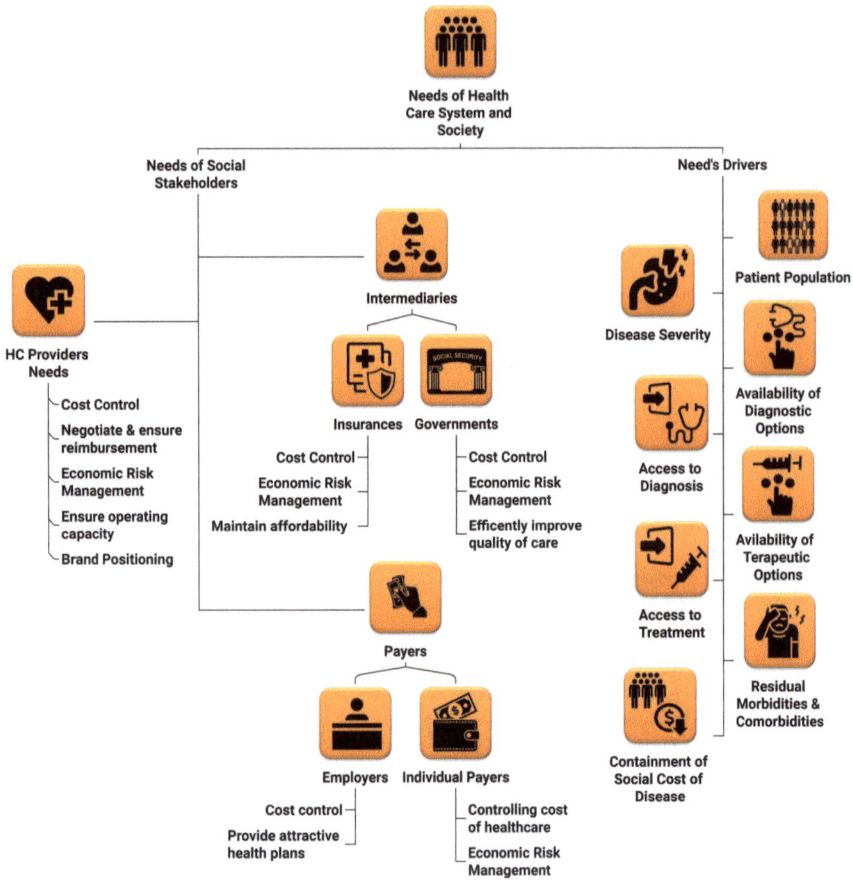

Fig. 7.11 The complete map of stakeholders' unmet medical needs

7.1.3 Key Questions Related to Assumptions and Decision-Making (Table 7.1)

Table 7.1 Market attractiveness and unmet medical needs: core assumption and decision-related questions

Market attractiveness and unmet medical needs	
Sub-cluster	Core assumption and decision-related questions
Market size	Is the market size (in the number of actual treatments or procedures) attractive?
	Is the market value (in the number of current sales) attractive?
	Which segmentation criteria can more precisely estimate the actual market size?
	Is the identified market sub-segment sufficiently large to maintain business interest?
Degree of rivalry	How many present and emerging competitors play in the same therapeutic or diagnostic arena?
	Is the current degree of rivalry acceptable for the new therapy's market entry?
	What degree of commoditization exists to increase rivalry and put pressure on prices?

(continued)

Table 7.1 (continued)

Market attractiveness and unmet medical needs	
Sub-cluster	Core assumption and decision-related questions
Patient needs	Can the degree of disease severity be considered a key component of patients' unmet medical needs?
	Are the morbidities or the mortality rates experienced with current therapeutic alternatives still high? Are they the key sources of unmet medical needs?
	What role do access to diagnostics and therapy play? Is access a substantial issue? How can it be measured?
	What is the change in the patient's quality of life and residual morbidities after the use of the currently available treatments?
	Does the disease cause significant impacts on the patient's social, professional, family, and economic life?
	To what extent is the safety of the existing treatments an unmet need?
	To what extent is the efficacy of the existing treatments an unmet need?
	To what extent is the treatment's efficacy an unmet need?
	To what extent is the safety of the existing diagnostic systems an unmet need?
	To what extent do the sensitivity and specificity of existing diagnostic systems address unmet needs?
	Would an earlier diagnosis offer the opportunity for beneficial earlier treatments?
	Has the patient's journey been drafted?
	What components of the patient's journey are more relevant to the unmet needs addressed by the innovative solution? Encounters? Tasks? Goals? Constraints? Actors? Emotional journey? Physical journey? Device touch points?
Needs of the healthcare system and society at large	What are the principal unmet needs of healthcare providers?
	What are the principal unmet needs of insurers?
	What are the principal unmet needs of the government?
	What are the principal unmet needs of employers?
	What are the principal unmet needs of individual payers?
	How big is the patient population that needs to be supported by payers and intermediaries? What is the economic burden that a cheaper therapy may relieve?
	Is the economic impact on society higher compared to other diseases?
	Is the economic impact on private insurance companies higher compared to other diseases?
	To what extent does the disease affect healthcare providers from an organizational and financial standpoint?
	Is residual morbidity a considerable healthcare issue? To what economic extent is it an issue?
	Is access to diagnosis or care a key determinant of societal needs? To what extent is access perceived as an unmet need?
	Is the lack of alternative diagnostic systems a social issue?
	Is the lack of earlier or more accurate diagnostic alternatives a recognized need?
	Is the lack of alternative treatments a social issue?

7.1.4 The Complete Map

Figure 7.12 summarizes the key elements relevant to determining and characterizing the market's attractiveness.

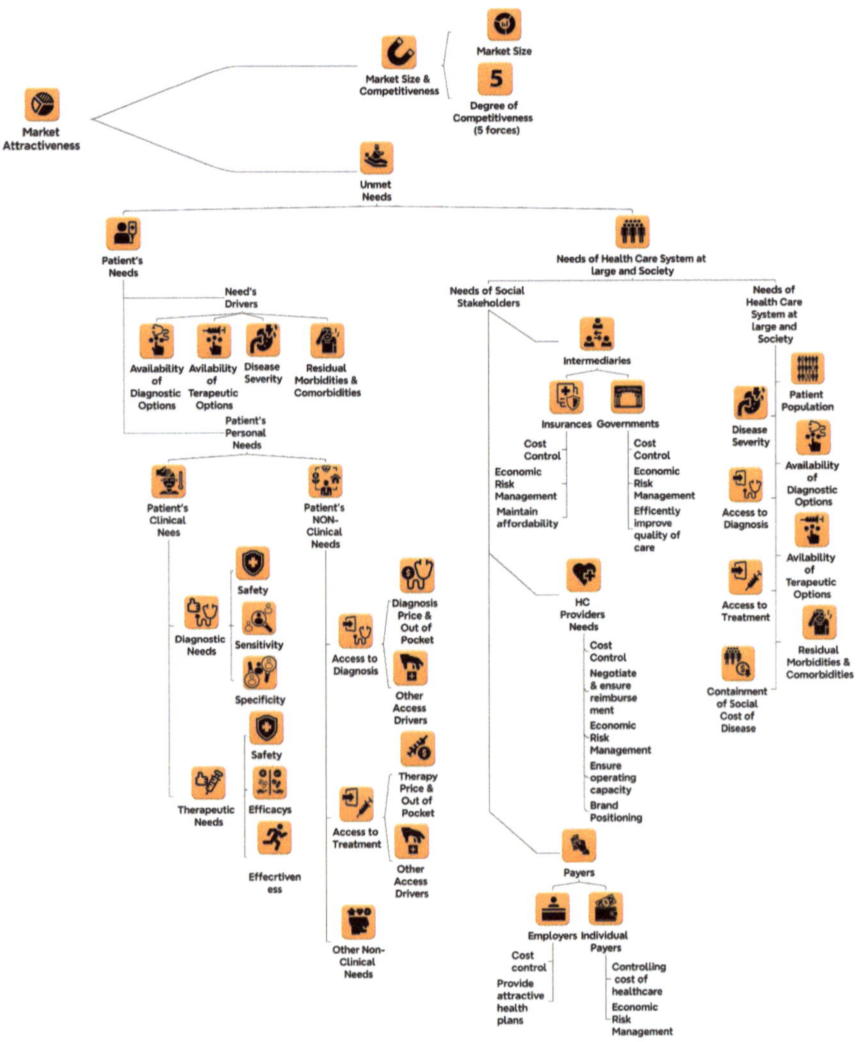

Fig. 7.12 Key elements relevant to determining and characterizing the market's attractiveness

7.2 Product Strength and Readiness

In medical technology startups, vague intentions for innovation are often the first source of decision-making biases and unsubstantiated claims about the value of an innovation. For this reason, the disciplined construction of a target product profile (TPP) is the first step in entrepreneurs' assessment of the value of their new ideas. TPPs are frameworks used to communicate, both internally and externally, a summary of the requirements for new products or services that satisfy essential unmet medical needs. Internally, the goal of TPPs is to align product development efforts with the intended features, including the testing of claimed therapeutic or diagnostic capabilities.

This chapter focuses on various components of the readiness and characterization of a new product in medical innovation, including its features, intended use, and degree of readiness, which strongly affect the entrepreneur's accuracy in predicting and proving the product's ability to deliver on the promised outcomes (Fig. 7.13).

7.2.1 Target Product Profile (TPP)

It can be challenging for entrepreneurs to efficiently and rationally assess how the new idea satisfies unmet medical needs unless these needs are very clearly defined. For this reason, entrepreneurs must adopt maps similar to the one discussed in Sect. 7.1 to focus on the most critical assumptions that must be made about unmet medical needs.

Without a compelling description of the product and its intended therapeutic or diagnostic targets, it is difficult to relate the innovation to a specific group of unmet needs or to evaluate the importance of these needs (Jambulingam, 2018); for this

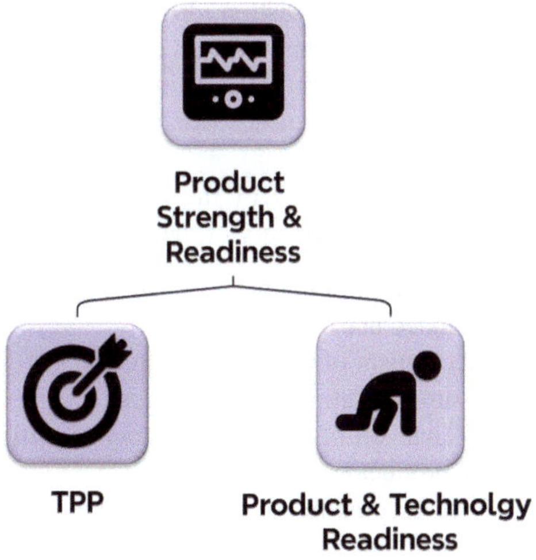

Fig. 7.13 The key problem clusters determining a product's strength and readiness

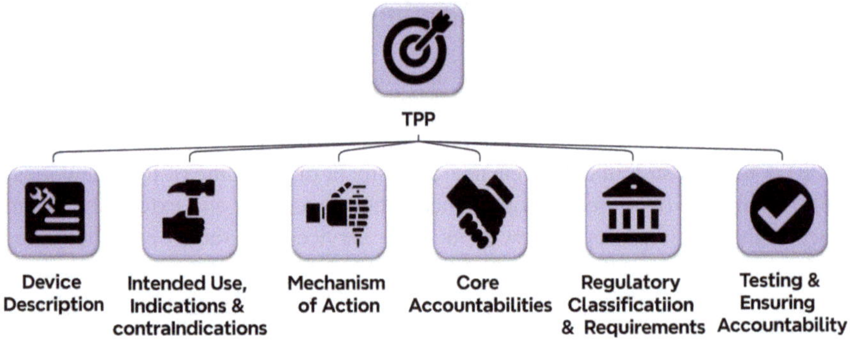

Fig. 7.14 The key problem clusters relevant to TPP

reason, the disciplined construction of the TPP is such a crucial first step for entrepreneurs.

Externally, the TPP supports discussions with future users, buyers, and concerned regulatory agencies. Internally, the TPP plays a crucial role in identifying new product ideas and their critical attributes before product development commences, ensuring that the final product version meets the needs of end-users (WHO, 2020). TPP is also essential in defining the commercial strategy, as the value proposition is based on key elements of the TPP. Without a well-defined TPP, it is virtually impossible to offer a coherent, sound value proposition and build a joint marketing and sales strategy.

An additional confirmation of the importance of TPP is that the FDA developed a TPP guidance document to facilitate effective interactions between the FDA and innovation companies, thereby potentially shortening the product development timeline and minimizing the risk of late-stage failures (Jambulingam, 2018). Indeed, a TPP provides a structure to ensure that the company clarifies all relevant medical, technical, and scientific aspects necessary to reach the desired commercial outcome (MaRs Discovery District, 2011). Moreover, the correct use of the TPP enhances alignment within the venture's organization and, consequently, it ultimately benefits the patient (Lambert, 2010). From a managerial standpoint, the TPP must be regularly reviewed to ensure that any changes in the strategic direction result in its revision.

As shown in Fig. 7.14, the TPP for medical products is a key strategic document that provides a summary of the following:

- Description of the product that is to be developed or under development
- Intended use of the product
- Mechanism of action of the product
- Contraindication of the product
- Mechanism of action, regulatory classifications and requirements pertaining to the product
- Core accountabilities in addition to studies and activities necessary to demonstrate the performance, safety, and efficacy of the product

7.2.1.1 Device Description

The device description focuses on the technology, design, hardware, and software components, their interactions, and the overall internal functionality of the product. If the device uses data or exchanges it with remote locations, individuals, servers, or an AI support platform, all these elements and their interactions constitute part of the TPP.

The correct description of the device and its components is also critical to the prediction of venture's future cash flow. Indeed, the TPP provides important information about the approximate cost of the final product in terms of the sum of the costs of its components. It may also deliver important information about the estimated cost of assembly. This section, therefore, is critical to predicting the Profit & Loss (P&L) lines related to the investments in product development and industrialization and to the cost of goods sold (COGS), which, together with price, defines the primary source of profit for the venture.

In summary, the main questions (and decisions) relevant to the product description are as follows:

- What technologies support the functioning of the device or the system?
- What software and hardware components are required?
- What are the predicted development investments?
- What are the expected industrialization investments?
- What overall assembly cost can be predicted?
- What manufacturing cost can be predicted?
- What type of function will link total cost and volume?
- What are the regulatory timeframes and the associated investments?
- Are the earliest price assumptions compatible with product's estimated costs?

7.2.1.2 Intended Use, Indications, and Contraindications

Intended use refers to what a device or software is designed to do. It corresponds to the purpose of the device or software, including which disease or injury it can diagnose, treat, or monitor. Indications for use refer to the patient and environmental circumstances in which the device or software can be safely and effectively utilized. Indications also require precise information about the intended patient population and medical conditions to be diagnosed, treated, or monitored. From the standpoint of a product idea's value prediction, the confines of the target patient population identified by the proper use of the device or software should correspond to the same, exact target population assessed in Sect. 7.1, concerning the determination of market attractiveness.

Contraindications refer to situations in which the device or software should not be used, as the risk of use outweighs any possible benefit. Examples that may, but not always, contraindicate the use of a device or software include the following (FDA Labeling Guidance #G91-1):

- Hypersensitivity to an ingredient of a permanently implanted device
- Substantial risk of being harmed due to the patient's age or sex
- Concomitant therapy, disease state, or any other condition
- Continued use in the face of an unacceptably hazardous adverse reaction

7.2.1.3 Mechanism of Action

The mechanism of action (MOA) is relevant when the medical device is intended to produce a therapeutic effect. The MOA pertains to how this effect is produced.

It is important to note how the MOA distinguishes a medical device from a medicinal product. In the case of medicinal products, the MOA is achieved through a pharmacological, immunological, or metabolic (Ph.I.M.) MOA. In the case of a medical device, the primary mechanism of action is achieved without the Ph.I.M. mechanism. In contrast, a Ph.I.M. mechanism may exist only as a secondary driver to assist the device in its primary function. These elements suggest that the primary MOA of a therapeutic medical device must be defined around non--Ph.I.M, chemical, or mechanical actions. The chemical MOA is intended to be the interaction of a substance with other substances in the body, such as the transformation of the initial chemical substances (the reactants) into different chemical compounds (the reaction products). To avoid falling into the area of pharmaceuticals, these chemical actions should not include the targeted interaction with a receptor and its signaling pathway. The physical MOA is intended as the interaction of a substance or material with other substances present in the body, such as to solely transform the surrounding environment or matter (Racchi et al., 2016). From the standpoint of a product idea's value prediction, the confines of the target patient population identified by the proper use of the device or software must correspond to the same, exact target population assessed in Sect. 7.1, concerning the determination of market attractiveness.

7.2.1.4 Core Accountabilities

Accountability refers to the ability of the product to actually deliver on all explicit and implicit promises. The explicit promises of new medical technologies often involve higher therapeutic efficacy, lower risks and fewer contraindications, improved diagnostic accuracy, improved usability, and earlier or improved access to therapy or diagnosis. Field experience with medical startups suggests that the venture's management must remember that these core promises are always accompanied by secondary commitments that are not quite negligible.

Secondary commitments include, for instance, a mean time between failures (MTBF) that is consistent with the patients' or users' expectations, sufficient resistance to dynamic shocks such as those experienced during transportation, the ability to maintain sterility while in storage, predictable and limited wear and tear to prolonged use, and an easy repair process independent of the geographical location where it is used. These are the areas of reliability, serviceability, usability, biocompatibility, shelf life, sterility, toxicity, and resistance to stress and wear (Fig. 7.15).

Cost-effectiveness is also a key capability of medical technology that, despite not belonging to the clinical domain, represents one of the most critical factors determining the success of an innovation. Indeed, the demand from scrupulous consumers and cost-pressured payers requires innovative healthcare products to not only be safe and effective but also exhibit cost-effectiveness in relation to existing

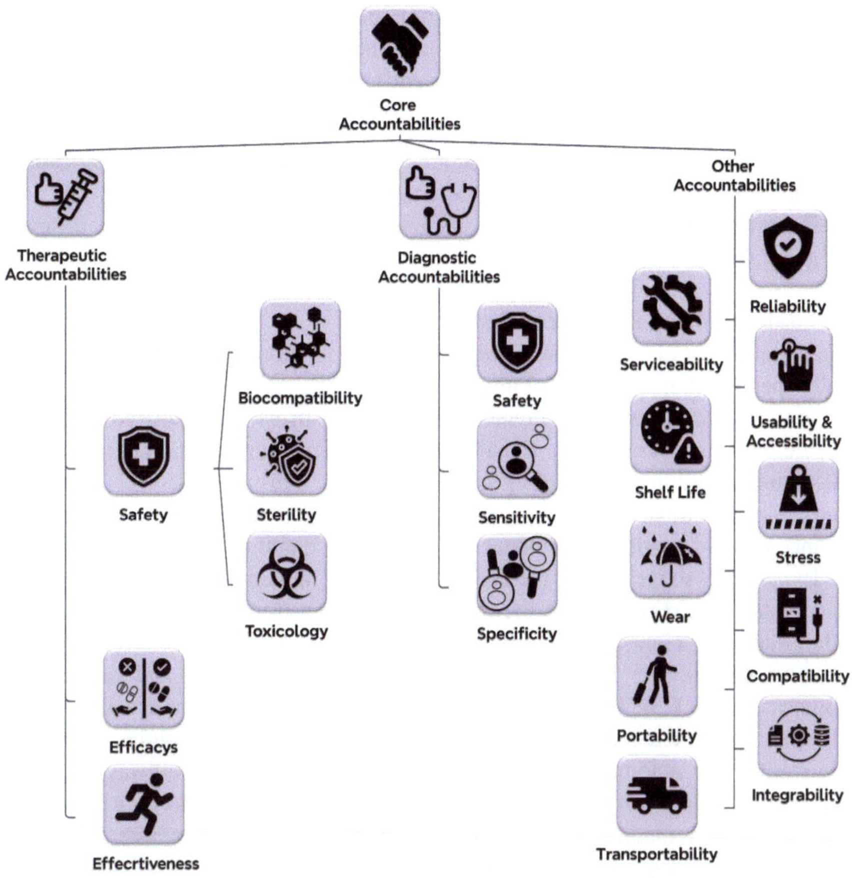

Fig. 7.15 The framework of core accountabilities for medical products or services

diagnostic or therapeutic solutions (Herzlinger, 2006). Section 7.3 focuses on value proposition attractiveness and analyzes the critical needs for cost-effectiveness.

7.2.1.5 Regulatory Classification and Requirements

The rapidly transforming landscape of the medical technology industry increases compliance risk for many innovative organizations. Only a timely and proactive approach can help innovators deliver value to patients while reducing regulatory risks. Section 7.5, which focuses on identifying, characterizing, and assessing the role of institutional stakeholders, addresses the regulatory aspects more extensively. In the context of medical devices, the regulatory classification system is being constantly revised to carefully reflect the possible health risks associated with the use of modern high-end technology. The general requirement consists of categorizing devices under the highest potential risk class, requiring medical device innovators to carefully review the latest rules and classify their products accordingly (Lukas et al., 2020).

It is important to remember that regulatory classification is necessary at the early stages of product development to understand the resources and the time required to receive marketing authorization in different target geographies. In particular, the entrepreneur must gain an early picture of the following:

- Minimum testing requirements that are typical of each regulatory classification
- Estimated investments related to these testing requirements
- Expected registration timelines characterizing the processes in the different geographies

Forces Influencing Regulatory Risk

Innovators must also face the unforgiving reality of pioneer entrants who statistically spend 34% more time (7.2 months) and about 7% more resources than follow-on entrants in regulatory approvals, making small firms less likely to enter new device markets (Stern, 2017). Existing rules are often complex and not necessarily clear, disproportionately affecting small enterprises. It results in several unintended and counterintuitive outcomes, including the diffuse perception of an underlying disincentive for small ventures developing genuine, quantum-leap innovations.

Experience suggests that the regulating bodies' intention to establish a robust and sustainable regulatory framework capable of ensuring a high level of safety while supporting innovation has not been realized yet. A failure to enact the desired changes may further jeopardize future innovations, outcomes, and care quality (Baines et al., 2023). These aspects of regulatory approval serve as crucial warning signs for entrepreneurs, who should not confidently assume that higher degrees of innovation will automatically lead to more rewarding market pathways. Instead, they must ask themselves whether, from a regulatory standpoint, they are pioneering an entirely new medical technology and whether the field of reference for their innovation is clearly and permanently regulated or is still in an evolutionary stage.

The following is the list of key questions (and decisions) about regulatory approvals that must be considered with regard to the TPP:

- Which class of reference best fits the innovative product?
- Is this class the same in all geographies? If not, which geographies will use which class?
- Which minimum testing requirements must the product undergo to be considered for registration?
- Which type of regulatory-driven testing will be necessary?
- What investment is required for regulatory-driven testing?
- Is the new product a "pioneer" from a regulatory standpoint?
- Is the regulatory framework consolidated or being defined?
- Can the compliance rules of reference be considered clear across geographies?
- What milestones can be predicted for submissions and registrations?

7.2.1.6 Testing and Ensuring Accountability

The need to test and, in turn, prove the ability to deliver on the new product's (or service) core accountabilities is as critical as the precise definition of the entire set

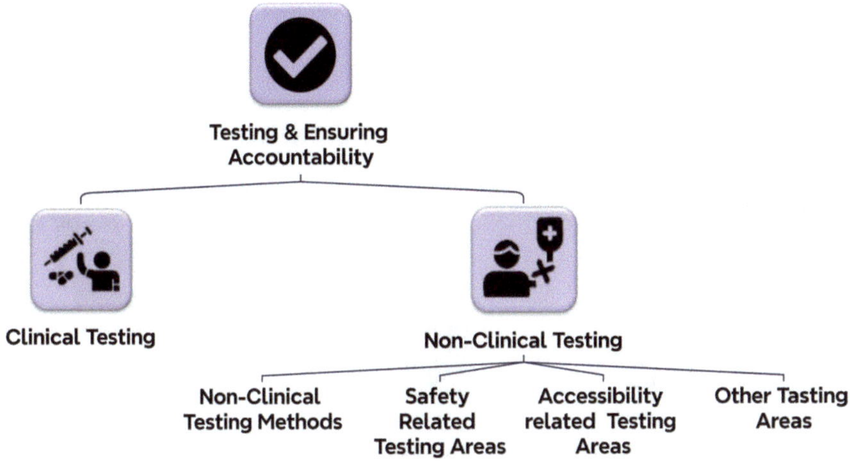

Fig. 7.16 Problem clusters relevant to test and ensure accountability of medical products or services

of promises. According to Fig. 7.16, this book proposes a separation of the testing activities into clinical and non-clinical areas.

The notion of proof of concept (POC) is key to both areas. In general, POC is evidence (not necessarily or always conclusive) that a business proposal, design idea, and so on will work, usually based on an experiment or a pilot project. In the pharmaceutical industry, POC may be defined as the earliest point in the drug development process at which the weight of evidence suggests that it is "reasonably likely" that the key attributes for success are present and the key causes of failure are absent (S. Cohen et al., 2010).

In the context of medical technology, POC involves building working prototypes of the necessary functionality and infrastructure of adequate quality to investigate the evidence for improving health in the daily life of a limited but relevant set of people serving as subjects for a suitable amount of time (Bardram, 2008). In short, in medical technology, POCs are often referred to as working prototypes (across product development and industrialization phases) that demonstrate the ability of the product to maintain performance promises.

Clinical Testing: Managerial and Business Implications

Clinical testing concerns aspects such as safety, efficacy, and effectiveness for treatments or those of safety, sensitivity, and specificity for diagnostics (Fig. 7.17). Leaving the point of clinical trial design and implementation from a medical standpoint to the extensive literature available, here we focus on the managerial implications of the design and of the implementation choices made for clinical trials.

In this context, broader clinical testing validity refers to a much larger area than the purely clinical one. Clinical trials supporting medical claims must be compelling in terms of the selected trial's endpoints (NCI, 2024a). However, the trial's endpoints must be significant for the patient's quality of life and meaningful for the

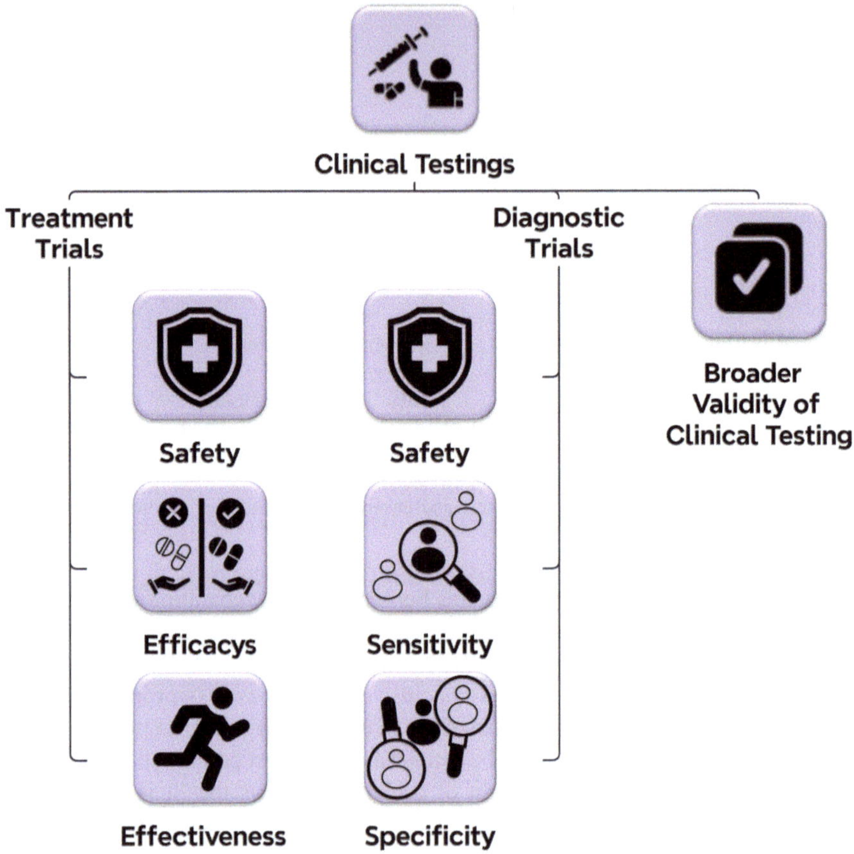

Fig. 7.17 The founding elements of clinical testing

physicians treating the target disease, and not merely satisfactory for the publisher, the researcher, or the research department seeking publications.

Formal statistical methods for analyzing clinical trial results are widely studied, honed, and accepted by the medical community. Unfortunately, the interpretation of trial results from the perspective of clinical relevance has not received similar emphasis. This lack of emphasis explains the common tendency to identify statistical significance with clinical importance and, conversely, to identify statistical insignificance with clinical irrelevance (Man-Son-Hing et al., 2002).

For instance, a medical rehabilitation device that improves the individual physical stability parameters (balance scores) measured with the Berg Balance Scale (BBS) for Parkinson's patients by 15% is undoubtedly a relevant statistical result if a strong p-value supports it.

However, in practice, it may not add any significant value to a patient's life unless this result is reflected in an actual, measurable decreased frequency of falls or loss of balance. Assuming that a statistically positive clinical trial result always translates into a tangible value for the patient may lead to future surprises. For instance, if, in real life, there is no actual recognition of the value to the patient, the adoption of the new rehab device may encounter significant and unexpected barriers.

Not even resorting to the concept of clinical association between clinical scales and predicted fall frequency or specifying that BBS scores are widely recognized as a reliable predictor of future falls is sufficient to compensate for the lack of a measured decline in patients' falls. In terms of burden of proof, the safer option always involves matching the endpoints with what matters to the patient and, in turn, to prescribing physicians. In this example about rehabilitation devices in Parkinson's patients, the entrepreneur should opt for more significant testing of the actual fall patterns in treated and control participants. In this context, the only challenge might come from the clinical trial's feasibility and cost. Below is a summary list of key questions (and decisions) concerning clinical testing:

- What are the trial's treatment or diagnostic endpoints?
- Does the clinical trial design faithfully interpret what matters to patients and physicians?
- What other endpoints can best prove the benefits for the patient?
- What other endpoints can make a difference from an adoption standpoint?
- What additional investment is necessary to prove broader clinical indications and claims?
- What signals are available to support the decision to invest in clinical trials? What additional signals may be necessary?

Non-Clinical Testing

Non-clinical testing encompasses all testing activities that exclude the actual clinical involvement of human beings (Fig. 7.18). Safety-related tests may involve sterility, toxicology, and biocompatibility assessments. Accessibility tests involve proving the solution's ability to be accessible in terms of usability, portability, compatibility, transportability, integrability, and even wearability. Other common non-clinical testing arcas include overall reliability, serviceability, shelf life, and resistance to wear and specific physical stresses. Typical non-clinical testing methods include well-known in vitro and animal tests as well as analytical validation, clinical association validation, and other real-life non-clinical tests.

Clinical association validation is another very important type of non-clinical testing. It is essential when, for instance, the functioning of the medical innovation is based on a predictive model, whether or not it is based on AI. A valid clinical association refers to how well the model's underlying assumptions are accepted and grounded in established scientific evidence. It should accurately reflect real-world healthcare scenarios and the conditions identified by the model-based solution. Establishing a valid clinical association is crucial to gaining early medical acceptance and support for clinical adoption.

Clinical association is indeed an important alternative way to support or even explain the results of clinical trials. Conversely, unless the tested clinical claims are clearly, undoubtedly, and redundantly proven by credible clinical trials, it is very

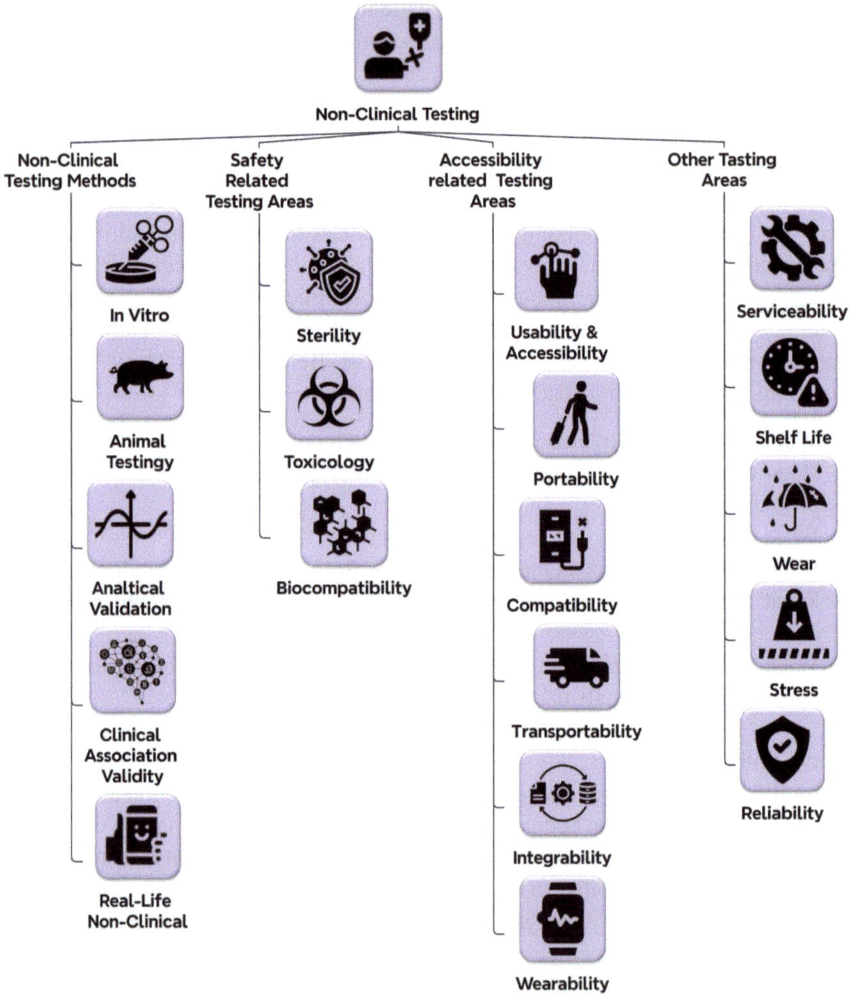

Fig. 7.18 The key problem clusters relevant to non-clinical testing

important for a medical entrepreneur to carefully weigh their investment in new medical products that base their value claims on predictive models, MOAs or diagnostic principles that result as counterintuitive for the scientific community.

The concept of valid clinical association is crucial for the software used by a medical device or a platform falling in the FDA's category called "Software as a Medical Device" (SaMD): it is used to refer to the extent to which the SaMD's output (concept, conclusion, measurements) is clinically accepted or well-founded based on an established scientific framework or body of evidence, and corresponds accurately in the real world to the healthcare situation and condition identified in the SaMD's TPP. A valid clinical association is an indicator of the level of clinical acceptance and of the degree of confidence that can be assigned to the clinical significance of the SaMD's output in the intended healthcare situation and clinical condition (FDA, 2017).

For instance, a European startup that claimed the ability to predict acute kidney failure in the ICU with a device and algorithm based on the derivatives of the patient urine flow was confronted with the skepticism by the scientific community: The flow of urine and its variations are historically not considered a reliable predictor of acute kidney failure. This startup based its predictive claim on retrospective studies that could hardly scratch the foundation built by years of well sedimented beliefs: This issue alienated investors and hindered the startup from reaching commercialization.

Analytical validation, instead, involves evaluating predictive performance, validity, and applicability of machine learning models. Is the model able to accurately and reliably process input data to produce precise output data? To evaluate this concern, it is essential to understand the integrity of the clinical data used in the model development process and ensure transparency in the reporting of how the training data is processed. Additionally, there should be a clear assessment of the algorithm's correctness, reproducibility, robustness, stability, and generalizability.

There are several other non-clinical, explicit, or implicit promises that have been demonstrated to be critical for the success of a medical venture. Surprisingly, many startups fail due to biased assumptions regarding the new product's ability to satisfy non-clinical requirements. For instance, in real life, accountability is compromised when a sophisticated device for post-surgery patient monitoring requires frequent servicing due to breakdowns after each transportation. Also, very basic usability elements can compromise accountability.

It is important to notice that usability testing must be performed in contexts that are as close as possible to the real-life conditions where the new technology is intended to be used. For instance, the usability of a newer infusion pump, which has already been clinically tested, should not be tested only in a laboratory or in a focus group setting. It can be tested in a hospital, bedside context, where the observed outcomes are limited to the quality of the user-device interactions. Real-life non-clinical testing refers to all tests that attempt to replicate real-life usage conditions without clinically involving the patient; this does not exclude the non-clinical involvement of a patient.

Usability testing is a rationally designed process focused on assessing whether a medical device meets its intended users' needs and preferences. Moreover, usability testing is used to determine whether a medical device is vulnerable to dangerous use errors that could lead to the user's or patient's injury or death (Wiklund et al., 2011).

There are several types of usability tests that medical entrepreneurs should consider. Here, we define four of the most commonly used test types (Carman et al., 2013). Formative Usability Testing involves evaluating an evolving design with the goal of identifying opportunities for improvement and confirming that the design is progressing in the right direction. The primary aim of formative usability testing is to refine the design of a medical device or identify potential opportunities for design pivots. Summative Usability Testing is conducted when a design is considered complete and nearly ready for industrialized production. The goal of this testing is to validate that the design meets the intended user requirements and facilitates safe and effective user interactions. Comparative Testing entails evaluating the current device in comparison to another device. This can involve older device versions or competitors' devices. The objective is to identify the relative usability strengths and weaknesses of the devices against one or more reference devices. This comparison can also be made against an ideal device, not actually existing, but defined by its Target Product Profile (TPP), which includes all established requirements. This process is often referred to as "benchmark testing," but it clearly falls within the category of comparative testing. An "Out-of-the-Box" test focuses on participants who begin with a sealed box or package, and their first task is to open it and see what is inside. Conducting a usability test this way not only reveals how participants interact with the packaging, but also how they handle the various items contained within the box, how they instinctively assemble the basic parts, or even power and turn on the medical device itself.

As a conclusion of this section about non-clinical testing, it is worth noting that a keen focus on innovative features can often lead to neglecting fundamental and implicit promises, such as reliability and serviceability. Non-clinical testing is, therefore, vital to prove the product's ability to deliver on promises beyond the core clinical claims and to avoid critical flaws in terms of implicit promises, such as usability and reliability.

7.2.2 Product, Technology, and Competence Readiness

The assessment of readiness involves both product and technology readiness, as well as the readiness of the required organizational competencies. A rational definition of the TPP presents a unique opportunity to evaluate the necessary competencies in the technological and medical domains (Fig. 7.19).

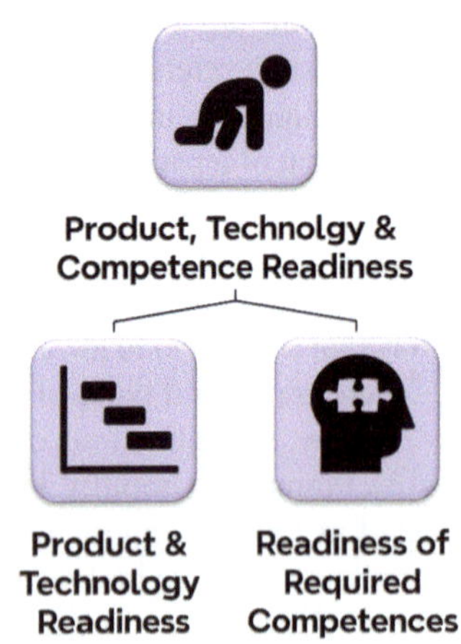

Fig. 7.19 The founding clusters of product's readiness

7.2.2.1 Product and Technology Readiness

Even when a startup has defined the fundamentals concerning the product, it is important to have constant clarity about technology readiness. Currently, many organizations use the technology readiness level (TRL) as a simple metric to indicate the maturity of the technology. Many medical startups enter the investors' world with ideas that are not yet expressed as a new product but are in less solidified forms of new technologies that can potentially be instrumental in tens or hundreds of medical applications. Despite this high degree of indetermination, many ventures jump from foggy product ideas into preparing and presenting business plans that, as a consequence, are unsubstantiated and financially inconclusive, thereby keeping most investors in a passive, waiting mode.

We have met several well-wishing ventures that are in a stagnant position, still awaiting an investor who does not need a compelling, bottom-up, product-based, and market penetration-based business and financial plan. Unless the new technology reveals very intuitive, straightforward, breakthrough possibilities of product declination, it is extremely difficult to narrow thc many possible uses down to a few precise use cases and rationalize their respective revenues and financial predictions so that they appear attractive to most investors. At this stage, startups are evaluating which specific declination of their technologies can be most fruitfully translated into a marketable product. This evaluation can be performed following the first few clusters of the K9-map, which are concerned with the early questions: Is there a medical need? Is there a market? Which product or service can deliver the optimal value proposition to these medical needs? Once these questions have been answered, the startup can move forward through the nine-cluster map and consider the overall degree of market competitiveness and the possible barriers to entry. The most common TRL frameworks are those used by the US Department of Defense Technology Readiness Assessment (Hook-Barnard et al., 2014) and the Human Factors and Ergonomics Society (HFES) (Salazar & Natalia Russi-Vigoya, 2021).

The US government, primarily through NASA and the Department of Defense (NASA, 2025), uses a 9-level scale to define Technology Readiness Levels (TRLs), ranging from TRL 1 (basic principles observed) to TRL 9 (technology proven in operational environment). These definitions are used to assess the maturity of a technology and can help with risk management, funding decisions, and communication between researchers, stakeholders, and investors. This scale is represented in Fig. 7.20.

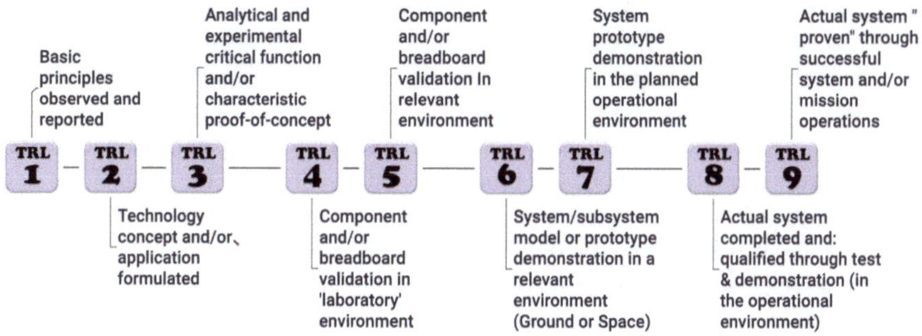

Fig. 7.20 TRL scale of NASA and of the US Department of Defense (NASA, 2025)

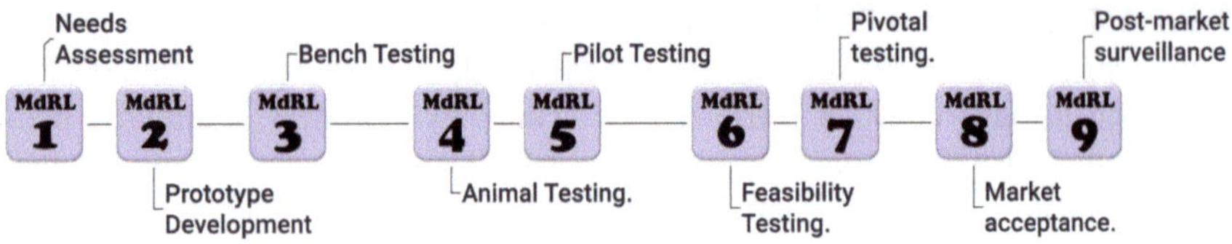

Fig. 7.21 The Medical Device Readiness Levels (MDRLs) scale. As defined by (Seva et al., 2023)

However, there are several newer readiness scales, such as the Medical Device Readiness Levels (MDRLs) scale (Seva et al., 2023). This scale proposes a particularly useful technology readiness framework for communicating and planning the development of Class III medical devices. The MDRL scale (Fig. 7.21) is based on nine sequential degrees of readiness: needs assessment, prototype development, bench-testing, animal testing, pilot testing, feasibility testing, pivotal testing, market acceptance, and post-market surveillance.

The authors (Seva et al., 2023) therefore suggest a MDRL scale that begins with Needs Assessment, where scientific and design principles are defined to solve a medical problem, focusing on safety, clinical value, and system integration. Next, Prototype Development transforms these concepts into a working model that demonstrates feasibility. User performance and integration issues are analyzed to refine the design. During Bench Testing, mechanical, electrical, and biological performance is evaluated in laboratory settings, including tests with animal or human tissues, to verify functionality and durability. Animal Testing follows, providing the first evidence of safety and performance in living systems, helping to identify and correct design weaknesses. In Pilot Testing, the device is used with a small group of healthy individuals to confirm safety and usability, leading to further design optimization. Feasibility Testing then measures clinical effectiveness on patients with the target condition, ensuring smooth interaction among users, clinicians, and systems. Next, Pivotal Testing evaluates the device in a larger patient population to confirm safety, efficacy, and statistical validity while resolving any remaining performance issues. After regulatory approval, Market Acceptance confirms that the device performs well in real-world conditions, ensuring comfort and user tolerance. Finally, Post-Market Surveillance monitors the device's performance and user satisfaction over time, ensuring continued safety and guiding future improvements.

7.2.2.2 Readiness of the Required Competencies

The precise definition of the TPP helps determine the required development activities and the needed competences and therefore it supports the decision about what type of professionals will perform them either internally or externally. In that case, the entrepreneur can better identify gaps in the available competencies and, therefore, make assumptions about the time and cost required for the acquisition and retention of potential new talents.

While the capabilities required must be analyzed within the area of technical and clinical competencies, a broader set of competences must be considered to keep future key stakeholders (users, patients, and so on) actively involved in the product development process. Therapeutic areas and indications enable the entrepreneur to quickly link the new technology to a specific group of healthcare stakeholders involved in the future use and acquisition of the technology. This link facilitates the making of essential assumptions about the competencies required by the future commercial organization.

7.2.3 TPP Provision of Additional Quality Signals for Financial Predictions

In general, approaching the decision-related problems of each of the nine clusters of the K9 map diligently provides the refined signals necessary for predicting the value of the venture. In this product-centric cluster, there is some obvious information that the decision-maker can collect and organize for future predictions. For instance, as already mentioned, the precise definition of the product's components can potentially lead to more accurate assumptions about the future manufacturing cost of the product. More accurate assumptions about product cost can lead to more realistic expectations about the variability of the product cost as a function of the volume sold.

From an investment and depreciation standpoint, the close look at product readiness suggested in this cluster helps refine the timing and amount of development and industrialization investments. Moreover, the precise definition of clinical and non-clinical testing required by a disciplined TPP definition certainly supports the better quantification of the need for investments in this area.

With respect to revenue prediction, a finer definition of intended use and indications can help further refine the boundaries of the target market and, therefore, the size of the achievable market. Moreover, a more precise definition of technology readiness and estimation of time to market helps place the milestones of revenue generation on the timeline. The organizational readiness of this cluster produces more refined information regarding future operating expenses. Analyzing the competencies required for the development of the new technology may better identify the gaps in the current competencies and, in turn, the potential additional future expenditures for the total compensation of the required talents.

7.2.4 Key Assumptions and Decision-Related Questions (Table 7.2)

Table 7.2 Product strength and readiness: core assumptions and decisions affecting the venture's value

Product strength and readiness	
Sub-cluster	Core assumptions and decisions affecting the venture's value
Device description	Which technologies support the functioning of the device or system?
	What software and hardware components are required?
	What are the core functionalities?
	What manufacturing cost can be predicted for the product?
	What investment must be budgetted for product industrialization?
Intended use, indications, and contraindications	What is the intended use?
	What are the indications?
	What are the contraindications?
	Is the intended use consistent with the total market assessments?

(continued)

Table 7.2 (continued)

Product strength and readiness	
Sub-cluster	Core assumptions and decisions affecting the venture's value
Mechanism of action	What is the mechanism that allows the therapeutic effect or the diagnostic performance?
Therapeutic or diagnostic accountabilities	What are the safety, efficacy, and effectiveness accountabilities?
	What are the safety, sensitivity, and specificity accountabilities?
	What are the side-effect accountabilities?
	What are the accountabilities regarding morbidities?
Other accountabilities	What are the explicit accountabilities (i.e., usability and integrability)?
	What are the relevant implicit accountabilities (i.e., reliability and portability)?
Regulatory classification and requirements	Is this medical innovation a regulatory pioneer?
	Is the reference regulatory context clear or an evolving target?
	What are the minimum testing requirements for the designated regulatory classification?
	What is the approximate time for regulatory approval? Is this time consistent with the expectations surrounding revenue milestones?
Clinical testing	What type of clinical tests is required?
	What therapeutic or diagnostic dimensions will be assessed?
	What is the trial's treatment or diagnostic endpoints?
	Does the clinical trial design faithfully interpret what matters to patients and physicians?
	What other endpoints can best prove the benefits for the patient?
	What other endpoints can make a difference from an adoption standpoint?
	What additional investment is necessary to prove broader clinical indications and claims?
	Are the planned clinical trials assessing endpoints unequivocally valid for patients and physicians?
	What timing and costs will be associated with the clinical tests?
Non-clinical testing	What type of non-clinical method will be required?
	Which tests will be required in the safety area?
	Which tests will be required in the accessibility area?
	What timing and costs will be associated with the non-clinical tests?
Product and technology readiness	What is the current TRL?
	What elements support the identified readiness?
	What is the estimated time to market?
	What are the competency gaps in the current organization?
	What type of talents is necessary to acquire in order to close any competence gaps?
	What are the costs associated with the attraction and retention of the identified talents?

7.2.5 The Complete Map (Fig. 7.22)

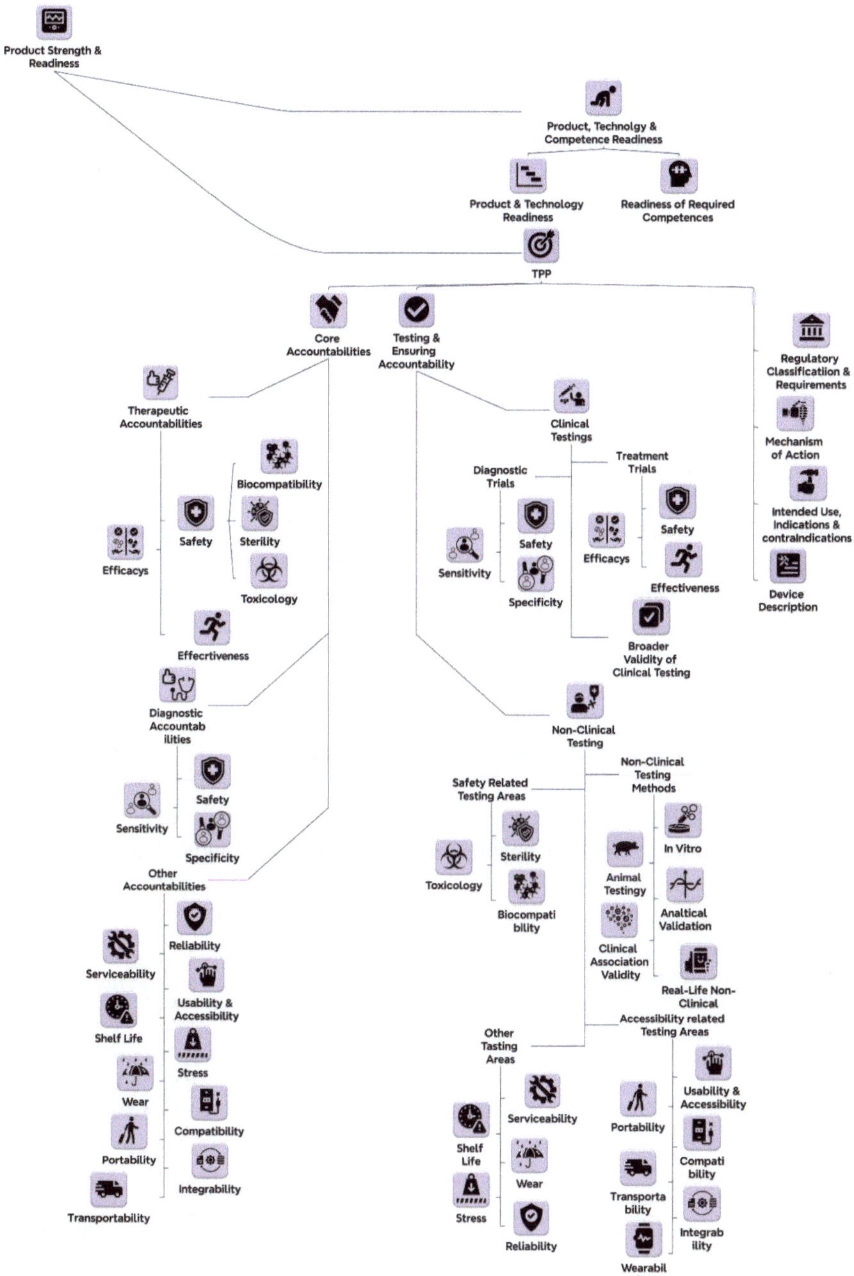

Fig. 7.22 The complete map of problem clusters relevant to product (or service) strength and readiness

7.3 Value Proposition Attractiveness

An articulation of the value proposition, which is the value created by technology, is the key starting point for the evaluation of a new idea in medical technology (Davey et al., 2011). The degree of attractiveness of a value proposition is determined by the balance between the ability to concurrently and uniquely satisfy a given set of unmet medical needs and the price proposed to buyers.

More often than not, young medical ventures fail to test, in due time, their assumptions about the ability of their innovative product to satisfy the unmet needs of their most relevant stakeholders. This bias, frequently associated with entrepreneurial optimism, can lead to a lengthy and expensive sequence of commercialization failures (Fig. 7.23).

Even more often, young medical ventures fail to make early, rational assumptions about the price level at which the innovative product (and its value proposition) will be commercialized. This failure leads to late, untested, and often biased price decisions that may result in costly commercial surprises.

Beyond being a core pillar of customer value, price is the single most critical factor in determining whether a product makes money. Nevertheless, it often remains a late consideration that is only made after the product's development is complete (Ramanujam & Tacke, 2016). Medical entrepreneurs must, therefore, simultaneously focus on testing their assumptions about value proposition fit to unmet medical needs and their pricing decisions. This chapter focuses precisely on these two aspects.

7.3.1 Value Proposition Fit to Unmet Needs

Prior to the late 1990s, rational and analytical discussions about value propositions were restricted to internal documents and processes within large consulting firms

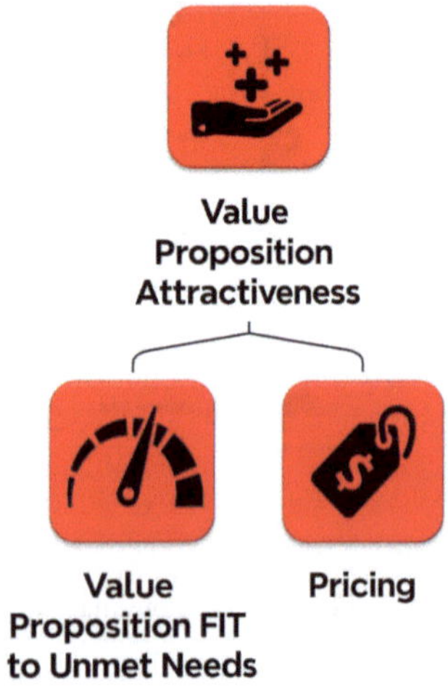

Fig. 7.23 The founding clusters of value proposition attractiveness

(Frow & Payne, 2011). The origin of the concept of value proposition can be found in consulting work performed by McKinsey & Co. in the 1980s; later, other authors defined value proposition as a statement of benefits offered (in exchange for a price) to a specific customer segment (Ballantyne et al., 2011).

The current literature identifies value proposition as a representation of the essence of strategy and the firm's most important organizing principle. However, research suggests that less than 10% of companies formally develop value propositions. While confirming this view, our experience taught us that few medical technology venture rigorously test their value propositions against the most critical medical needs.

Figure 7.24 highlights the key aspects necessary to determine the degree of fit of the value proposition to unmet medical needs. In essence, the map highlights the need for a precise comparison between targeted unmet needs and the main components of value proposition. Since the sub-cluster related to unmet medical needs is treated in depth in Sect. 7.1, this section focuses just on the building blocks of the value proposition: value to patients and other healthcare and social stakeholders.

7.3.1.1 Value Proposition

In medical technology, the leading approach to defining value assessment involves the identification of a few broad "value drivers" that should be incorporated into the assessment process. These value components are not limited to clinical efficacy; instead, they also include the impact of new technology on a broader set of value-based performance metrics and reimbursement models. They also extend the

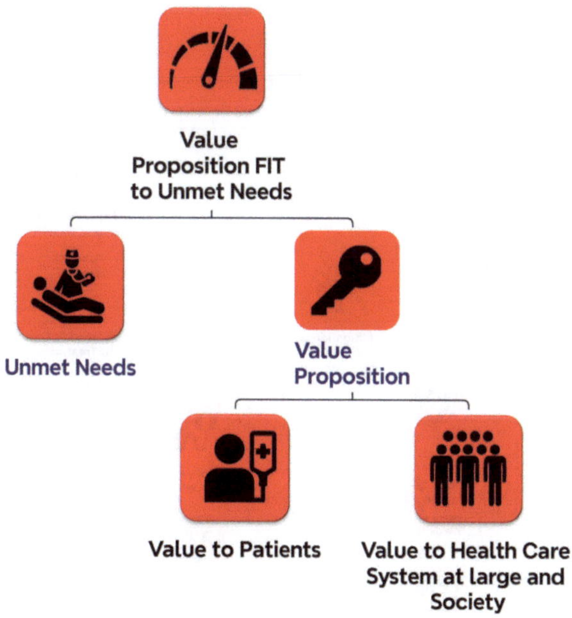

Fig. 7.24 The founding clusters of the value proposition fit to the unmet medical needs

meaning of value assessment to the wider socio-economic and political context in which patients are situated (Deloitte & AdvaMed, 2017). These components are as follows:

- Clinical impact on the patient
- Non-clinical impact on the patient
- Healthcare revenue- and cost-related impact
- Public (societal) impact

The value frame suggested by this chapter is based on these four components.

Value for the Patients

The clinical impact on the patient is the degree of clinical utility and health outcomes associated with using a new medical technology. This utility typically comes from the value of enhanced accessibility to earlier and more accurate diagnostics and to safer, more efficacious treatments. It is important to note that the clinical impact on patients is defined by not only the technology's ability to diagnose, prevent, or mitigate the severity of the disease but also its capacity to reduce residual morbidities or comorbidities.

The non-clinical impact on the patient includes benefits for the patient (often with reflections on caregivers), such as improved accessibility to diagnosis and/or therapy and consideration of other non-clinically relevant issues, such as the patient's personal experience with the new treatment or diagnosis, the impact on the patient's family and social life, the patient's access to work, and the patient's finances. Figure 7.25 offers a visual description of the critical value components for the patient.

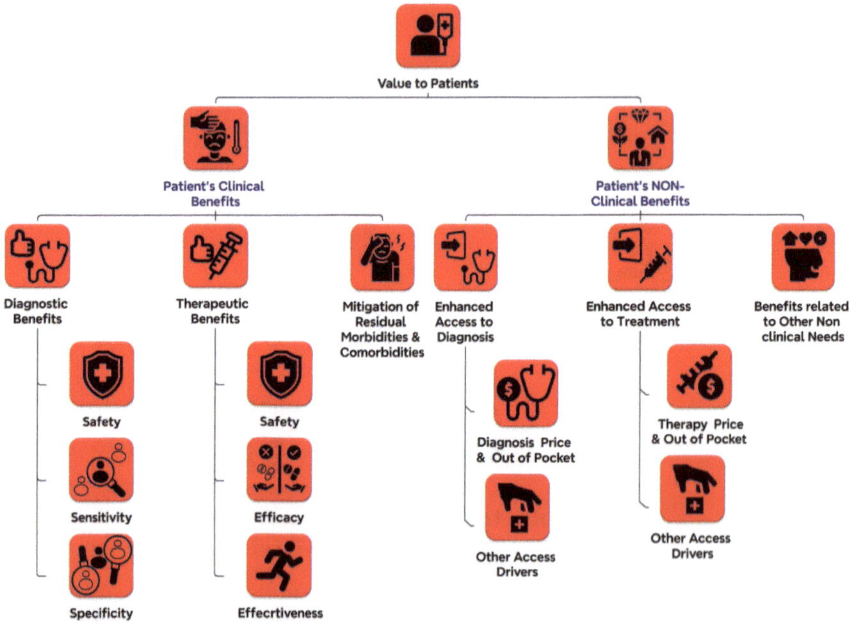

Fig. 7.25 The value proposition framework for patients

The clinical benefits of therapeutic or diagnostic technologies are based on their value pillars: safety, sensitivity, and specificity for diagnostics and safety, efficacy, and effectiveness for therapies. Even if, for a patient, effectiveness is much more important than efficacy, the map includes the concept of efficacy so that it ties in with the clinical testing required for the new therapy. As mentioned, another critical element defining a clinical value for a patient is the mitigation of residual morbidities.

Non-clinical benefits include essential elements such as improved access to therapy or diagnosis, which is limited by affordability (price), often linked to lack of reimbursement, and many other access-limiting factors, such as the actual availability in the patient's country or at a reasonable distance from the patient. Other non-clinical benefits include the patient's experience with a particular therapy or diagnostic procedure, the ability to navigate therapeutic processes, the impact on their finances, and the impact of the new technology on the support that a patient receives from the caregivers.

It is important to bear in mind that the function of this map is simply to help the medical entrepreneur ensure that all the building blocks of the value-to-patient have been considered while designing the value proposition of the new technology. Each cluster of the map corresponds to a group of critical questions the entrepreneur must answer. For instance, the cluster related to "therapeutic benefits" must trigger questions such as the following:

- Is the new technology able to safely improve efficacy?
- If yes, to what extent? What specific elements of efficacy are improved?
- What are the actual meaningful consequences of higher efficacy in patients' lives?
- Are side effects a threat to safety? To what extent? What impact do diminished side effects have on patients' lives?
- What additional advantages are offered to patients due to diminished comorbidities?

Value for the Healthcare System and Society at Large

In addition to patients, other major stakeholders (characterized in greater detail in Section 7.5) have a vested interest in the availability and in the value proposition of new technologies: Entire healthcare systems (both public and private) are affected by the employment of new medical technologies. Therefore, healthcare providers, intermediaries (such as insurance companies and the government), and payers (such as taxpayers, patients, and employers) have their respective sets of expectations concerning the adoption of new medical technologies, and they also critically contribute to the decision of new technology adoption. Not only are they interested in better diagnostic and therapeutic performances and in the reduction of morbidities and comorbidities, but they are also interested in improving access to diagnosis and treatment while controlling the overall healthcare costs.

Medical technology manufacturers are painfully aware that healthcare systems aim to achieve therapeutic, diagnostic, and access goals while relentlessly and tightly controlling the increasing costs of healthcare delivery. New medical technologies have a significant impact on the general public and societies concerned with the cost and added value of public and private healthcare, individual health risks, and the ability of society to function despite being affected by high-prevalence diseases such as diabetes, CVDs, and respiratory infections.

It is, therefore, clear why the group receiving benefits is not merely limited to patients but includes a much broader group of stakeholders. Indeed, in the life sciences industry, value must be assessed by considering each group as a component of a more complex stakeholder framework and with its own respective view of the new technology value. For instance, individual patients seek quality of life, patient associations seek affordable access to innovative treatment, regulators are concerned with safety and efficacy, and payers focus on managing financial risk. On the other hand, it is essential to remember that paying attention to the complete set of stakeholders often leads to the tendency to overlook the fact that the actual "consumer" is the patient.

Medical technology innovators frequently witness how some stakeholders overlook or even ignore the patient-centered aspects of value propositions. Instead, they focus on creating and delivering their value propositions for one another (Villela, 2013). Medical entrepreneurs can use the map shown in Fig. 7.26 to ensure that all of the stakeholders' value drivers are considered. Of all the eight value drivers shown, health economics deserves additional consideration due to its complexity and its growing role in the decision of stakeholders to adopt new medical technologies.

7.3.1.2 Health Economics and Health Technology Assessment

The optimum use of resources for the care of the sick and the promotion of health defines the field of health economics (Mushkin, 1958). Health economics is a branch of economics concerned with efficiency, effectiveness, value, and behavior in the production and consumption of health and healthcare (M. Smith, 2021). Health economists are concerned with the organization of the market for health services and the net yield of investment in people's health.

Health technology assessment (HTA) comprehensively evaluates the costs, benefits, and risks associated with healthcare technologies. It uses scientific evidence to inform policymakers about healthcare investments' medical, social, economic, and ethical implications. HTA also supports informed decisions pertaining to the prioritization of healthcare investments, treatments, and public health measures. It is an evidence-based tool used to inform funding and coverage decisions by national and regional healthcare systems. HTA can have an impact on whether and to what extent a new medical technology should be adopted and how much funding and coverage should be allocated (Fontrier et al., 2024).

HTA is the "systematic evaluation of properties, effects, and/or impacts of health technology. Its main purpose is to inform technology-related policy-making in healthcare, and thus improve the uptake of cost-effective new technologies and prevent the uptake of technologies that are of doubtful value for the health system" (WHO, 2015). Entrepreneurs, manufacturers, and vendors bringing new technologies to the healthcare market are under increasing pressure every day to provide a

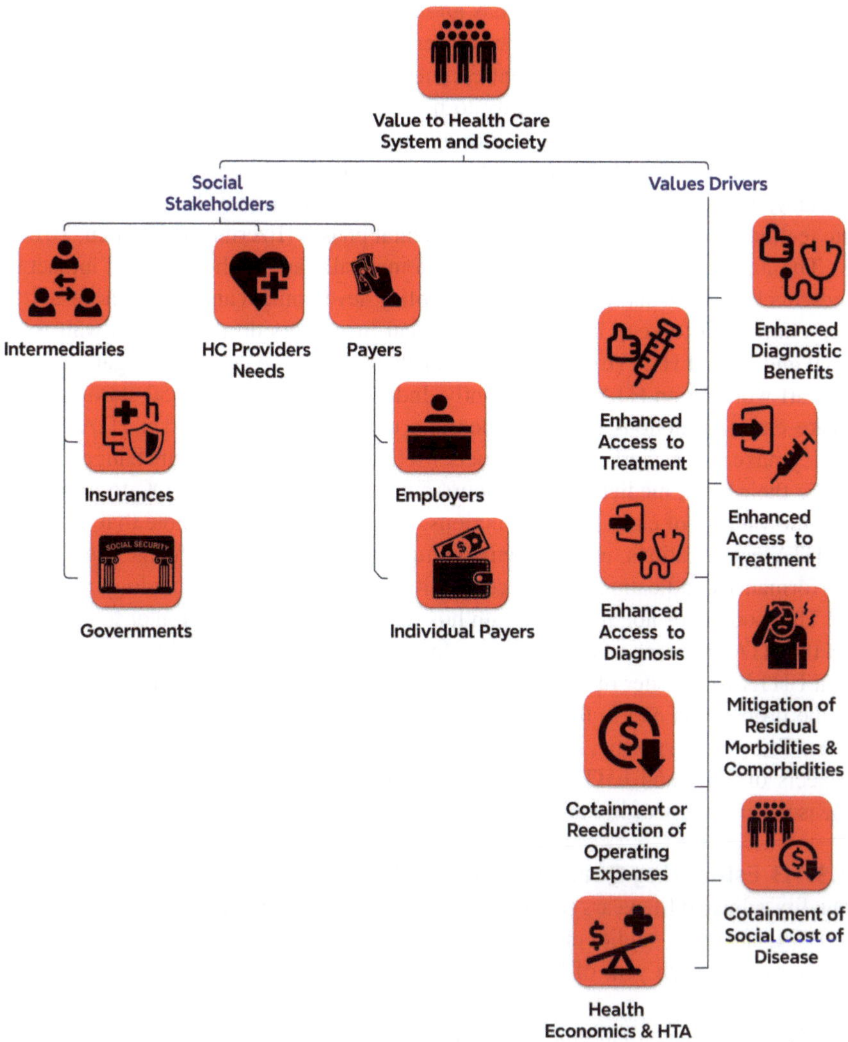

Fig. 7.26 The value proposition framework for the healthcare system and society

fact-based demonstration of the positive net value offered to patients and society after subtracting the price asked for the new technology.

Several countries do not strictly implement rigid and evolving utility schemes, but most adopt HTA to inform decision-makers. In these latter cases, utilitarian considerations strongly bind innovators and policymakers. In the UK, for instance, the National Institute for Health and Clinical Excellence (NICE) established formal processes of technology economic evaluation that became the funding criteria defining the acceptability of prices for new drugs and new devices (Girling et al., 2010). It is important to highlight that HTA recommendations are not necessarily legally binding. In most countries, HTA has an advisory role in the decision-making process without holding any regulatory power.

Therefore, the most critical entrepreneurial assumptions about a new medical technology must concern its ability to meet fundamental HTA prerequisites. This prediction must be made long before authorities or insurance companies conduct mainstream HTA evaluations. For this reason, early health technology assessment (EHTA) has been introduced. It represents an emerging form of HTA that can evaluate technologies that are still in developmental phases (Ijzerman & Steuten, 2011).

The main objective of EHTA is to systematically assess the potential impact of these technologies in order to anticipate policy development and market access. It is very important to keep in mind that medical innovators do not have to wait for a final and formal HTA (from authorities or their respective delegated organizations) to start working on HTA independently. Indeed, early-stage medical startups must make early predictions about the possible outcomes of post-market HTA. While policymakers use marketable products (defined and practically unchangeable) to make their HTA, medical entrepreneurs must use their TPP and POC to make and test the validity of early HTA predictions. It is also important to note that, contrary to HTA organizations, entrepreneurs can pivot and, therefore, change the clinical performances, indications, and accessibility of a new medical technology.

This book does not aim to explain how HTA and EHTA function or to elaborate on the HTA approach of each country. Instead, it merely seeks to highlight the fact that EHTA is a cluster of assumptions and decisions critical to the definition of the value proposition and to the ability to meet HTA requirements in the target countries. To be more specific, it is important to keep in mind one of the most significant aspects of HTA and EHTA: the use of quality-adjusted life years (QALYs) as a measure of quality of life and its monetary value.

7.3.1.3 Estimating Quality of Life and Its Monetary Value

Quality-adjusted life years (QALYs) have become the dominant outcome of interest in HTA, and many studies employ a cost-per-QALY analysis. A QALY is calculated by measuring each year of life, that is, by multiplying each year by its respective health state.

The health state is summarized in a parameter that varies between 0 (= death) and 1 (= perfect health). In this way, a period of life (for example, the time spent after a life-saving surgical procedure) is evaluated by embedding both the length and quality of life spent, instead of purely basing the evaluation on duration (Scuffham et al., 2008). Quantitatively, a QALY is represented by the size of the area of the rectangle built in a survival/quality of life Cartesian chart. Moreover, the application of QALYs extends beyond individual treatments to broader public health interventions, such as vaccination programs. For instance, the cost-effectiveness of pneumococcal vaccination can be evaluated using QALYs to determine the health benefits relative to the costs incurred (Mangen et al., 2015, 2017).

This method facilitates a nuanced understanding of the effectiveness of healthcare interventions, as it accounts for both survival and the quality of life experienced during that time (Fig. 7.27). QALYs are widely used in various health economic

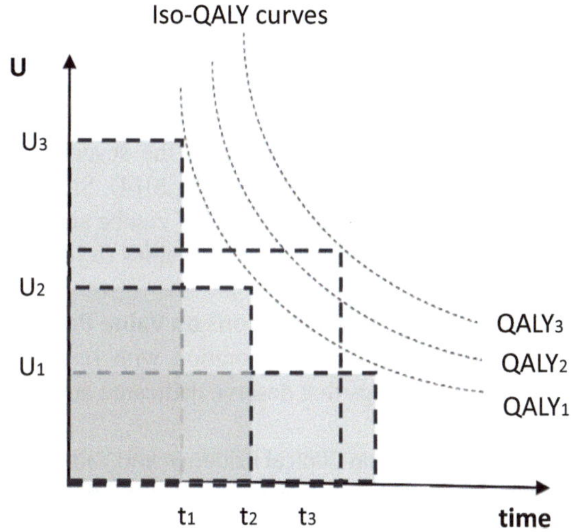

Fig. 7.27 QALY accounts for both survival and the quality of life experienced during that time. This chart shows how iso-QALY curves can be obtained

evaluations, including cost-utility analyses, where they serve as the primary outcome measure. The dollar value of a QALY is a critical concept in health economics, particularly in the context of cost-effectiveness analysis and HTA. The monetary valuation of QALYs varies significantly across different healthcare systems and contexts, often reflecting the societal willingness-to-pay thresholds for health gains. In the USA, the commonly referenced willingness-to-pay threshold for a QALY is between US$100,000 and US$150,000 (Chen et al., 2023). Conversely, in other countries such as Spain, the threshold is typically set at 25,000 € to 30,000 € per QALY (Vallejo-Torres et al., 2017).

In the UK, the threshold for the incremental cost-effectiveness ratio, which usually lies in the range from GBP 20,000 to GBP 30,000 (Girling et al., 2015), reflects different economic contexts and healthcare priorities. The threshold for the incremental cost-effectiveness ratio is often indicated with the acronymic ICER or the Greek letter "λ." In the business words of each country and in every healthcare system, the value of λ represents the monetary value that society is willing to pay in exchange for one additional year of perfectly healthy life for a patient.

The valuation of a QALY (and, therefore, of λ) can also be influenced by the method of estimation. Various approaches, including contingent valuation and time trade-off methods, have been employed to derive monetary values for QALYs. These methods often yield different results due to their inherent assumptions and the populations studied (Cameron et al., 2018; Aviles-Blanco, 2021). For example, contingent valuation studies may report higher willingness-to-pay (WTP) for QALYs associated with severe health states compared to mild ones, indicating that the perceived value of health improvements can vary based on the severity of the condition under treatment (Bobinac et al., 2012; Tilling et al., 2014).

The choice of specific estimation methods can also impact the valuation of QALYs. Various approaches, including contingent valuation and time trade-off methods, have been employed to derive monetary values for QALYs. These methods often yield different results due to their inherent assumptions and the populations studied (Cameron et al., 2018; Aviles-Blanco, 2021). For example, contingent valuation studies may report higher WTP for QALYs associated with severe health states compared to mild ones, indicating that the perceived value of health improvements can vary based on the severity of the condition under treatment (Bobinac et al., 2012; Tilling et al., 2014). Similarly, analyses of cancer treatments have shown that the cost per QALY can be significantly greater, reflecting the high costs associated with advanced therapies (Ryen & Svensson, 2015).

Other Important Observations on Value Proposition

Field experience and collaboration with several medical startups have provided additional learnings that deserve dedicated and even redundant mention.

Alignment Between Clinical Evidence and Value Proposition

Clinical trials should emphasize endpoints that reflect what is fundamentally important to patients, such as quality of life, symptom relief, and functional improvement. As already noted in Sect. 7.2, it is not uncommon for medical ventures to design clinical trials and the respective endpoints in a manner that is scientifically and academically impeccable but lacks a clear association with the value proposition deliverables. For instance, in a real-life case, the therapeutic value proposition of therapy was focused on slowing the progression of multiple sclerosis (MS): The entrepreneurs-academicians decided to use research endpoints associated with the radiological changes observed over time through multiple MRIs. The promising results observed in such controlled trials supported significant investments to prepare for the commercial launch. Unfortunately, a later trial that was based not on radiological observations but on actual, measurable clinical improvements in patients, such as mobility and frequency of MS relapses, did not support the existence of a statistically significant relationship between therapy and positive patient outcomes. In this case, the misalignment between the initial clinical trial endpoints and the actual patient needs cost tens of millions to investors and caused a late, forced exit decision for the venture.

Designing a clinical trial that addresses the actual needs of patients, rather than just concentrating on academically or intellectually significant endpoints, requires several strategic approaches. These strategies should prioritize patient engagement, relevance, and real-world applicability. Here are some key methods that can be applied in this process, along with potential references to support them.

Engaging patients early in the clinical trial design process is essential for understanding their needs and preferences. Successful collaborations between sponsors and patient groups help align therapeutic developments with what patients truly need, enhancing the potential effectiveness and acceptance of treatments. By incorporating patient feedback, researchers can identify meaningful outcomes that resonate with patients' experiences, thus improving the relevance of these outcomes (Caffery et al., 2023; Bloom et al., 2018).

Including patient-reported outcome measures (PROs) in clinical trials provides direct insights into how treatments impact patients' daily lives. PRO measures are increasingly recognized as valid endpoints for clinical trials, allowing researchers to capture user-centric data on health-related quality of life and treatment experiences (Gipson et al., 2018; McNiff et al., 2024). This inclusion ensures that trials focus on outcomes that are important to patients.

Taking into account the contextual factors surrounding the use of specific medical technologies allows trials to better reflect real-world scenarios. Various methodologies can be employed to assess the relevance of treatments for different patient populations, potentially leading to more clinically applicable trial results. Creating a multi-stakeholder environment that includes patients, healthcare providers, regulatory bodies, and industry representatives enhances the trial design process. This approach allows researchers to gather insights from multiple perspectives and ensures the trial addresses broader healthcare needs.

Establishing mechanisms for continuous feedback from participants during trials enables mid-course corrections. This iterative process facilitates adjustments that align ongoing research with actual patient needs, making trial results more applicable and insightful.

Value Proposition Traps for New Diagnostics

The identification of rheumatoid arthritis (RA) at its first presentation and treatment at an earlier stage can significantly affect the course of the disease, prevent the development of joint erosions, and even retard erosive progression. Early diagnosis and treatment may, therefore, affect disease outcomes and even lead to a state of remission (Heidari, 2011). This concept is crystal clear in the global medical community; therefore, any test for RA that allows earlier and more accessible intervention is likely to receive the synergic support of several stakeholders. Another case of positive impact from early diagnosis is obtained from a startup I recently assessed: shortening the waiting times from weeks to minutes for the completion of histological tests on potentially oncological patients. Patients can get to know earlier if they are actually affected by cancer, and if they are, the early cancer characterization will allow for an earlier beginning of treatments.

However, an earlier diagnosis does not always offer significant and positive changes for the patient's journey. An instance may occur when there is no therapeutic advantage related to early diagnosis.

What about a very early diagnosis of autism? What if a new AI-based system could perform an early, fast, MRI-based diagnosis of autism at a very early stage (<1 year)? Are there early treatments that, at this stage and age, could actually affect the course of the disease? Unfortunately in this case, there are no pharmacological treatments able to significantly impact the future of these young patients. Treatment for children under 1 year old diagnosed with autism focuses on early intervention using play-based and parent-guided strategies to improve social communication and overall development. Interventions, such as those in programs like ESDM (Early Start Denver Model), teach parents how to support their baby's development of social skills and language by embedding learning opportunities into daily routines and play. These treatments are helpful but unfortunately the extent of their efficacy is still debated and certainly not resolutive.

Therefore, even if the new diagnostic technology demonstrates both early access and very reassuring sensitivity and specificity, only the unavailability of efficacious, early treatments capable of changing the patient's journey can support the enthusiasm of the innovator. If, instead, the new test is intended for the large-scale screening of children in an effort to minimize the number of late detections, in this case, the mission of the value proposition becomes more complicated, as it implies some kind of larger public interest in the health screening decision, which is not easy either to influence or to forecast. In other words, while fighting for early diagnosis, entrepreneurs in this field must pay the same attention to the actual therapeutic implications for patients. With a more "extreme" expression, one could argue that new diagnostic technologies, capable of detecting previously undetectable diseases, may offer a compelling value proposition only if the medical community can take meaningful action to cure newly diagnosed patients.

Quantum Physics of Value Proposition

While focusing on the design of a value proposition, certain field learnings must be shared. These learnings suggest that the post-market adoption rates and their related revenue dynamics in medical innovation are not a linear function of the strength of the value proposition.

When the novel value proposition of a medical innovation has a limited impact on a patient's life, we often witness the absence of a substantial market response or, in the best cases, a very slow response, leading to adoption curves that are much longer than what any startup can financially sustain. The fact that a medical innovation represents an improvement over existing medical solutions does not automatically make it a successful innovation. Similar to the discrete energy levels of quantum physics, a medical innovation must support value propositions that represent a quantum-leap improvement over the existing standard of care. In young, early-stage medical startups, value propositions claiming marginal improvements on patients' lives or improvements in areas that do not lead to meaningful impacts on patients are often doomed to failure. The only exception to this rule is provided by new technologies that, while not offering an enhanced effect on patients' lives, allow significant savings to the health care delivery organizations.

Big MedTech giants such as Medtronic can afford to bet on marginal medical innovation, but only because these large, global companies enjoy a pre-existing, strong portfolio and commercial franchise that allows any innovation to reach millions of patients and buyers worldwide on the very day of market launch. When analyzing marginal-improvement ideas, medical entrepreneurs need to remember that startups neither can count on established global franchises nor can they afford a significant amount of time for a delayed market response.

7.3.2 Pricing

If value is the difference between benefits and costs, a value proposition statement constitutes the articulation of the measurable value that an organization or individual will obtain by accepting an offering at a specific price. In other words, a value proposition statement must include three key product elements: What the product can do for the customer, how it can help the customer succeed, and how much the customer must pay for it (Barnes et al., 2009).

As shown in Fig. 7.28, pricing considerations are made using several points of reference, such as the product's value proposition, HTA, existing or predicted reimbursement price, competitive pricing, the idea of "willingness to pay," the product's manufacturing cost (COGS), the sought market penetration and its speed.

It is important to note that these seven areas are partially overlapping and not always independent. For instance, reimbursement decisions are normally driven by HTA, and HTA is based on the value proposition of the new medical technology. Despite these partial overlaps, we see how, in pricing decisions, it is feasible to start by approaching them separately.

7.3.2.1 Price and HTA

In the context of medical innovation, particularly in the realm of medical devices, one of the simplest processes for estimating the devices' price level is the headroom approach. This approach is instrumental, as it relies on assessing a value-based price ceiling for the new device. It can be used to refine this price ceiling at different stages of the development cycle. Such price ceilings roughly assess the maximum market price opportunity for new products.

7.3.2.2 The Headroom Method

Owing to its adaptability, the headroom approach can be used in the earlier stages of development as a "reality check" regarding the viability of a new device in a specific healthcare market. The headroom method can also be used to support product development decisions using a pivoting approach, and it can contribute to the pricing strategy in the case of a high degree of uncertainty in reimbursement

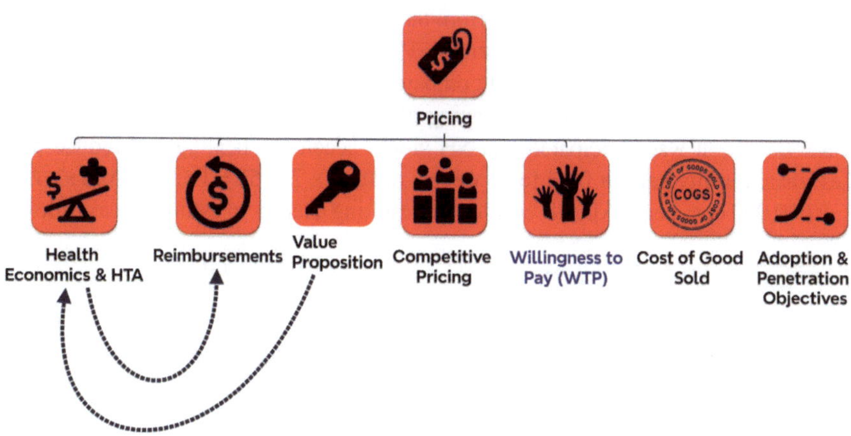

Fig. 7.28 The founding problem clusters of Pricing

prediction. Using this decision framework, entrepreneurs can anticipate the value of new pivoting options and justify the costs of newly required clinical trials or usability tests.

The framework operated by NICE in England balances the health and social care cost (HSCC) of a medical intervention against the health benefits, measured in QALYs. This approach assumes that the newly proposed technology brings two types of values (Girling et al., 2015):

- Savings in the HSCC
- Improvement in the patient's QALYs for a specific timeframe.

In this context, for a new device, the term "(commercial) headroom" constitutes the net benefit that the healthcare provider would recognize if the device were supplied free of charge to the health service (Girling et al., 2015). This value for the healthcare system is calculated as the sum of the monetary value of the savings resulting from the use of the new technology (i.e., supporting shorter hospital stays), which we indicate here with "ΔHSCC," and the monetary value derived from the patient's improved QALYs (ΔQALYs) for a certain number of years (Δt), calculated using a threshold for the incremental cost-effectiveness ratio "λ," which fits the geographical and healthcare system context. Therefore, we can calculate the price ceiling, for instance, in the UK context:

$$\text{Monetary Value} = \text{Price Ceiling} = \Delta\text{HSCC} + \lambda^{*}\Delta\text{QALY}(\Delta t)$$

The components of this expression are defined as follows:

$$\lambda \sim \text{GBP}20{,}000$$

$$\mathbf{\Delta HSCC} = \text{New HSCC} - \text{Current HSCC}$$

$$\mathbf{\Delta QALYs} = \text{NewQALYs} - \text{CurrentQALYs}$$

Let us analyze an example taken from the literature on health economics (Girling et al., 2015). Stapled hemorrhoidectomy (SH) was introduced in the UK at the end of the last century. Its principal benefit compared to the existing standard of care was a shorter hospital stay (an estimated saving of 2 days, at GBP300 per day, in 90% of cases), while it was anticipated that SH would accelerate healing and reduce pain (estimated utility gain of 0.24 for 2 weeks, or 0.24 × 2 ÷ 52 = 0.009 QALYs). At a GBP20,000 cost-per-QALY threshold, the resulting headroom estimate is as follows:

$$\text{Price Ceiling} = \left(2^{*}300^{*}90\%\right) + 20{,}000^{*}0.009 = \text{GBP}720$$

In 2007, SH (and the related stapling device) was recommended by NICE and introduced at GBP420, leaving the public buyer a utility quantifiable in GBP720 − GBP420 = GBP300.

7.3.2.3 Price and Reimbursement

In the medical technology industry, reimbursement refers to the amount of money a third party pays a healthcare provider for the costs of a medical device or procedure. The third party can be a private or public insurance company or, in some countries, the government. The reimbursement pricing for medical technologies, particularly medical devices, is a complex and evolving decision area that is influenced by key elements such as HTA, other economic evaluations, and the regulatory environment across different countries.

The increasingly tighter integration of HTA into reimbursement decisions has been driven by its ability to systematically evaluate medical technologies based on the economic value of their clinical effectiveness and cost-effectiveness (Konishi & Yasunaga, 2023; Sorenson et al., 2013). While pharmaceuticals typically undergo a more straightforward pricing and reimbursement process, health technologies and medical devices are confronted with challenges due to their diverse classifications and rapid technological advancements (Tsai et al., 2022). This complexity demands tailored reimbursement policies that can adapt to the unique characteristics of medical devices, ensuring that they are evaluated not only for their clinical benefits but also for their economic impact on healthcare systems.

Countries such as Taiwan and Japan have implemented specific frameworks for the reimbursement of medical devices, emphasizing the role of HTA in these processes. In Taiwan, for example, the reimbursement coverage decisions for medical devices are influenced by economic evaluations of the devices' value relative to the existing alternatives (Tsai et al., 2022). Similarly, Japan has established stringent reimbursement pricing controls that require a thorough evaluation of innovation in order to ensure that new devices are appropriately valued in the market (Tamura et al., 2019). These practices highlight the importance of early dialogue between manufacturers and reimbursement agencies to ensure the alignment of expectations and streamline the assessment process (Backhouse et al., 2011). Moreover, the reimbursement landscape is further complicated by the presence of multiple stakeholders involved in the funding and decision-making processes. In Australia, for instance, the funding arrangements for high-cost medical devices are intricate, involving multiple purchasers and shared responsibilities, which can lead to disparities in access and reimbursement levels (Saing et al., 2019). This situation underscores the necessity for clear and fair criteria for reimbursement coverage, particularly with the emergence of new technologies that promise improved patient outcomes (Lee & Salole, 2014). The different types of reimbursement currently used by the government and insurance companies to compensate healthcare-delivery organizations constitute an additional source of complexity. These reimbursement practices are often not centered on the use of a specific product or technology. Let us shortly summarize the four most common reimbursement practices:

- **Fee for service** (a fee is paid for each service provided to a patient):
  - Hospitals maintain what we often call a "chargemaster," which is a list of all of the services that the hospital can provide and the amount that the hospital charges for each service.

  - Cost-based reimbursement: Hospitals present intermediaries with their costs for taking care of a patient, and the payment is based on the cost.
- **Per diem** (payment is made per patient day):
  - Hospitals and intermediaries set a fixed amount that the hospital will be paid for a day the patient spends in the hospital, regardless of the hospital's charges or the cost incurred for caring for that particular patient on that specific day.
  - Per-diem payments can vary:
    By the type or complexity of services
    By the day of the hospital stay
    Carve out: Some specific services are paid for separately from the per-diem amount.
- **Diagnosis-related group** (DRG): Hospitals are generally paid for the inpatient care they provide. In the DRG system, hospitals are paid a flat amount for each specific discharge diagnosis, implying payment per hospital stay. The DRG system has a method for assigning patients to groups based on their diagnoses. This grouping is done by an algorithm, which is often referred to as a grouper. Each DRG is associated with a specific amount of payment allowed. One common way to associate each DRG with an allowable payment amount is to assign it a weight, reflecting the complexity or expected cost for the hospital in caring for a patient in the group. Conversion factors are utilized to calculate the weights.
- **Global budget model**: It is a way of paying for inpatient care, and it can easily be extended to cover the outpatient care provided by the hospital. Under this system, the hospital is paid a fixed amount for a period of time often one year, for taking care of a known or predicted population and for providing a defined scope of services.

  One general aspect of global budgets is that they often work best when they can capture a large share of a hospital's business, possibly even the entirety of it. Global budgets are harder to use when there are many intermediaries, each making their own deal with the hospital, although that is not to say that this way has not been attempted in various situations.

7.3.2.4 Early Price Decisions as Monetization of the Value Proposition

HTA and its application of the headroom method are often focused on the value of clinical benefit and the operational savings of a new technological solution. However, HTA may overlook other aspects of the value proposition that entrepreneurs must account for. For instance, in competing healthcare settings, such as in the US healthcare market, hospitals may desire a certain new technology for reasons that have little to do with HTA. Certain more advanced hospitals may seek the adoption of certain new technologies, as these technologies may seemingly be able to present an image of innovation and, consequently, create competitive advantages over other hospitals that are perceived as less innovative. Estimating a dollar value for these types of features is a challenge, but some draft estimations must be prepared. This positive component of the technology's value proposition clearly does not constitute a part of HTA.

Another example of a value driver not included in HTA is the perceived value of particularly attractive business models that buyers view as very convenient. For instance, payment models based on per use or per procedure are sometimes preferred, as they move the budgetary risk from the buyer-user to the manufacturer. If the risk-mitigation value of these payment models is not thoroughly considered in HTA, the medical entrepreneur must account for their values and seek its specific price-related recognition.

On the contrary, there are often elements of the new technology that may represent sources of negative components in the calculation of the delivered value: These are sometimes overlooked or underestimated in HTA. For instance, a piece of new equipment may create compatibility issues with existing equipment and require the hospital to support additional costs for the creation of a compatible environment. Therefore, after completing the estimations of the advantage reflected in the headroom or in any other HTA study, it is important to return to the value proposition and to all its components beyond HTA, in order to identify all positive or negative elements that contribute to the monetary value of the new technology.

Figure 7.29 presents an example of how a predicted price ceiling (PPC) can be estimated by attributing a specific value to each feature of the value proposition (VF*i*, with $i = 1, \dots n$).

A chart, such as the one represented in Fig. 7.30, can be created, for instance, for a new disposable connector used to perform the sterile connection of a syringe to a pump-led infusion line. A new connector may decrease the probability (Δp) of the patient's bacterial contamination, which, if experienced, can add several days to the patient's hospital stay. The value (VF1) of this feature can be obtained by multiplying the economic value of the average length of the prolonged stay in case of infection by Δp. The use of the connector may generate savings in terms of decreased nurse time due to its easier connectivity features (VF2). Moreover, the buyer may perceive value in adding such modern technology to their

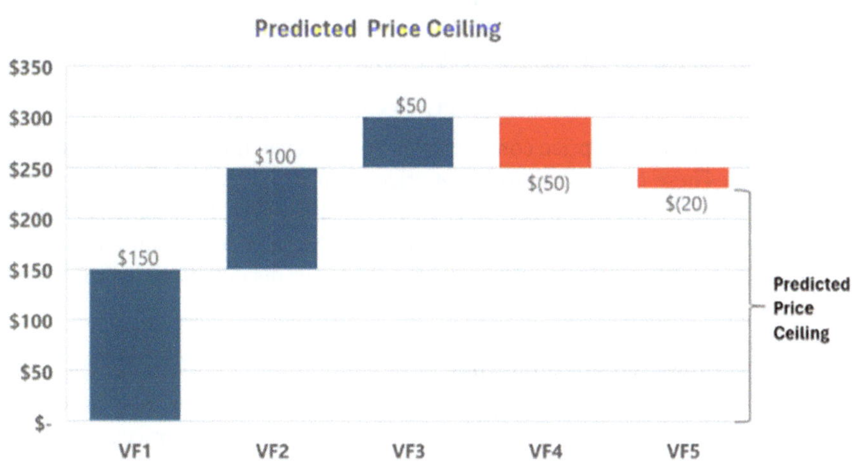

Fig. 7.29 An graphical example of Price Ceiling Calculation

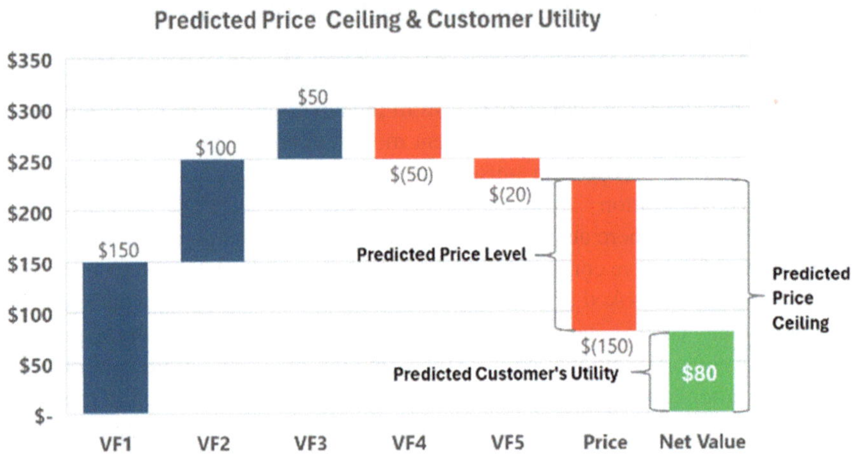

Fig. 7.30 The chart shows how the Predicted Price Level, applied to the Predicted Ceiling Price affects customer's utility

facilities, as they create an image of aggressive safety measures (VF3). On the negative side, the entrepreneur must account for the need to re-train all the nurses to use a new adaptor (VF5) as well as the need to replace some injection pumps that are incompatible with the new connector (VF4), representing hidden costs for the hospital.

The PPC for the new device can be calculated as follows:

$$\mathrm{PPC} = \sum_{i=1}^{n} \mathrm{VF}_i = \$230$$

PPC is the price level at which the manufacturer can completely erase the net customer's utility linked to the new product's adoption. Therefore, to ensure that a measurable, residual customer utility (PCU) can be delivered, the predicted price level must be smaller than the PCC (Fig. 7.30).

$$\mathrm{PPC} - \mathrm{PPL} = \mathrm{PCU} > 0$$

This focus on internal price consistency is crucial, even if the exact price has not yet been set. These considerations about price, focused on its consistency with the value proposition, must be made early in the process, even if at an approximate level. Many startups wait too long to determine the optimal market entry price, allowing the market to present them with potentially disappointing or disruptive surprises.

7.3.2.5 Competitive Pricing Issues

After completing the internal analysis of price-value consistency, it is crucial to assess whether the set price range aligns externally with the known prices of alternative technologies or services. In some cases, it is sufficient to confront all competitive prices head-to-head, but in most cases, it is necessary to compare the net value that each competitor delivers to its clients. Section 7.4 focuses more specifically on competitive pricing and the relative attractiveness of each player. While designing a value proposition, the entrepreneur must ensure that the set price range is compatible with those of the existing alternatives.

7.3.2.6 Willingness to Pay

The most successful product innovators begin by determining what the customer values are and how much they are willing to pay. Then, they design their products around these inputs and put in place a clear monetization strategy that they follow through with (Ramanujam & Tacke, 2016).

Entrepreneurs must have a conversation about the "willingness to pay" with customers early in the product development process in order to accurately prioritize the most valued product features and understand whether customers will pay for the new product.

The "willingness to pay" cluster, therefore, reminds us that, independently of the entrepreneur's internal considerations on pricing, value proposition, HTA requirements, positioning vs. competitive prices, the cost of the product, the target market penetration speed, and the reimbursement price do not necessarily offer an accurate prediction of what buyers will actually accept and buy.

Indeed, payers, healthcare providers, and intermediaries take into consideration their perceptions of utility and their multiple, often competing, priorities (Fig. 7.31).

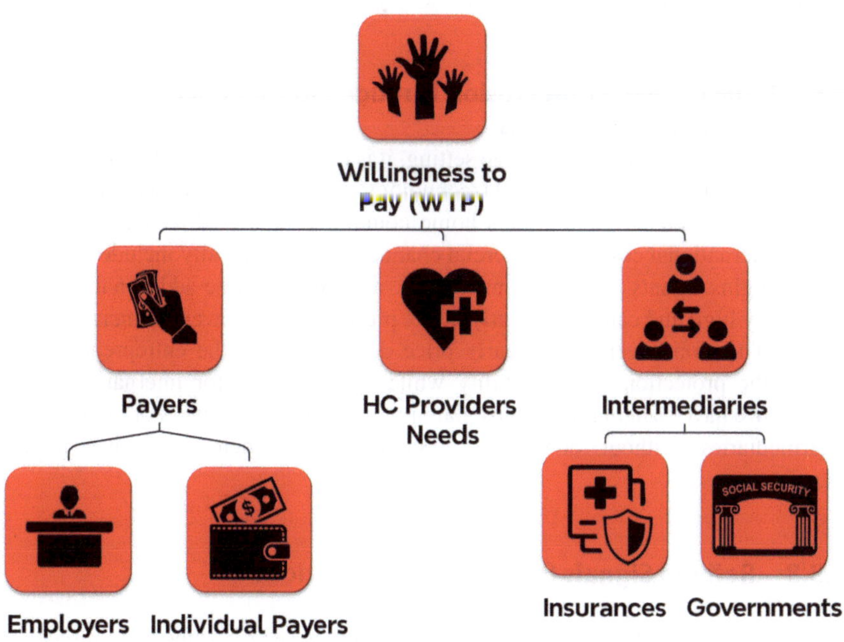

Fig. 7.31 The founding problem clusters relevant to the prediction of the willingness to pay

With reference to a real case, let us imagine launching an unbeaten technology that, in a safe but expensive manner, inactivates platelets for transfusion to oncologic patients, removing all of the receiving patient's viral and bacteriological risks.

Irrespective of the relatively high price of the required equipment, it appears to be a golden value proposition for hospitals with advanced oncology departments. Indeed, sales conversations with various hospitals' hematologists confirm the enthusiasm about the reduction in transfusion risk for immunosuppressed patients. At a later stage, when the general management becomes involved in the buying decision, the commercial talks become significantly more difficult. Cross-departmental hospital management does not set its priorities with a disease-focused approach. They weigh the safe-platelet investment against a larger context of investment opportunities.

In many cases, they prefer to invest in other departments with known structural issues instead of perfecting the platelet quality in the oncology department. One hospital declined the offer of the platelet inactivator and chose instead to invest in increasing the number of beds in the cardiology department, which did not meet the patient demand previously. Another hospital also declined the offer and bought two new MRI machines, as the radiology department was perceived as its operational bottleneck. These examples reinforce the claim that all key stakeholders involved in the buying-decision process must recursively run decisions about acceptable price levels. Therefore, medical entrepreneurs must disconnect from common product-centric or disease-centric approaches and gain insights into the depth and breadth of decision criteria used by the larger group of key stakeholders involved.

7.3.2.7 The Issue of Profit Erosion Through International Commercial Channels

While focusing on the issue of price setting, it is essential to remember that international commercialization of medical technology bears significant costs. For instance, the prices applied in the venture's home country do not need to account for the whole international cost of commercial channels, which typically includes distributors' margins, agents' commissions, freight, and duties. These additional costs are often very high and can significantly affect profitability and even financial sustainability. Therefore, even at this early price decision stage, the entrepreneur must ensure the protection of profitability while setting prices for international end-buyers. We have seen many startups planning to use multiple layers of international intermediaries without adequately accounting for their respective margin requirements.

7.3.3 Refined Signals for Financial Predictions

While focusing on the value proposition, the entrepreneur is somehow "forced" to make explicit decisions and assumptions about the attractiveness and price level of the value proposition (Fig. 7.32). This disciplined focus on making decisions and assumptions on the value proposition helps the entrepreneur become more specific about the building blocks of the overall value and of their relative weights. This degree of detail is helpful in comparing the product's (or service's) value proposition with that of competing players. In other words, the work done on this cluster feeds the next cluster with detailed information and tested assumptions about the new idea's ability to uniquely satisfy the unmet medical needs identified in the first cluster of the K9 Map.

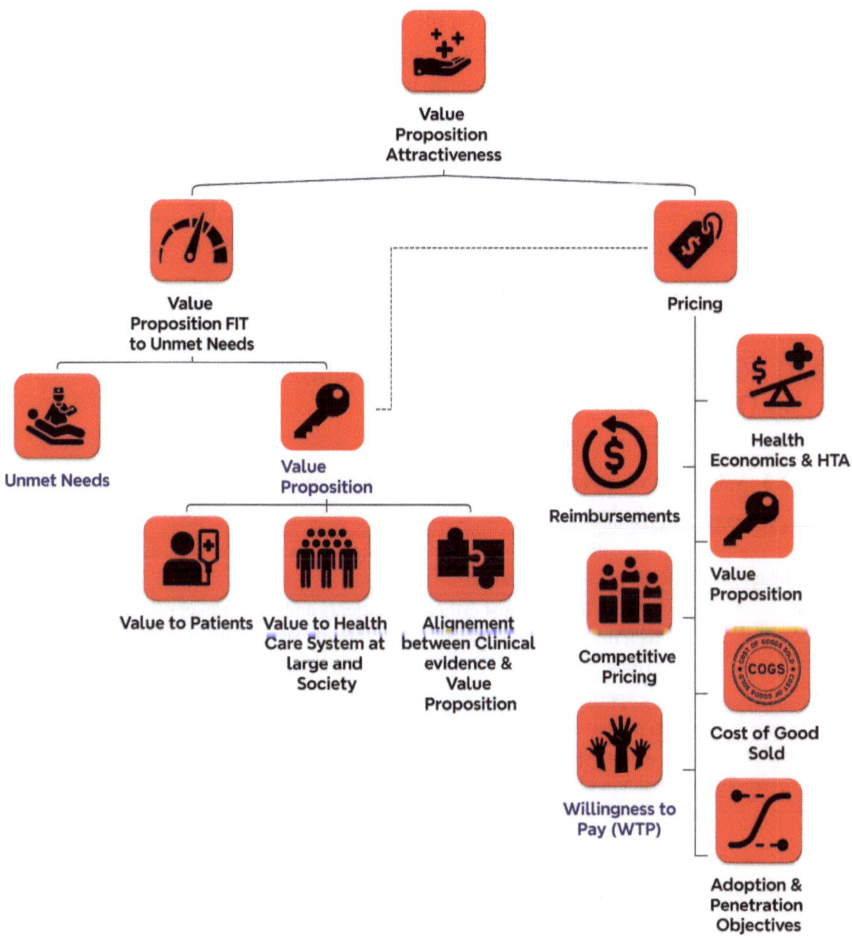

Fig. 7.32 The founding clusters driving value proposition attractiveness

In other words, this cluster represents the first opportunity for entrepreneurs to start making predictions about future acceptance, adoption, and, in turn, market penetration rate and speed. These evaluations support some basic penetration assumptions that feed forecasting models such as the Bass diffusion model (Dodds, 1973; Sillup, 1992), which is a classic marketing tool for predicting the market shares of several newly introduced products. Regardless of whether the entrepreneur uses the Bass diffusion model or any other forecasting approach, the degree of attractiveness of the value proposition remains a key determinant of market penetration dynamics. Later on, other clusters will allow us to refine these predictions more precisely. For instance, the competitiveness cluster will suggest approaches to preliminary adoption predictions based on the comparison among competing solutions and their respective offerings. The status quo cluster will provide information on the likelihood of securing government or insurance reimbursement, which can likely boost adoption. The segmentation, targeting, and positioning cluster will further sharpen these assumptions within each target segment. Instead, the analytical focus on making decisions and assumptions about the price level narrows the range of possible prices. This decreased uncertainty is not only valuable for analyzing the product's competitiveness but also essential for starting earlier revenue and margin predictions.

7.3.4 Key Assumptions and Decision-Related Questions (Table 7.3)

Table 7.3 Value proposition attractiveness: key assumptions and decision-related questions

Product strength and readiness	
Sub-cluster	Key assumptions and decision-related questions relevant to value propositionattractiveness
Value proposition fit to unmet needs.	Which needs are the most critical to patients and other stakeholders?
	How well does our value proposition align with satisfying the stakeholders' most critical needs?
	Which experiment can be performed to test assumptions about this degree of fit?
Value proposition	Which patient needs are targeted by the value proposition?
	What are the specifically targeted clinical needs? What are the specifically targeted non-clinical needs?
	What are the stakeholders involved in the buying decision?
	What specific stakeholder needs does the value proposition target?
	Are the targeted medical needs critical for patients or for any other relevant stakeholder?
	Which tests support the ability to meet the value proposition commitments?

(continued)

Table 7.3 (continued)

Product strength and readiness	
Sub-cluster	Key assumptions and decision-related questions relevant to value propositionattractiveness
Pricing	What are the key references used to set our price level?
	Is the set price level sufficient to cover the product cost profitably?
	What experiment can help testing the degree of price acceptance?
	How does our price level compare to that of our competitors?
	Do we expect a reimbursement? If "yes," how much? What can be done to obtain reimbursement?
	What kind of adoption curve will the set price allow?
	What maximum market share can be achieved with the set price level?
	Is the draft pricing accounting for export-related costs such as freight and duties?
	Is the draft pricing properly accounting for the margins of international intermediaries?
	What tests can be done to verify the accuracy and the viability of international end-user (end-buyer) pricing?
Health economics	Does the innovation comply with the minimum cost-advantage requirements of health economics?
	What are the HTA requirements and standards of the target countries?
	Are there QALY advantages for patients in relation to the new technology?
	What is our estimation of λ?
	What type of role does HTA play in the target geographies?
	How will the innovation perform in post-market HTA?
	What evidence can be gathered to estimate EHTA?
	What experiments can help support the EHTA prediction?
Willingness to pay	What is our prediction of the stakeholders' willingness to pay?
	What experiment can be conducted to support this prediction?
	Which stakeholders are most involved in the buying decisions?
	What are the potential buying motivations of each stakeholder?
Other questions	Given the answers to the above questions, does the value proposition represent a significant innovation or merely a marginal improvement?
	Are the testing activities actually aligned with the promises of the value proposition?

7.3.5 The Complete Map (Fig. 7.33)

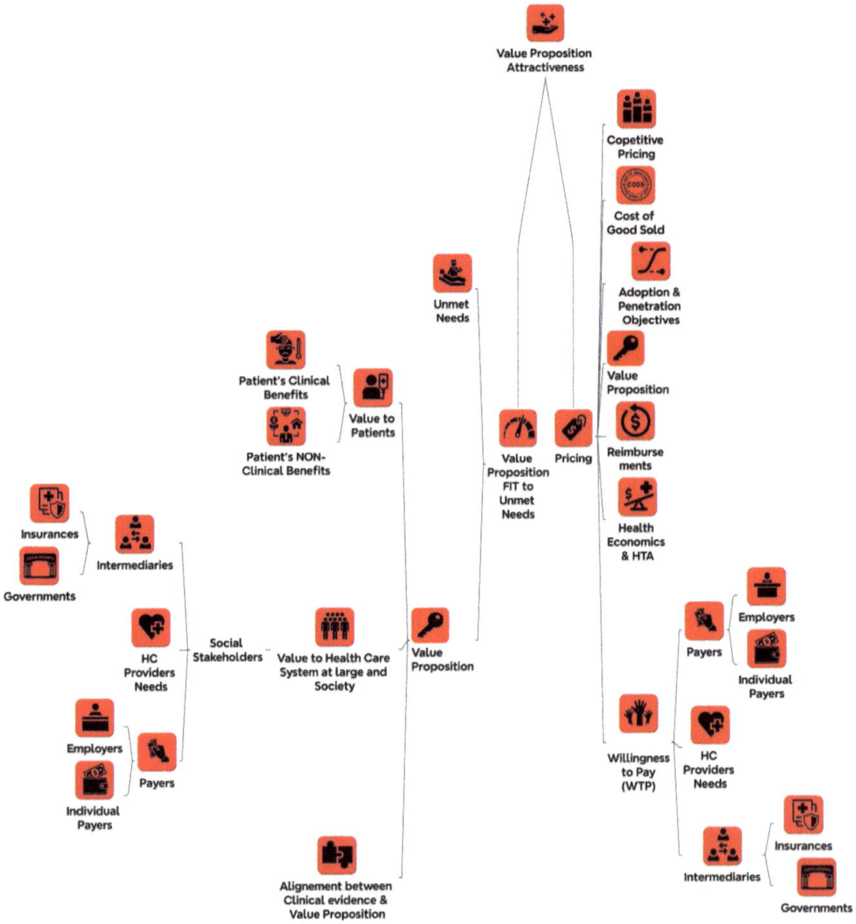

Fig. 7.33 The complete map of problem clusters relevant to the degree of value proposition attractiveness

7.4 Competitiveness

The emotional engagement of medical entrepreneurs and of the venture's management in developing their innovative solutions often weakens their ability to perform objective and disciplined evaluations of the competitive environment. Indeed, most managerial efforts are dedicated to the acceleration of product development and the sharpening of functional aspects. Most medical innovation pitches I've encountered lack a sincere and disenchanted evaluation of existing alternatives to the proposed innovation. This common flaw is often accompanied by a lack of attention to the emergent competing technologies, which demands a prospective analysis of new patents and newly started clinical evaluations. In these cases, when explicitly asked about the sustainability of their innovative solutions, most startup management teams answer using anecdotal information.

Medical technology, similar to other industries, does not have a universally accepted definition of competitiveness. Some organizations define competitiveness as the superior ability to persuade customers to buy their products. In contrast, others define it as the innovative ability to improve products and services, generating a superior offering. Section 7.1 analyzed competitiveness from the perspective of the overall market's degree of rivalry and commoditization. We clarified that, to mitigate investment risk, early-stage medical startups should be cautious when approaching markets with high competitive intensity, especially if they are not developing quantum-leap innovations.

This chapter, instead, is more concerned with the head-to-head degree of competitiveness of a specific innovative product or service against each existing and emerging competitor. In particular, the attention is on a new product's degree of competitiveness against three potential sources of competition: the current standard of care, the direct competition (current or emerging), and the indirect competition (Fig. 7.34). In parallel, this chapter addresses some fundamental issues related to the sustainability of competitive advantage.

The degree of competitiveness of an individual product in a target market is an essential co-determinant of its likelihood of adoption and, therefore, of its success. Moreover, the same degree of competitiveness contributes to predicting the return on investment. We have all seen many medical startups make biased assumptions about their product's degree of competitiveness, which leads to biased predictions about the overall investment risk and the potential return on investment. Once again: To mitigate investment risk, medical startups should avoid developing and launching new products without first making a rational, disenchanted, or even cynical assessment of their actual competitive edge.

7.4.1 Current Standard of Care

One of the most important competitors for innovative products in life sciences is the current standard of care, which can be defined as a diagnostic and treatment process that a clinician should follow for a specific type of patient, illness, or clinical circumstance (Shiel, 2020). In other words, the standard of care consists of guidelines based on scientific evidence and collaboration between medical professionals, specifying the appropriate choices and, therefore, guiding physicians in making decisions about all current preventive, diagnostic, and medical treatments. The current standard of care represents the first, most obvious competitor for any innovative therapy.

For instance, prostatectomy is a type of surgery in which a surgeon removes all or part of the prostate in order to treat prostate cancer or benign prostatic hypertrophy. Historically, the only standard of care for prostatectomy was an open or

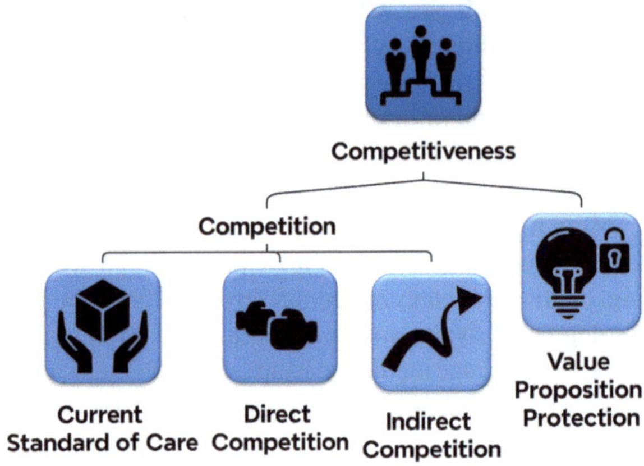

Fig. 7.34 The key clusters relevant to the degree of competitiveness of a medical product or service

laparoscopic surgical procedure, manually performed by a physician. For years, robotic surgery has challenged this standard of care, providing superior performance. In other words, it has been demonstrated that, compared to patients opting for traditional open surgery, those patients who undergo robot-assisted radical prostatectomy experience less blood loss, less pain, and shorter hospital stays. At present, many patients actively seek the robotic option as their preferred procedure.

Like in the case of robotic surgery, medical startups must analyze the current standard of care, including all involved products or services, and then demonstrate that the new product (or service) is superior to that standard.

In other words, the medical entrepreneur must adopt a position in consideration of the following questions:

- Does the current standard of care represent a direct competition to the new medical technology? If yes, what part of the new product's value proposition is explicitly challenged?
- What are the unmet needs addressed by the current standard of care? How can the new idea better address those needs?
- What are the leading healthcare providers' costs and savings from the new idea compared to the current standard of care?
- What are the main barriers to exiting the current standard of care for healthcare providers?

7.4.2 Existing Products: Direct Industrial Competitors

Competition against similar products or treatments (usually offered by other companies) is often based on a head-to-head, feature-by-feature comparison, where the different products' physical and functional characteristics, as well as the benefits to users and patients, are evaluated. This comparison is performed both by competitors to prove (or confirm) the superiority of their product and by customers (hospitals and physicians) to autonomously assess their best therapeutic options (Burns et al. 2007, 2007).

The analysis of the challenges associated with competitive assessment in the domains of MedTech and HealthTech, as well as the analysis of real-life examples, constitutes a great tool to represent the complexities faced by medical entrepreneurs.

An example of feature-by-feature product comparison is the one typically adopted for stents, which are devices used in percutaneous coronary intervention, critical to the standard treatment for patients with significant coronary artery disease or acute coronary syndromes. A metallic drug-eluting stent comprises a metallic platform and a drug, usually released from a polymer coating on the same metallic platform. A wide range of stents, with differences in material and design as well as in the type of polymer and drug, is currently available. Competitors usually compare head-to-head, feature-to-feature stents, challenging each other on features like frame design, materials, thickness, polymer degradation time, and the benefits of the eluted drug (Parker & Iqbal, 2020).

Medical entrepreneurs in this specific field must use this type of straight comparison to identify (and, possibly, visualize) elements of product superiority or inferiority. Although entrepreneurs may find differences in the competing products' physical, biological, and pharmacological properties, the vital question is whether these differences affect actual clinical outcomes for patients. Medical entrepreneurs are, instead, often trapped in the belief that a superior (or different) set of technical features naturally and automatically translates into superior outcomes for patients or for other relevant stakeholders.

But what truly matters are the different patient outcomes related to the use of the various products. This is usually achieved through randomized clinical trials comparing the various elements of clinical outcomes. The most common goal is to prove the superiority or non-inferiority of the clinical outcomes related to the product concerned.

When comparing different product characteristics and performances, product positioning maps are powerful tools for competitive analysis, offering a visual representation of the competitive landscape. They aid strategic decision-making and provide valuable insights into comparative dynamics and customer preferences (D'Aveni, 2007). These maps help visualize the competitive landscape by spatially illustrating the relationships among brands and products (Ringel, 2023). Essentially, a positioning map is a visual tool utilized to analyze how the new medical product or service compares against its competitors in customers' minds. A typical example is shown in Fig. 7.35, a chart developed by a young and successful venture commercializing bionic prosthetic hands. These maps plot the innovative product and the key competitors on a graph, with the axes representing important product attributes such as price, quality, usability, or expected clinical outcomes. The intersection of the latitudinal and longitudinal lines marks a product's location (positioning). Numbers and letters are often added to the maps on the grid, showing greater detail.

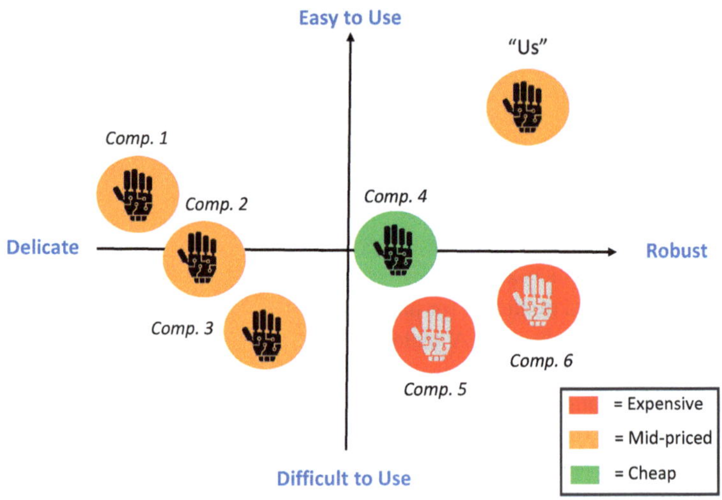

Fig. 7.35 An example of a medical product's positioning map

The problem with product positioning maps is that every startup has one, and almost all startups position their product in the same top-right position. Neither my colleague nor I have ever seen a startup present a positioning chart in which its product is not placed on the top-right part of the map. Most medical startups, at best, concede a half-win to another competitor in one dimension but always exchange it for a significant advantage in the other dimension. A good example is provided by "Company 6" in Fig. 7.35.

The assumptions and the decisions that need to be addressed while comparing a new product or service with that of the direct competition in a positioning map are as follows:

- Do the chosen dimensions correspond to clear and measurable parameters that are rationally selected from among the ones that are the most significant to buyers and patients?
- Are the assessed positioning coordinates obtained by formulating explicit, testable assumptions?
- Which experiments were performed to test the position of each of the competitors? Which other experiments may be needed?

7.4.3 Emerging Competition

The competition within the emerging medical technology sector is characterized by rapid innovation. Various factors, including the increasing prevalence of startups, advancements in technology, and the evolving needs of healthcare systems, shape this highly dynamic environment.

Despite venture capital's increasingly prudent approach to investing in medical technologies, the MedTech and HealthTech industries are continuing to experience growth, driven by the proliferation of innovative technologies and the integration of AI and the Internet of Things (IoT) into healthcare solutions. The number of global patent applications related to medical devices has tripled over the past decade, indicating a surge in innovation (Alexander et al., 2019). Furthermore, the development cycles of these technologies have halved in recent years, facilitating quicker market entry (Alexander et al., 2019).

In summary, the era when an entrepreneur's competition was a well-known, fixed contextual parameter is far gone. Medical entrepreneurs today must allocate part of their resources to constantly monitoring emerging competitors. It can be achieved through the recursive interrogation of global companies' databases, such as those provided by PitchBook or CrunchBase, by accessing the US Patent Office's (USPTO) public data, or by consulting databases that report ongoing or completed clinical trials, such as www.clinicaltrials.gov.

The assumptions and the decisions that need to be addressed concerning emerging competition are as follows:

- Does the emerging competition threaten our venture's idea's unicity? What hypothesis can be set and tested to support an assumption about this threat?
- How many startups globally label themselves as players in the same therapeutic or diagnostic area as our new product? At what stage of product development are these companies currently positioned?
- How many companies are conducting clinical trials in the areas near the one that concerns the new product idea? How many of these are trials, and in which phases are they?
- Are there technological alternatives to replicating or copying the deliverables of the new product without infringing on the patents obtained?
- Which companies hold patented technologies in the same therapeutic or diagnostic area?

7.4.4 The Competitiveness of the Business and Commercial Model

Competitiveness cannot be determined by focusing only on product features, without assessing the strength of the competition's models for commercialization and monetization. While Sect. 7.7 focuses, with significant depth, on the decision-related problems surrounding commercialization and monetization, here we focus on the roles of these issues as complementary elements to a product-focused competitive analysis.

Figure 7.36 highlights the three main components of a business model: the commercial model, the monetization model, and the organizational competencies. A monetization model articulates how a business converts its value proposition to financial income, which is crucial for long-term sustainability and competitiveness

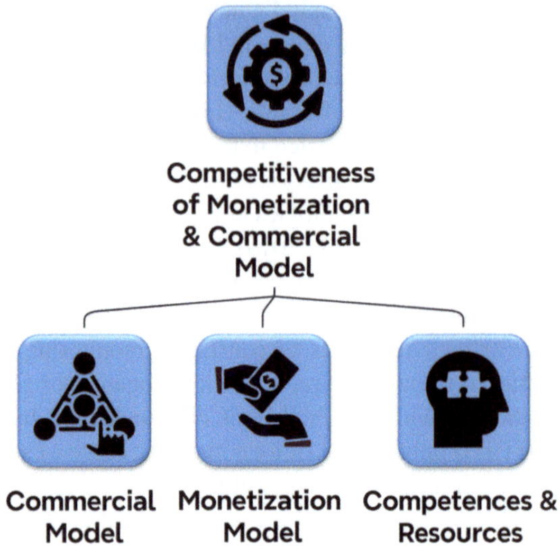

Fig. 7.36 Three essential complementary elements to a product-focused competitive analysis

(van Angeren et al., 2022). Monetization refers to the methods and processes by which a business converts its resources—such as products, services, and IP—into revenue streams. In contrast, a commercial model focuses more on the actual go-to-market approach and includes elements such as market strategy, customer interaction, sales channels, and commercial organization.

Each competitor's monetization and commercial model, as well as the competencies built and acquired to support these models, must be assessed, as they may contain critical competitive advantages that are neither fast nor easy to emulate. As already mentioned, only new medical solutions with non-marginal, quantum-leap innovations can count on some degree of relief from the competitive pressure of global commercial organizations adopting proven monetization pathways. These considerations represent a crucial warning for the many medical startups that want to believe that product-based competitive edges are:

- enough to justify substantial adoption rates and the measurable reshuffling of relative market shares and
- sufficient to drive changes in medical practices.

Only global companies can rely on hundreds of product managers and specialists distributed worldwide, striving to influence the adoption of new medical technologies and meet growing demand. Young medical startups must win their battles one by one, hospital by hospital, country by country, unless they are soon acquired by one of the global market players. Therefore, the high-level assumptions and the decisions that need to be addressed with respect to the business and commercial model are as follows:

- Do competitors benefit from the strengths of more prominent, mature commercial organizations?
- Is the monetization model of our venture's innovation as attractive as that of the main competitors?

- How efficient and profitable is the commercial model and organization of the new product compared to the competitors?
- Is there a meaningful gap in commercial competencies between my venture's organization and the competition?

7.4.5 Indirect Competition

According to Porter's five forces model, substitute products are different from the concerned product but can still satisfy the same needs of customers (Porter & Canada, 1985). Indirect competition deserves particular attention in the life sciences industry. The following examples illustrate how important it is for medical entrepreneurs to consider threats from competitive fields that are apparently distant from the ones they are currently focusing on.

For instance, there are several pharmaceutical, nutraceutical, and MedTech approaches to dealing with nocturnal enuresis in children. However, urologists know that human biology can resolve these conditions autonomously, even if in uncertain time-frames (Cendron et al., 1999). Therefore, all the medical devices that support children and their families in managing or mitigating this problem compete against normal human development, diminishing the intensity of the unmet medical need.

While predicting future revenues and designing clinical trials, one of our startup clients, which focuses on enuresis devices addressing behavioral approaches, had to seriously consider the "competition" from human biology, which can resolve the issue over a more extended period. If the natural solution is relatively rapid and painless, it is not easy to justify the adoption of new therapies.

Another example of indirect competition comes from therapeutic or preventive approaches that are unrelated to the specific disease in question. For instance, scientific findings have proven that particular diets can contribute to the prevention of medical conditions that lead to the implantation of stents. Indeed, substantial evidence indicates that diets comprising nonhydrogenated unsaturated fats, with whole grains as the primary form of carbohydrates, abundant fruits and vegetables, and sufficient omega-3 fatty acids can offer significant protection against coronary heart disease (Hu & Willett, 2002).

Continuing with examples from the stent market, a large international study led by Stanford and New York University found that invasive procedures are no better than medications and lifestyle changes at treating severe but stable heart diseases (Maron & Harrington, 2019). In this case, players and innovators investing in the design or launch of a new stent type should consider the competition from new pharmaceuticals that offer the same results to patients without the need for surgery.

While considering indirect competition, the medical entrepreneur must answer the following key questions:

- Can preventive procedures or practices significantly alleviate the unmet medical needs targeted by our invention? If yes, to what extent, and which preventive procedures or practices are available?

- What other procedures or products can address the same unmet medical needs through off-label use?
- Are there non-medical processes that can alleviate the target unmet medical needs?

7.4.6 Competitiveness of the Portfolio in Medical Innovation

It is essential to note that medical entrepreneurs frequently face the challenge of entering the market with a single product. If managing customer relationships for a single product is complicated, creating new relationships with customers from scratch represents a significant challenge. Let us take hospitals, for example.

Medical inventions often respond to a very narrow area of the needs of patients, physicians, and hospitals. However, hospital buyers make purchasing decisions focusing on the broader needs of the hospital, in order to guarantee the proper delivery of the entire set of healthcare services rendered to patients. Therefore, the competitive stance of small startups selling a very narrow product portfolio represents a competitive issue when competing with companies capable of offering a broader portfolio, which is more likely to draw buyers' attention and recognition.

Unless the medical venture has identified a few multinational companies that are interested in their new medical technology as a good fit for those companies' broader portfolios, entrepreneurs must choose between two options: identify adjacent products that can enhance the existing portfolio or seek a good portfolio fit with downstream distribution channels. While focusing on the subject of portfolio-driven competition, the medical entrepreneur must answer the following key questions:

- Is the direct competition more or less advantaged by their portfolio?
- Will the narrow portfolio create an issue in market entry?
- For what established corporation could the new technology represent a compelling portfolio addition?
- What other product could be a smart portfolio fit when added to my new technology?
- What type of medical distributors can effectively integrate the product or service, minimizing marginal commercial costs and maximizing exposure to potential buyers? Does this type of distributor actually exist?

7.4.7 Patents vs. Protection of the Value Proposition

This section does not focus on the legal and technical aspects of patenting, such as the requirements of "patentability," "novelty," "non-obviousness," and "enablement" (Mayfield, 2016). Instead, it explores the fundamental, underlying business question pertaining to IP protection in a medical startup: Can the venture protect the idea's value proposition in real therapeutic and diagnostic practice?

There is a recurring misconception about IP protection for medical technology. It is based on the immensely popular and biased belief that patenting one product's technological component is sufficient to protect the entire value of the new idea. Let us use an example to be very specific about this critical issue. In the context of epidural anesthesia, developing an epidural catheter that informs the anesthesiologist when the catheter tip has reached the epidural space can add significant and unique value to the safety of epidural procedures (Arendt & Segal, 2008).

Suppose a new venture develops a catheter that, through an electromagnetic sensor, can provide the anesthesiologist with information about the catheter's position and its reach into the epidural space. This venture has obtained a patent for using an electromagnetic sensor to deliver information on the position of an epidural catheter. What we have seen happening very often is that ventures like this one assume that their patent protection guarantees uniqueness. After repeatedly claiming this assumption internally and externally (without actually testing it), the entire venture's organization gradually develops a false sense of security about the sustainability of its competitive advantage.

In reality, these assumptions are often affected by fundamental biases. For instance, in this venture's case, the patent is limited to using an electromagnetic sensor to inform the anesthesiologist about the catheter's position; it does not cover the broader idea of informing the anesthesiologist about the catheter's tip position while inserting it. Companies commercializing devices that perform the same duty using other technologies already existed. Indeed, competitors commercializing similar solutions that use acoustic waves or optical sensors have soon become an overwhelming barrier for this venture.

In these cases, to avoid a misleading sense of security, the venture should prioritize an investigation into the presence of market players (or new market entrants) offering the same medical value added with alternative technologies.

Again, the focus must be on protecting the foundations of the entire value proposition, not on protecting one of the many possible technical or clinical ways to deliver the same value. When broader protection cannot be achieved, ventures must recognize this limitation and rationally reflect on and account for the reduced protection in their predictions of future competitive scenarios. In these cases, it is essential:

- to acknowledge the actual degree of protection while formulating competitive arguments leading to the prediction of market shares and
- to budget for possible additional patenting activities, providing wider and stronger protection.

In short, a rational, disenchanted analysis of the actual degree of value proposition protection is critical, as it helps correctly frame future penetration expectations and more accurately predict the future expenses for obtaining and maintaining additional patents. Therefore, while focusing on this topic of value proposition protection, the medical entrepreneur must answer the following key questions:

- Is there an IP protection strategy for the new product?
- Is the IP protection strategy capable of protecting the entire value proposition? If not, how can a lower degree of protection influence the predicted competitive scenarios?
- What other protective measures can be adopted to better protect the value proposition?

7.4.8 Key Assumptions and Decision-Related Questions (Table 7.4)

Table 7.4 Competitiveness: core assumptions and decisions affecting the venture's value

Competitiveness	
Sub-cluster	**Core Assumptions and Decisions Affecting the Venture's Value**
Current standard of care	Is the current standard of care a substantial competitor?
	What evolutions of the current standard of care are possibly threatening the competitive advantages of the new innovation?
	What health economic advantages are offered by the use of the new technology compared to the current standard of care? How can we prove or test our hypotheses of an economic advantage?
Direct competition	How many products directly compete with the new innovation?
	Which products compete more aggressively?
	What dimensions of the value proposition of the new product are currently unmet by the competition?
	How much patent-based IP protection time is left for the competitors?
	How much patent-based IP protection time is left for the new technology?
	To what extent does the new product display superiority in the identified key competitive dimensions?
	Are the strengths of the new product focused on truly critical unmet needs? How are unmet medical needs being prioritized and ranked? How can the validity of this ranking be tested?
Indirect competition	Can preventive procedures or practices significantly alleviate the unmet medical needs targeted by the new invention? If yes, to what extent, and which preventive procedures or practices are available?
	What other procedures or products can address the same unmet medical needs through off-label use?
	Are there non-medical alternatives that can alleviate the targeted unmet medical needs?
Emerging competition	How many and which companies are targeting similar therapeutic or diagnostic solutions or technologies?
	In which clinical development phases do emerging competitors currently stand?
	What known clinical results already support emerging competitors? How many and how compelling are these supports?
	What time-based advantage vs. Emerging competitors?
	Is emerging competition a substantial risk that needs to be monitored?
Portfolio competitiveness	Is the direct competition more or less advantaged by their portfolio?
	For which established player can the new product be a portfolio fit?
	What other product can represent an effective portfolio addition?
	What type of medical distributors can effectively integrate the new product?
Value proposition protection	Is there an IP protection strategy for the new product?
	Is this strategy capable of protecting the entire value proposition?
	If not, how can a lower degree of protection influence the competitive scenarios?
	What other protective measures can be taken to better protect the value proposition?

7.4.9 The Complete Map (Fig. 7.37)

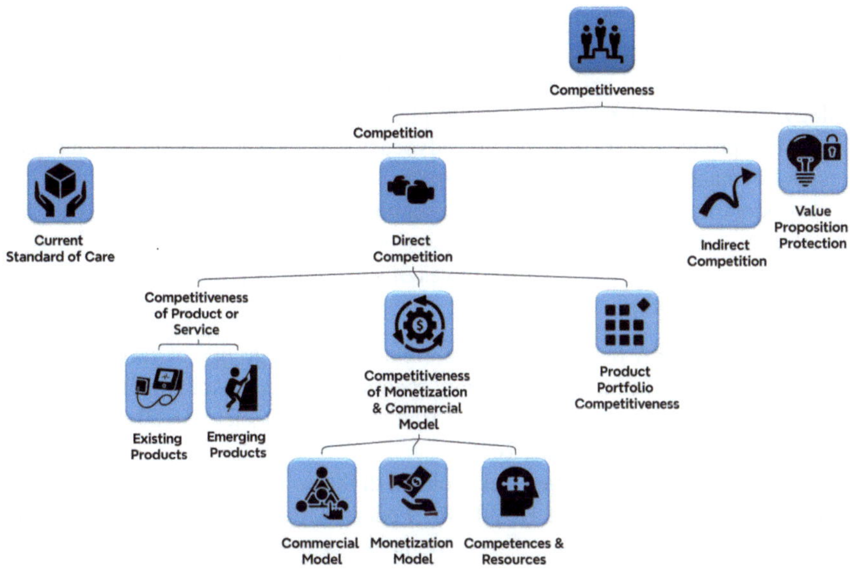

Fig. 7.37 The full map of decision-related clusters relevant to the determination of the degree of competitiveness

7.5 Suitability for the Status Quo

In the life sciences business, the customer landscape is rapidly changing. Both public and private payers are becoming increasingly more involved and more selective in awarding reimbursements every day. Hospital networks are consolidating into larger groups with increasing negotiation power, while physician practices are being acquired by large healthcare organizations that naturally assume control over procurement decisions and can influence therapeutic choices. These trends are changing the various subjects (stakeholders) involved in the decision-making process for adopting new medical products. The medical sales stereotype, once consisting of a single sales representative selling to a doctor, has significantly changed into a broader set of mutually influencing subjects capable of determining the actual adoption of a new product (Abdullah et al., 2013).

A stakeholder is any individual, organization, or group that "can affect or is affected by the achievements of the organization's objectives" (Friedman & Miles, 2006). In the life sciences industry, from the entrepreneur's (or the manufacturer's) perspective, "stakeholders" mainly refer to the most relevant individuals, groups, or organizations that influence the decision-making process of adopting and purchasing the proposed device or therapy.

Several stakeholders must be considered in the medical device sector throughout the development of the new product or service, including patients, clinicians, manufacturers, regulators, and healthcare providers (Laurisz et al., 2023). Other scholars, specifically concerned with the identification of key stakeholders in medical technology, suggest similar lists, including: physicians, KOLs, hospital procurement, public and private payers, patients and advocacy groups, HTAs, regulatory authorities, policymakers, and policy-influencers.

The identification of specific key stakeholders in life sciences depends on the specific disease or therapeutic area of concern, as well as on the particular geographical region. Here are a few examples of stakeholders relevant to different diseases in various regions:

- In the area of RA therapy, there are eight identified groups of stakeholders in Portugal: general practitioners (GPs), rheumatologists, hospital managers, hospital pharmacists, budget holders, the Rheumatology Society, and the RA Patient Association (Sendyona et al., 2016).
- In Switzerland, the identified stakeholders in the context of chronic diseases are the Cantonal Ministries of Health, the Federal Office of Public Health, health insurance companies, the Swiss Medical Association, Swiss Societies of Medicine (general-internal medicine, endocrinology-diabetology, pneumology, and cardiology), Cantonal Societies of Medicine, companies and physicians with experience in developing chronic disease management projects, Institute of Nursing, Swiss Society of Pharmacists, home healthcare centers, and patient organizations (Lauvergeon et al., 2012).
- For the US-based hemophilia market, the list of stakeholders able to impact a hemophilia product includes the American Society of Hematology, the American Thrombosis and Hemostasis Network (ATHN), a large network of pharmacies (like CVS or Walgreens), the Hemophilia Alliance, the Hemophilia Federation of America, the National Hemophilia Foundation, the New England Hemophilia Association, the New England State Consortium Systems Organization (New England Medicaid Programs), a group of manufacturers (such as Novo Nordisk, Roche, Shire, and Genentech), the United Hemophilia Foundation, and the World Federation of Hemophilia (ICER, 2018).

In order to rationally approach this environmental complexity, entrepreneurs must identify all relevant stakeholders, understand their respective needs and their reciprocal relationships and influencing powers. Working with market stakeholders and potentially integrating them into the product development process is a vital competency necessary for medical entrepreneurs (Symmank et al., 2015).

In other words, stakeholders serve as one of the most important sources of information for entrepreneurs while testing the hypotheses that support expectations (and/or claims) about the attractiveness of the value proposition, market penetration, and future revenues. This chapter characterizes the seven most common stakeholder groups relevant to medical innovation: healthcare providers, manufacturers and service providers, patients, society, intermediaries, and regulatory agencies (Fig. 7.38).

7.5.1 Patients

In recent years, the role of patients as market stakeholders in healthcare has gained significant attention, reflecting a paradigm shift from the traditional view of patients as passive recipients of care to their role as active participants in their health management. This transformation is primarily driven by the concept of patient empowerment, which emphasizes the importance of patients assuming an active role in their healthcare decisions and management. Patient empowerment not only enhances self-management but also contributes to improvements in health outcomes and overall well-being. Indeed, research indicates that when patients are involved in value co-creation alongside healthcare providers, the sustainability of healthcare systems is positively impacted (Popa et al., 2022; Russo et al., 2019).

Historically, patients were viewed merely as consumers of healthcare services; however, contemporary perspectives advocate for their involvement in shaping healthcare delivery (McDermott & Pedersen, 2016). This shift is supported by evidence suggesting that patient engagement leads to better healthcare outcomes and

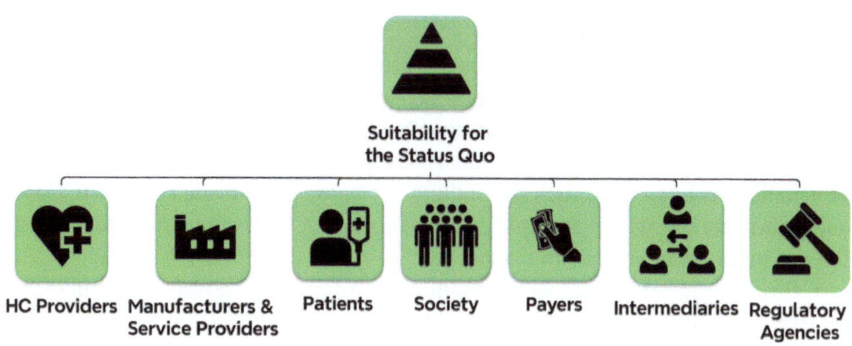

Fig. 7.38 A summary map of the most relevant stakeholders in medical technology innovation

enhances the quality of care. Furthermore, the integration of digital health technologies has facilitated this transition, allowing patients to access health information and actively participate in their treatment decisions (Preko & Budu, 2024).

The implications of viewing patients as consumers extend beyond individual care experiences to encompass broader systemic changes within healthcare organizations. For instance, the adoption of patient-centered care models necessitates that healthcare providers recognize and respond to patients' beliefs, values, and preferences, which are critical for effective engagement (Harrison et al., 2019; Batalden et al., 2016.) This engagement is not merely restricted to the provision of information; it involves fostering a collaborative relationship in which patients feel (and are) empowered to voice their needs and preferences (Lopatina et al., 2019). Such collaborative dynamics can enhance patient satisfaction, which is increasingly recognized as a key metric for assessing healthcare quality (Naidu, 2009).

In conclusion, recognizing patients' centrality as market stakeholders is reshaping the landscape of healthcare delivery. Providers can enhance service quality and patient satisfaction by empowering patients and involving them in the co-creation of healthcare value.

It is essential to underline the fact that patients often become the actual users of medical technology, which requires the definition of a specific, related stakeholder: the patient as the user.

Moreover, in the context of certain diseases, patients often form associations to seek stronger representation in society and the healthcare system. These associations represent another important type of stakeholder. In many cases, patients rely heavily on and collaborate with their caregivers, who constitute the most important intermediaries between the patient, the healthcare system, and the rest of the world. Figure 7.39 shows these small groups of non-medical stakeholders with strong relationships with patients.

7.5.1.1 Patients as Users

For decades, some patients have been the primary users of new medical technologies as well. For instance, peritoneal dialysis (PD) is a treatment option for individuals with kidney failure, which involves peritoneal dialysis machines and a series of steps that the patient performs at home. Other examples can be found in the area of wearable technologies, such as continuous glucose monitoring systems (CGMS) and smartwatches, which exemplify how patients can leverage medical technology for self-monitoring and health management.

CGMS, for instance, has been recognized as a significant advancement in diabetes care, allowing patients to track their blood glucose levels in real time, which can lead to better glycemic control and reduced complications (H.-S. Kim & Yoon, 2020). Patients' role as users in the wearable medical industry is expected to be prominent, with studies indicating that 98% of patients find wearable technology

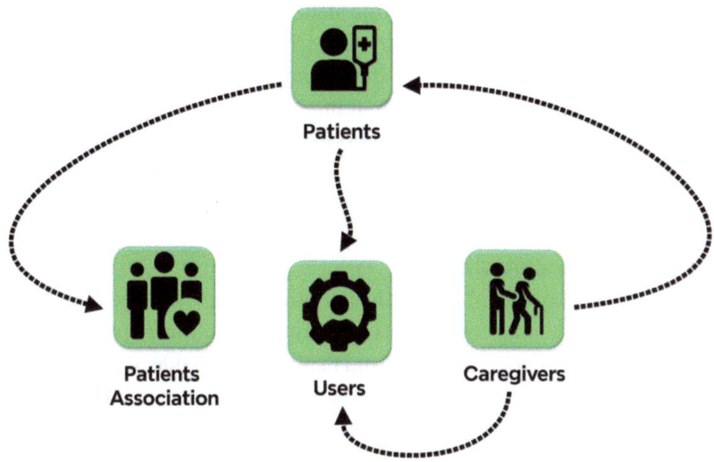

Fig. 7.39 The relationships between Patients, Patients as users, Patient's Associations, and Care Givers

convenient, highlighting a preference for noninvasive and user-friendly health monitoring solutions (Raja et al., 2019).

7.5.1.2 Caregivers

Caregivers play a crucial role as stakeholders in healthcare, particularly in the context of chronic illness and palliative care. Their involvement contributes to the monitoring and enhancement of the quality of care provided to patients and significantly impacts patients' decision-making processes. This important role encompasses emotional support, information-sharing, and active participation in treatment decisions, which are essential for the effective delivery of patient-centered care.

Family caregivers are often the primary source of support for patients, especially those with illnesses such as cancer and dementia. They are integral to the decision-making processes pertaining to treatment options and care pathways. In this context, caregivers' perspectives are vital in identifying the unmet needs within the healthcare system.

The importance of caregivers as stakeholders is further underscored by their role in facilitating communication between patients and healthcare providers. Caregivers often act as intermediaries, conveying critical information about patients' medical histories and preferences, which can significantly influence treatment decisions (Elliott et al., 2014). Their involvement in care transitions is particularly crucial, as they help ensure continuity of care and adherence to treatment plans (Salikhanov et al., 2023).

The importance of the caregivers' role necessitates that medical entrepreneurs focusing on innovations that concern the caregivers' involvement actively involve them in all stages of product or service development, as the caregivers are known to be involved in the patient's decision-making, therapy management process, and therapy compliance. Initiatives that incorporate caregiver feedback into service design have been shown to enhance the relevance and applicability of healthcare solutions (Wu et al., 2023).

7.5.1.3 Patient Associations

Patient associations play a vital role in the adoption of new therapeutic and diagnostic procedures by advocating for patients, influencing healthcare policies, and facilitating communication between patients and healthcare providers. Their involvement is essential in ensuring that patient's needs and preferences are prioritized in the healthcare system, particularly in the context of chronic diseases and emerging therapies.

One of the primary functions of patient associations is to advocate for better healthcare policies and practices. These associations work to highlight unmet needs and drive the development of patient charters that articulate the rights of patients and the standards of care they should expect. For instance, patient advocacy groups have been instrumental in developing charters for chronic diseases, which serve to inform policymakers about the specific needs of patients and improve access to care (Bonella et al., 2016). This advocacy is crucial in ensuring that new therapeutic procedures are developed and made accessible to all patients, as disparities in access can significantly affect treatment outcomes (Bonella et al., 2016; Selgas et al., 2017).

Moreover, patient associations facilitate shared decision-making (SDM) between patients and healthcare providers, which is vital for the successful adoption of new therapies. SDM involves patients' active participation in their healthcare decisions, which can lead to improved satisfaction and health outcomes (DeMeester et al., 2016; Baig et al., 2016). The role of patient associations in promoting SDM is particularly important for marginalized groups, such as racial and ethnic minorities, who may face additional barriers to effective communication with healthcare providers (Baig et al., 2016). By advocating for culturally competent care and the inclusion of diverse patient voices, these associations help ensure that new diagnostic and therapeutic options are tailored to meet the needs of all patient populations.

In addition to advocating for policy changes and promoting SDM, patient associations play a significant role in educating both patients and healthcare professionals about new treatments and diagnostic procedures. They provide resources and support that empower patients to understand their conditions and treatment options, which are essential for informed decision-making (Ronnebaum & Schmer, 2015; Choi, 2015). This educational role is complemented by their efforts to mobilize healthcare professionals to engage in advocacy, thereby creating a more patient-centered healthcare environment (Lewinski & Simmons, 2018). For instance, the involvement of nurses in patient advocacy has been demonstrated to enhance trust in the nurse–patient relationship and improve patient outcomes by ensuring that patients are well-informed about their healthcare options (Ronnebaum & Schmer, 2015).

Furthermore, the impact of patient associations extends to the broader healthcare landscape by influencing health policy at various levels. These associations can mobilize grassroots-level movements that lead to significant changes in healthcare delivery systems, as evidenced, for instance, by the coordinated efforts in Spain to reform renal replacement therapy through a multi-stakeholder advocacy group (Ronnebaum & Schmer, 2015). This example demonstrates that patient associations not only advocate for individual patients but also contribute to systemic changes that benefit the entire patient community.

A few examples of patient associations are the Acoustic Neuroma Association, Accelerated Cure Project for Multiple Sclerosis, Alliance for Aging Research, Alzheimer's Association, and American Brain Tumor Association. It is important to note that these associations normally include a global organization linked to multiple regional and country chapters.

In conclusion, patient associations are pivotal in adopting new therapeutic and diagnostic procedures. They advocate for patient rights, promote SDM, educate stakeholders, and influence health policy to prioritize patient needs in the healthcare system. Their polyhedric role underscores the importance of patient engagement in the healthcare process and highlights the necessity for ongoing collaboration between patients, healthcare providers, and policymakers.

In summary, while focusing on the topic of patients as stakeholders, the medical entrepreneur must focus on the following questions:

- To what extent do patients individually represent an active party in the stakeholders' map?
- If patients play a critical role in the decision to adopt a medical technology, in what manner does this role manifest?
- With which other stakeholders and in what way do patients interact while influencing adoption decisions?
- If patients play a critical role in adoption decisions, can the market be considered a consumer market? To what extent?
- Do caregivers play an essential role in adoption decisions?
- In what way do caregivers influence decisions pertaining to adoption?
- If patients cannot play a critical role individually, are there patient associations that can assume this role?
- If yes, are patient associations open to a dialogue with the venture regarding the adoption of the new medical technology?
- In what way do patient associations influence decisions pertaining to adoption?

7.5.2 Payers

Payers are individuals or entities that pay in exchange for healthcare services. While we may be inclined to identify health insurance or the government as payers, actual payers are those who pay for insurance and pay taxes to the government to provide healthcare coverage. Insurance companies and the government act only as intermediaries between patients and healthcare providers to mitigate individual financial risk.

As illustrated in Fig. 7.40, all individuals who are almost certain to become patients at some point pay taxes to the government, which provides social healthcare, and/or pay insurance premiums to private insurance companies, which offer private healthcare coverage.

Even if employers are frequently classified as payers, one may argue that, in reality, they pay for healthcare coverage as part of their workers' compensation. According to this perspective, employees' monetary compensation is diminished by the provision of health care insurance services, and working individuals can be viewed as the actual health insurance payers. On the other hand, in countries where employers pay and decide on coverage and health insurance plans, they play a significant role in establishing the level of patients' financial protection and the portfolio of reimbursable therapies and diagnostic procedures.

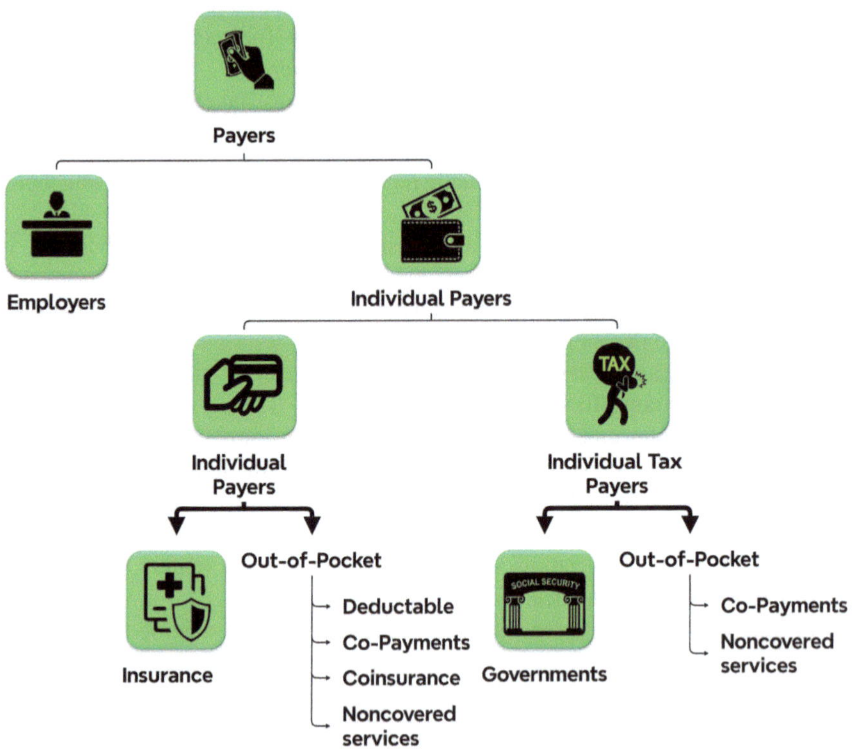

Fig. 7.40 The health care payers and the payment types

Patients' out-of-pocket costs are those medical expenses that they are responsible for, which are not covered by their health insurance plan.

Furthermore, even if private insurance covers patients, they are still exposed to different kinds of disbursements:

- Deductibles: These constitute the amount that a patient pays before insurance covers a service or a prescription.
- Co-payment: It is a fixed amount that a patient pays for a covered service after the deductible has been paid.
- Co-insurance: It is the percentage of a covered service a patient pays after subtracting the deductible.
- Non-covered services: These constitute the costs of services that are not covered by the patient's insurance plan.

High out-of-pocket costs can have serious consequences for patients, forcing them to forgo medications or opt for less effective treatments. Patients covered by public health insurance plans are usually less financially vulnerable and may be responsible only for copayments and not for the cost of non-covered services.

On the topic of payers, the medical entrepreneur must answer some important questions, such as the following:

- Do employers represent important stakeholders who are driving the adoption of specific new therapies or diagnostics?
- What role do employers play in the election and adoption of new therapies and diagnostics under their employees' plans?
- What is the role of employees in selecting insurance plans that cover specific services?
- Are employees likely to actively elect certain insurance plans because they cover special therapeutic or diagnostic procedures?
- To what extent are patients willing to pay (or co-pay) for certain new therapeutic or diagnostic procedures or for the use of medical devices? Which tests can support or weaken a hypothesis concerning the predicted patient's willingness to pay?
- In the context of the disease in question, are there patient associations that lobby for the inclusion of specific, unique new treatments or diagnostics under public or private insurance plans? Which tests can help answer this question?

7.5.3 Society

In contemporary healthcare systems, society emerges as a pivotal stakeholder, influencing the dynamics of health delivery, policy formulation, and ethical considerations. The COVID-19 pandemic strongly underscored the interconnectedness of health outcomes, societal well-being, and the collective opinion about the acceptability of the cost of health. It suggests that holistic healthcare evaluation mechanisms such as HTA should incorporate broader societal values, ensuring a certain degree of alignment between the need for innovation in healthcare and societal needs and expectations (Chan et al., 2023).

Moreover, the idea of healthcare as a fundamental human right is gaining traction, particularly in educational settings. Collective opinion appears to be increasingly supportive of an understanding of healthcare access and equity among medical students, reinforcing the notion that healthcare should be viewed as a societal obligation rather than a privilege (Sahanaa et al., 2023). This idea points to the need for a government–society partnership model to rejuvenate primary healthcare systems, emphasizing collaboration as a means to address resource limitations and improve service delivery (Kasi, 2023).

The economic implications of healthcare decisions also reflect societal stakes. For instance, the burden of diseases such as sickle cell disease extends beyond individual patients, impacting families and communities both economically and socially (Silva-Pinto et al., 2022). This interconnectedness necessitates a comprehensive approach to healthcare that considers societal impacts and the broader economic consequences of health conditions on society (Canavan et al., 2014).

In conclusion, the role of society as a healthcare stakeholder involves certain ethical, economic, and educational dimensions. As healthcare continues to evolve, integrating societal perspectives will be crucial in shaping effective, equitable, and sustainable healthcare systems.

While focusing on the topic of societal impact, the medical entrepreneur must answer the following questions:

- Will the new technology, its adoption, and its cost have a sensible impact at the societal level?
- If there is a significant impact at the societal level, is it related to concerns about the increased social costs of health? To what extent?
- If there is an impact at the societal level, is it related to the collective positive feelings about the diagnosis or cure of a high-severity disease? To what extent?
- Are there patient associations that may influence the collective opinion? To what extent? How long may the influencing process take?

7.5.4 Intermediaries

To manage individual health risks and avoid placing the entire responsibility of paying for potentially long and costly therapeutic paths on one individual, society has established the use of intermediaries. These entities pull the risk of large groups of individuals together to make the overall risk manageable. These intermediaries collect funds from a group of people, pool them together, and use them to pay for healthcare delivery for the individuals who are covered (Fig. 7.41).

There are two main types of intermediaries: insurance companies and the government (Thomas, 2008). Insurance companies are private companies that issue insurance policies, which are contracts that provide payments against the policyholder's medical bills in exchange for the regular payment of insurance premiums. These policies remove the risk that their holders may face high medical bills that

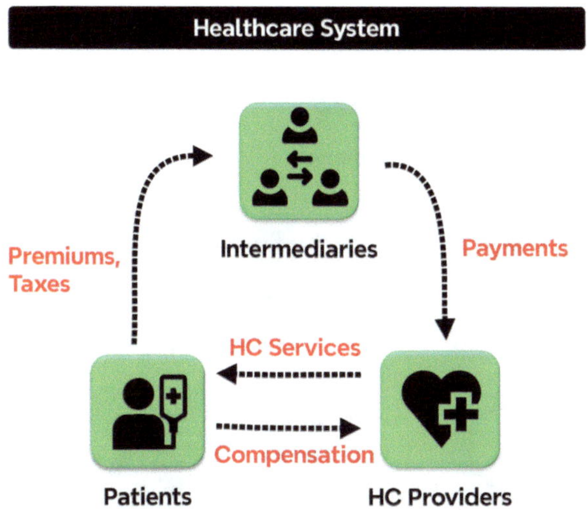

Fig. 7.41 The streamlined relationships between Patients, HC Providers, and Intermediaries

they are required to pay in full individually. Government payers are also referred to as public intermediaries. The government agrees to cover healthcare costs for certain groups, collecting funds from the population to finance these costs through the tax system. In some cases, these funds may be structured as a premium.

As mentioned in Sect. 7.3, most developed countries, particularly those with a predominance of government-managed healthcare, use HTA programs to broadly evaluate health technologies and guide coverage and pricing decisions at the country level. In these countries, HTA bodies (such as NICE in the UK) play an essential role in deciding new technology adoption and pricing. In contrast, the USA does not have a national HTA program to evaluate health technologies and broadly guide coverage and pricing decisions. However, several organizations in the USA, including the Department of Veterans Affairs and the Centers for Disease Control and Prevention, engage in HTA activities that are relevant to their respective areas of focus (Mulligan et al., 2020). Therefore, private insurance reimbursement decisions in the USA constitute a complex interplay of negotiations between healthcare providers and insurers, influenced by various factors, including service type, patient demographics, and regulatory frameworks (Fig. 7.42).

For instance, in the USA, the negotiation for reimbursing an optical imaging agent (5-ALA) that allows neurosurgeons a precise visualization of certain brain cancers was achieved sequentially. Initially, neurosurgeons requested and obtained access to this breakthrough medical product for their hospitals. Hospitals initially purchased this device at the hospital's expense, without an immediate change in the DRG reimbursement related to this type of brain surgery. In future interactions with insurance companies, hospitals requested and negotiated an increased DRG reimbursement for the surgical removal of brain cancers due to the inclusion of several new technologies (including this new optical imaging agent) and their respective prices. To date, private payment coding for 5-ALA reimbursement depends on the

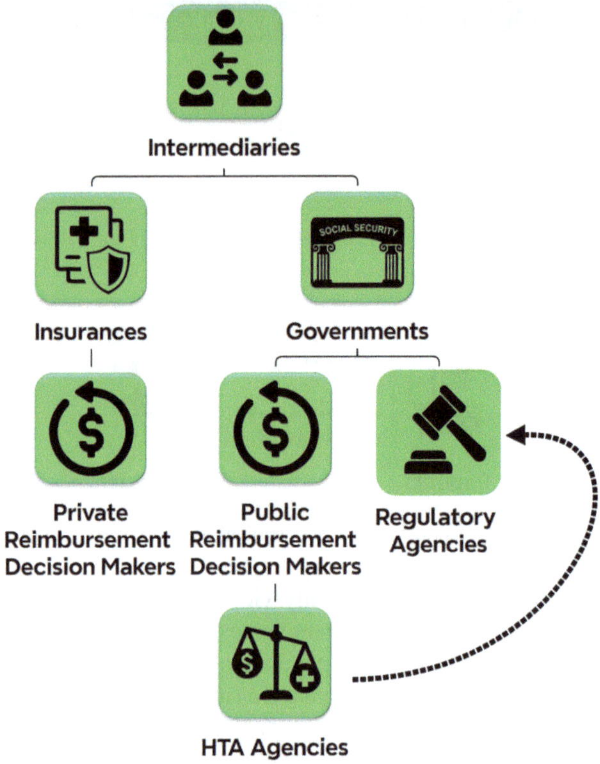

Fig. 7.42 The main health care intermediaries and their relationships with regulators and reimbursement

insurance involved, while the public reimbursement price varies depending on the negotiated payment rates with different Medicare carriers.

While focusing on the topic of healthcare intermediaries, the medical entrepreneur is required to make significant assumptions and decisions by answering questions such as the following:

- Is private insurance likely to reimburse the use of the new medical technology? How can private reimbursement be predicted or influenced?
- Are governmental entities likely to reimburse the use of new medical technology? How can public reimbursement be predicted or influenced?
- What decision-making mechanisms are used by private and public intermediaries with regard to reimbursement and pricing?
- What is the predicted timeframe to obtain private or public reimbursement?
- Can decision-making be influenced to maximize adoption and optimize reimbursement price? How?
- What is the role of hospitals in reimbursement negotiations with private and public intermediaries?

- What is the role of physicians in triggering reimbursement-decision processes?
- What role do patient associations play in influencing public or private reimbursement? Can this reimbursement be ethically influenced?

7.5.5 HealthCare Providers

The healthcare system comprises diverse providers, each specializing in different aspects of care. Understanding the different types of healthcare providers is crucial for effectively navigating healthcare services. As illustrated in the top-right side of Fig. 7.43, healthcare providers can be broadly categorized into primary, secondary, and tertiary care providers, alongside various specialized practitioners. It is a broad classification that focuses on the degree of specialization of the services provided. Primary care providers (PCPs) are typically patients' first point of contact. They include family physicians, internists, pediatricians, and general practitioners who provide comprehensive care and preventive services and manage chronic conditions. Primary care is essential in managing overall health and coordinating care among various specialists. Secondary care providers include specialists who focus on specific areas of medicine, such as cardiologists, dermatologists, and orthopedic surgeons. These providers often require a referral from a PCP. Tertiary care providers offer highly specialized services, often in hospital settings, and include neurosurgeons, oncologists, and other specialists who deal with advanced medical conditions. Access to tertiary care is usually determined by the healthcare infrastructure and the availability of specialized services in a region (Lamptey et al., 2017). Moreover, alternative and complementary healthcare providers, such as chiropractors, physical therapists, and acupuncturists, offer additional treatment options outside the domain of conventional medicine. Figure 7.43 outlines a classification of providers based on the physical identification of the individuals or the organizations involved in the delivery of healthcare. Individual healthcare providers are typically physicians, nurses, and other medical professionals such as chiropractors and osteopaths. Organizations include all non-individual practices, such as hospitals, physicians' practices, and facilities that make spaces available to a wide spectrum of individual providers.

7.5.5.1 Individual Healthcare Providers as Users

Individual healthcare providers may be or become technology users. For instance, nurses are the main users of infusion pumps, surgeons use robotic surgery equipment, and, more in general, they play an important role in making decisions on the adoption of new technologies. Even if users are not always the final decision-makers for new technology adoption, they often play an important role in the decision-making process, and they always play a key role in the actual usage intensity and frequency (or even permanent dismissal) of new medical technologies. If users do not "like" a system, they can become a barrier to its usage and its ability to render the desired impact in healthcare delivery. This might happen independently by the role of the same users in the initial adoption decision. To maximize the probability

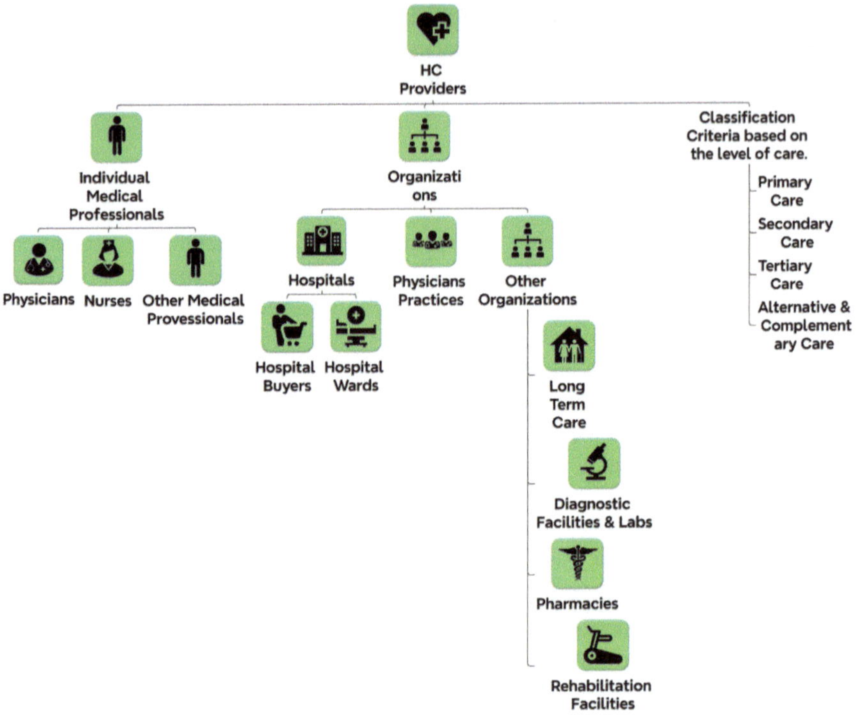

Fig. 7.43 A high-level framework of healthcare providers

of achieving broad user satisfaction, user-centered design must be incorporated early and carried throughout the device development process. It can be a challenge, particularly for smaller companies that, in many cases, lack the necessary knowledge and expertise (Martin et al., 2012).

While focusing on users, entrepreneurs can take advantage of the rich information accessible through "usability testing" (already described in Sect. 7.2) which is one of the most popular and customer-intimate ways of collecting and analyzing, in an orderly manner, useful information (signals) from users in the marketplace in the MedTech and HealthTech industries. Usability testing is a rationally designed process focused on assessing whether a medical device meets its intended users' needs and preferences. Moreover, usability testing is used to determine whether a medical device is vulnerable to dangerous use errors that could lead to the user's or patient's injury or death (Wiklund et al., 2011).

7.5.5.2 Individual Healthcare Providers as Key Opinion Leaders

Medical Key Opinion Leaders (KOLs) play a crucial role in the healthcare landscape, particularly in the development, dissemination, and acceptance of new medical practices and innovations. They contribute invaluable insights that shape research agendas and clinical trials, guiding pharmaceutical and medical technology companies during the design stages of new products. This ensures that the technologies developed address real clinical needs and relevant medical challenges.

KOLs also facilitate effective scientific communication between manufacturers and healthcare providers, which is essential during the commercialization phase. Companies rely on KOLs to educate healthcare professionals and stakeholders about new medical technologies. Their expert opinions provide reliable information and help address healthcare professionals' concerns regarding new treatments. By presenting unbiased product comparisons, KOLs enhance the credibility of the information shared, leading to greater acceptance and endorsement of new technologies among practitioners.

Furthermore, KOLs are instrumental in the market acceptance and integration of medical technologies. They often participate in advisory boards and conferences, providing platforms for sharing best practices and experiences related to new products. Their influence spans various domains, including clinical guidelines, the adoption of drugs or technologies, and patient engagement in clinical trials. KOLs also lead the development of clinical guidelines for diagnostic and therapeutic protocols, influencing the inclusion of new therapies in standard medical practices and facilitating their acceptance and utilization. Through their involvement in scientific associations, KOLs play a guiding role in the healthcare community.

The identification of relevant KOLs and of their respective roles in the scientific community is therefore an essential duty of medical ventures. Mapping KOLs by region and understanding their relationships and mutual influencing mechanisms is crucial to optimizing the venture's interaction with the scientific community. KOLs' segmentation also plays a critical role for innovative ventures in planning a rational approach to the overall interactions with these leaders. KOL segmentation in the life sciences involves the disciplined and rational identification and categorization of Key Opinion Leaders (KOLs) based on their individual influence, expertise, and specific roles within various therapeutic areas.

KOLs' segmentation helps innovative ventures in developing dedicated and targeted engagement strategies and optimizing their scarce resource allocation effectively. Common approaches to KOL segmentation include differentiating between global thought leaders and regional influencers, as well as delineating clinical roles, such as distinguishing between clinical leaders and network connectors. It is therefore essential to identify segmentation criteria relevant to the venture's approach to adoption and adoption's acceleration. It is important to identify those KOLs' segments that are not only the most credible, but also more likely to respond promptly to the proposed innovations and more likely to accelerate the inclusion of other relevant KOLs.

7.5.5.3 Scientific Associations

Scientific associations play often an important role in the adoption of medical technologies through their roles in promoting innovation diffusion, establishing HTA methodologies, setting implementation standards, addressing safety concerns, and providing education and training. Their influence ensures that new technologies are not only adopted but also integrated into clinical practice to foster patient safety and care quality.

Scientific associations contribute significantly to the advocacy for new medical technologies and to the formation of internal KOLs networks characterized by common interest in certain new technologies.

For instance, US startups working on the next needleless, wearable glucometer might be looking at creating and strengthening collaborations with major international organizations like the European Association for the Study of Diabetes (EASD) and the International Diabetes Federation (IDF), along with prominent national bodies such as the American Diabetes Association (ADA) in the USA, the Società Italiana di Diabetologia (SID) and the Associazione Medici Diabetologi (AMD) in Italy, and the German Diabetes Association (DDG) in Germany.

Unlike these associations, other relevant scientific associations might not be characterized by an interest in a specific pathology or disease. For instance, Organizations such as the European Federation of Medical Informatics (EFMI) and the International Medical Informatics Association (IMIA) are critical in supporting the roadmap for technology adoption across approximately 60 countries (Hofdijk et al., 2014). Their advocacy role and efforts are crucial as they support the provision of policies that favor the integration of innovative medical technologies into healthcare systems. They also provide a collective voice for healthcare professionals advocating for modernization in practice.

Entrepreneurs must therefore identify scientific associations that best understand the value of their innovations and determine if and how these innovations can be applied to the safe delivery of health care services.

7.5.5.4 Group Purchasing Organizations (GPO)

Hospital buying groups, also known as group purchasing organizations (GPOs), play a crucial role in the healthcare procurement landscape by leveraging collective purchasing power to negotiate better prices and terms for medical supplies and services. For this reason, it is important for medical technology innovators to regard GPOs as key stakeholders. One of the primary advantages of GPOs is their ability to aggregate demand from multiple hospitals, which significantly increases their bargaining power with suppliers. By negotiating bulk purchasing agreements, GPOs can secure lower prices for medical devices, pharmaceuticals, and other essential supplies, thereby reducing the procurement costs for individual hospitals (Rajib Lochan Saha et al., 2019; Rajib L. Saha et al., 2010). This negotiation capacity is critical in an environment where healthcare expenses are always on the rise and hospitals are under pressure to manage their budgets efficiently (Lee et al., 2023).

According to the American Hospital Association, 68% of all hospitals used a GPO for their primary purchasing needs in 2000, and this percentage rose to an estimated 98% by 2014 (Bruhn et al., 2018). Furthermore, it has been reported that 72–80% of healthcare expenditures in acute-care settings are routed through GPOs, underscoring their integral role in the healthcare supply chain (Hu & Schwarz, 2011). At present, over 95% of US hospitals engage with GPOs for their purchasing needs, highlighting the widespread reliance on these organizations to manage procurement in an efficient and cost-effective manner (Dean et al., 2024; Bruhn et al., 2018).

Medical entrepreneurs must therefore determine which GPOs are most relevant to their specific medical products or services and identify target GPOs in the different target countries. It is also critical to produce a version of the value proposition tailored explicitly to the target GPOs and to present a clear business case showing how the new product or service will lead to significant savings for GPO members through quantity discounts and other efficiencies.

While focusing on the topic of healthcare providers, the medical entrepreneur must answer the following questions:

- Which healthcare providers are most involved, individually or as organizations, in delivering care relevant to the innovative medical technology?
- How do individual healthcare professionals influence the decisions of larger organizations such as hospitals?
- Which other professionals are involved in the adoption decisions of large healthcare organizations?
- Do target hospitals have an internal adoption approval process? If yes, how do they work?
- Are there healthcare professionals who are the actual or potential users of the new proposed technology? Who precisely is the intended user of the technology?
- What are the most important unmet needs of users targeted by the invention? Which needs are more critical? How can the accuracy of these assumptions be tested?
- Will the product address the users' most critical needs? How can this assumption be tested?
- Are there KOLs that can play a role in the broad adoption of the new technology?
- Are there scientific associations that can play a similar role? If yes, which ones are followed, and what is the influencing model?
- Are GPOs involved in purchasing decisions? Which degree of influence do GPOs exercise in the category of products or services to which the invention belongs?
- What respective organizational competences are required to establish fruitful working relationships with healthcare professionals, healthcare organizations, KOLs, scientific associations, patient associations, and GPOs? What competence gap must be bridged in the venture? How long will this process take, and what are the costs involved?

7.5.6 Regulatory Organizations

Regulatory and registration issues are always crucial while launching an innovative medical product and attempting to predict its commercial value. Medical product regulations, the related product registration processes, and their requirements play a vital role in the protection of public health. On the other hand, they significantly influence the design, development, and commercialization of new medical technologies (Pietzsch et al., 2013). It is easy to explain in a few words the critical role of regulatory compliance: no product registration means no sales.

The first level of complexity for medical innovators comes from regional differences in regulations. Indeed, each country or region follows its own regulatory agency, which has different processes and poses various problems for companies.

This condition implies a significant degree of complexity pertaining to the different regulatory risks that new technologies are required to face in other regions. For instance, in a 2010 AdvaMed study, the FDA was unfavorably compared to European regulatory authorities over many aspects, such as predictability, reasonableness, transparency, and the rating of the overall process experience (Makower, 2010). The regulation and approval of medical devices differ across major jurisdictions, although all aim to ensure safety, efficacy, and quality before products reach the market. The United States, the European Union, Japan, Australia, and Canada each maintains distinct systems reflecting their regulatory traditions and institutional frameworks.

In the USA, the regulation of medical devices falls under the Federal Food, Drug, and Cosmetic Act, enforced by the Center for Devices and Radiological Health (CDRH) within the Food and Drug Administration (FDA). The American system adopts a three-tier risk classification, distinguishing among Class I (lowest risk), Class II (intermediate risk), and Class III (highest risk) devices. Approval processes correspond to these classes: Class I devices are subject to general controls; Class II devices require a pre-market notification, which is also known as the 510(k) process; and Class III devices require premarket approval (PMA), a very rigorous evaluation of safety and effectiveness.

In the European Union, medical devices are regulated by the Medical Device Directive (MDD), under which Notified Bodies (NBs)—independent, authorized private organizations—carry out regulatory control and are empowered to issue the CE mark. The EU uses a four-class scheme: Class I (which includes subclasses Is and Im), Class IIa, Class IIb, and Class III. Each class represents increasing levels of risk and therefore, stricter regulatory scrutiny. Manufacturers must obtain CE marking to demonstrate that their devices are safe and perform as intended for their specific medical purposes.

In Japan, regulatory oversight is provided by the Pharmaceuticals and Medical Devices Agency (PMDA). Devices are divided into four groups: general medical devices (Class I), controlled medical devices (Class II), and specially controlled medical devices (Classes III and IV). The approval process is progressive: Class I devices require only a notification; Class II devices need a device certificate; and Classes III and IV require formal device approval from the PMDA. The Japanese approach emphasizes documentation and certification proportional to risk.

Australia regulates medical devices through the Therapeutic Goods Administration (TGA), which applies a system similar to that of the European Union. Indeed, devices are classified as Class I, Class I supplied sterile, Class I incorporating a measuring function, Class IIa, Class IIb, Class III, and Active Implantable Medical Devices (AIMD). Manufacturers are required to provide evidence that both the device and its manufacturing process meet to the requirements of therapeutic goods legislation. This ensures compliance not only with design and performance standards but also with quality assurance systems.

In Canada, medical devices are supervised by the Therapeutic Products Division (TPD) of Health Canada, under a four-tier classification system where Class I represents the lowest risk and Class IV the highest. Approval requirements increase with risk: Class I devices are subject only to post-market monitoring; Class II devices require evidence of safety, efficacy, and quality as well as post-market surveillance; and Classes III and IV undergo premarket assessments for both safety and efficacy.

FDA is well known for representing the precious "key" to the biggest pharmaceutical and medical technology market (USA) and for enforcing the most demanding regulatory procedures. Indeed, it takes an average of 8 years to obtain FDA clearance from the moment of the first test. Only one out of six pharmaceutical compounds succeeds in passing the FDA's rigorous assessments of safety and efficacy for the proposed drug and indication. Analyses of clinical trial data from 2010 to 2017 reveal four key reasons contributing to the 90% clinical failure rate in drug development: insufficient clinical efficacy, unacceptable toxicity, poor drug-like properties, inadequate commercial needs, and suboptimal strategic planning (Sun et al., 2022). Generally, drug development follows a classical process. It takes an exact and disciplined quality system to drive and capture the design history, the validation tests, and their results for the FDA. A 2010 Stanford study estimated the average cost to bring a 510(k) a low- to moderate-risk product from concept to market at $31 million, of which more than 77% was spent on registration and FDA-related issues (Makower, 2010).

7.5.6.1 HTA Agencies

Registration bodies and HTA agencies work together outside the USA and often hold overlapping responsibilities. In the list of countries without a formal, central HTA system published by the WHO, the only notable absence is the USA. Outside the USA, most public entities deciding on reimbursement and reimbursement prices use the support of HTA agencies for decision-making. Private insurers tend to align with public entities' reimbursement and pricing practices in the long term. Therefore, from an HTA perspective, the impact of the innovation must be explicit while designing a new technology and its value proposition.

Tables 7.5 and 7.6, published by the European Commission (Chamova, 2017), illustrate the broader roles of HTA bodies in different EU countries with respect to specific regulatory functions.

While focusing on the topic of regulatory approvals, the medical entrepreneur must answer the following questions:

- In which category or class will the new medical technology fall?
- What clinical requirements are set for the main world regions?
- What is the expected cost of regulatory-driven clinical trials?
- What is the expected cost of the regulatory filing?
- What is the timing of clinical trial execution and approval?
- Which regions are the most attractive from a regulatory standpoint?
- What is the expected global calendar for regulatory approvals?
- What is the role of HTA or HTA agencies in the regulatory approval?
- What HTA-focused evidence can be provided as support for the adoption of the product/service?

Table 7.5 The broad roles of HTA bodies in different EU countries with respect to specific regulatory and reimbursement functions

	Type of HTA functions	Countries
One national HTA body	One national HTA body, having only one role: HTA recommendations	Bulgaria, Denmark France, Poland, Slovakia, UK
	One national HTA body with specific regulatory functions	Finland, Hungary
	One national HTA body with pricing and/or regulatory functions	Latvia,Luxembourg, Malta, Netherlands
	One national HTA body with regulatory + pricing and/or regulatory functions	Cyprus,Czech Republic, Portugal
Two or more national HTA bodies	Two or more national HTA bodies, at least one of which has regulatory functions	Spain
	Two or more national HTA bodies, with at least one having pricing and/or regulatory functions	Austria, Belgium, Croatia, Germany, Ireland, Sweden
	Two or more national HTA bodies,combining regulatory and pricing and/or imbursement functions	Estonia, Italy, Lithuania, Romania, Slovenia, Norway

Table 7.6 The broad roles of HTA bodies in different EU countries

HTA body functions	Countries
Quality standards	Austria, Croatia, Cyprus, Denmark, France, Germany, Ireland, Italy, Netherlands, Slovakia, Spain, United Kingdom, Norway
Development of clinical guidelines	Austria, Belgium, Cyprus, Estonia, France, Germany, Hungary, Ireland, Italy, Latvia, Luxembourg, Netherlands, Spain, United Kingdom, Norway
Healthcare promotion	Austria, Belgium, Bulgaria, Denmark, Estonia, France, Italy, Poland, Portugal, Slovenia, Spain, United Kingdom, Norway
Horizon scanning	Austria, Belgium, Estonia, Italy, Netherlands, Portugal, Slovakia, Spain, Sweden, United Kingdom, Norway
Registries	Austria, Belgium, Bulgaria, Denmark, Estonia, Ireland, Portugal, Slovenia, Spain, Sweden, United Kingdom, and Norway
Education	Austria, Belgium, Bulgaria, Croatia, Denmark, Estonia, Germany, Hungary, Ireland, Italy, Malta, Poland, Spain, United Kingdom, Norway
Early dialogs and scientific advice	Belgium, Estonia, Finland, France, Germany, Ireland, Netherlands, Portugal, Slovenia, Spain, Sweden, United Kingdom, Norway

7.5.7 Key Assumptions and Decision-Related Questions (Table 7.7)

Table 7.7 Fit to status quo: core assumptions and decisions

Fit to Status Quo	
Cluster	Core Assumptions and Decisions to Be Made
Patients	To what extent do patients individually represent an active party in the stakeholders' map?
	If patients play a critical role in the decision to adopt, in which way is this role exercised?
	With what other stakeholders and in which way do patients interact while influencing adoption decisions?
	If patients play a critical role in adoption decisions, can the market be considered a consumer market? To what extent?
	Do caregivers play an essential role in adoption decisions? In what way do caregivers influence decisions about adoption?
	If patients cannot play a critical role individually, are there patient associations that fill this role?
	If yes, are patient associations open to a dialogue with the venture regarding the adoption of the new medical technology?
	In what way do patient associations influence decisions about adoption?
Payers	Do employers represent important stakeholders trusted to adopt specific new therapies or diagnostics?
	What role do employers play in the adoption of new therapies and diagnostics in their employee plans?
	What is the role of patients in electing insurance plans that include certain specific services?
	Are patients likely to actively elect certain insurance plans because they cover special therapeutic or diagnostic procedures?
	To what extent are patients willing to pay (or co-pay) for certain new therapeutic or diagnostic procedures or for the use of medical devices? Which tests can support or weaken a hypothesis about the predicted patient's willingness to pay?
	In the context of the disease in question, are there patient associations that lobby for the inclusion of specific, unique new treatments or diagnostics under public or private insurance plans? Which tests can help answer this question?
Society	Will the new technology, its adoption, and its associated costs have a discernible impact at the societal level?

(continued)

Table 7.7 (continued)

7.1.1

Fit to Status Quo	
Cluster	Core Assumptions and Decisions to Be Made
	If there is a significant impact at the societal level, is it related to concerns about the increased social costs of health? To what extent?
	If there is an impact at the societal level, is it related to the collective positive feelings about the diagnosis or cure of a high-severity disease?
	Are there patient associations that may influence the collective opinion? To what extent? How long can the influencing process take?
Intermediaries	Are private insurance companies likely to reimburse the use of the new medical technology? How can public reimbursement be predicted or influenced?
	Are governmental entities likely to reimburse the use of the new medical technology? How can public reimbursement be predicted or influenced?
	What decision-making mechanisms are used by private and public intermediaries when it comes to reimbursement and pricing?
	Can decision-making be influenced to maximize adoption and optimize reimbursement price? How?
	What is the role of hospitals in reimbursement negotiations with private and public intermediaries?
	What is the role of physicians in reimbursement-decision processes?
	What role do patient associations play in influencing public or private reimbursement? Can this role be ethically influenced?
Healthcare providers	Which healthcare providers are most involved, individually or as organizations, in delivering care relevant to the innovative medical technology?
	How do individual healthcare professionals influence the decisions of larger organizations such as hospitals?
	Who else is involved in the decision-making process for large healthcare organizations regarding adoption?
	Does the target hospital have an internal approval process for adoption? If yes, how does it work?

(continued)

Table 7.7 (continued)

Fit to Status Quo	
Cluster	**Core Assumptions and Decisions to Be Made**
	Are there healthcare professionals who are the actual or potential users of the new proposed technology? Who precisely is the intended user of the technology?
	What are the most important unmet needs of users targeted by the new invention? Which needs are more critical? How can the accuracy of these assumptions be tested?
	Will the new product address the users' most critical needs? How can this assumption be tested?
	Are there KOLs that may play a role in the broad adoption of the new technology?
	Are there scientific associations that can play a similar role? Which ones are followed, and what is the influencing model?
	Are GPOs involved in the purchasing decisions? If yes, in what countries or regions?
	What degree of influence do GPOs exercise in the category of products or services to which the invention belongs?
	What criteria do GPOs use to include new technologies in their portfolio?
Regulatory organizations	In which category or class does the new medical technology fall?
	What clinical requirements are set for the main world regions?
	What is the expected cost of regulatory-driven clinical trials?
	What is the expected cost of the regulatory filing?
	What is the timing of clinical trial execution and approval?
	Which regions are the most attractive from a regulatory standpoint?
	What is the expected global calendar for regulatory approvals?
	What is the role of HTA or HTA agencies in the regulatory approval?
	What HTA-focused evidence can be provided as support for the adoption of the product/service?

7.5.8 Complete Map (Fig. 7.44)

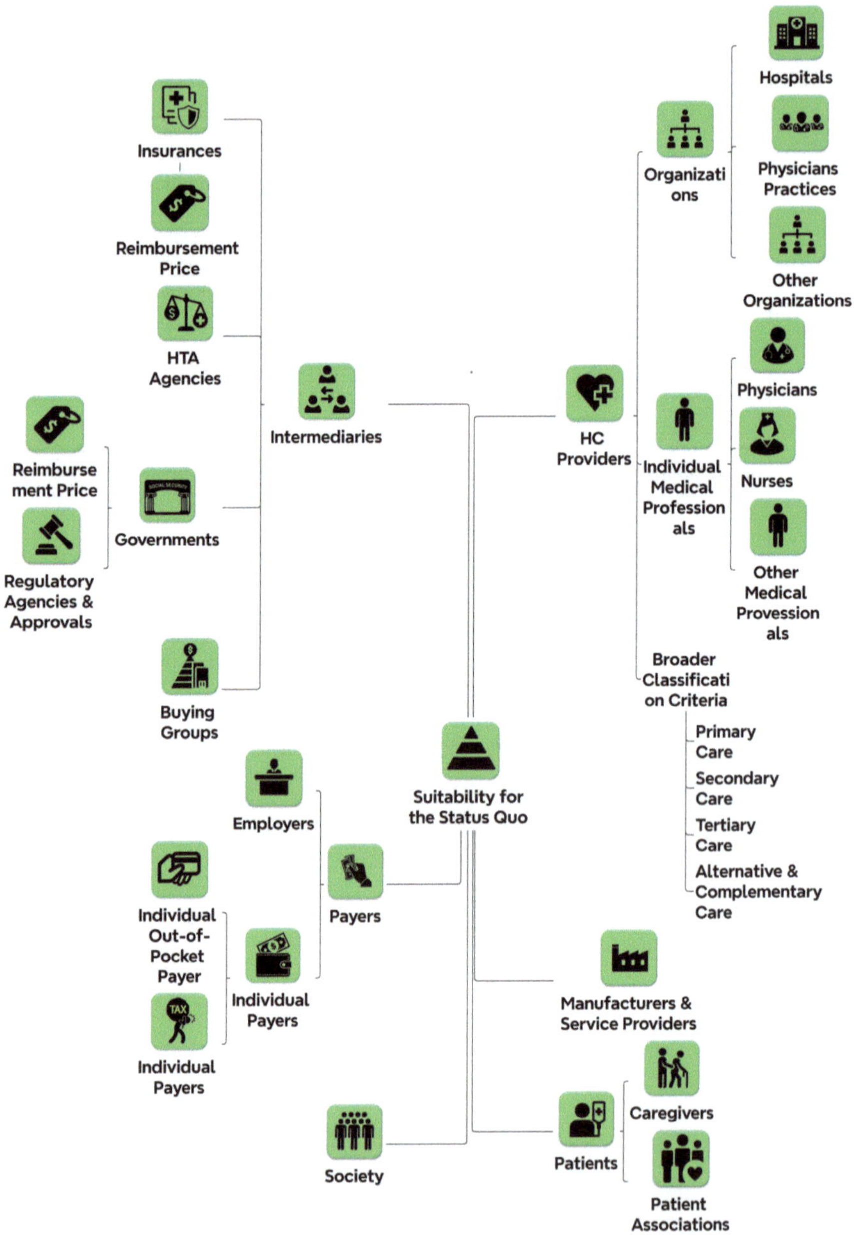

Fig. 7.44 The complete map of problem clusters relevant to the degree of status quo alignment of a new medical technology or service

7.6 Attainability of Target Segments

Marketers must genuinely understand and map each market segment's needs and behaviors to perform correct segmentation (Goyat, 2011). Marketing researchers seek to produce segmentation processes that are sufficiently general to be applicable across industries, but some industries hold distinctive features that require industry-specific knowledge (Eliashberg et al., 2006). This consideration applies to the life sciences industry, which is characterized by a set of unique challenges posed by the vast and diversified group of stakeholders involved in demand creation and purchasing decisions (Stremersch & Van Dyck, 2009).

In established life sciences corporations, the primary function of segmentation is optimizing marketing resources to maximize each segment's competitive advantage, growth and profitability. In early-stage medical ventures, the emphasis of market segmentation is less on the optimal allocation of commercial resources and more on the research and identification of substantial target segments that are suitable for the profitable and energetic adoption of the innovative solution. Only once these segments are identified, ventures can start planning an appropriate commercial strategy (Tynan & Drayton, 1987).

Therefore, before approaching key assumptions and decisions about business and commercial models, entrepreneurs must fine-tune their market segmentation and elect (decide) the segments' attractiveness criteria. In this phase, entrepreneurs must focus on the sequential prioritization of geographical areas and identify and prioritize smaller sub-segments characterized by similar behavior of stakeholders in each target region. The degree of attractiveness and the adoption dynamics of a new medical technology can be better assessed and predicted within well-characterized, homogeneous market segments. Homogeneous contexts support more granular predictions of the degree of competitiveness of the medical innovation and its price. It is precisely for this reason that Section 7.5 is fully dedicated to a deeper understanding of the key stakeholders and their respective roles: defining target segments means identifying groups of stakeholders with virtually homogeneous unmet medical needs, similar purchasing-decision processes, and aligned incentives.

After identifying the target segments, it is crucial to focus on how to position the new medical technology within each segment. This step is vital for assessing the potential for commercial success, as well as the pace and depth of market penetration. In summary, the sequence "segmentation, targeting, and positioning" represents a familiar strategic approach for life sciences marketing and is one of the most widely applied marketing processes. This chapter focuses on the application of these three sequential concepts in medical technology innovation (Fig. 7.45).

7.6.1 Segmentation

In many cases, the highest level of segmentation is represented by the geographical one.

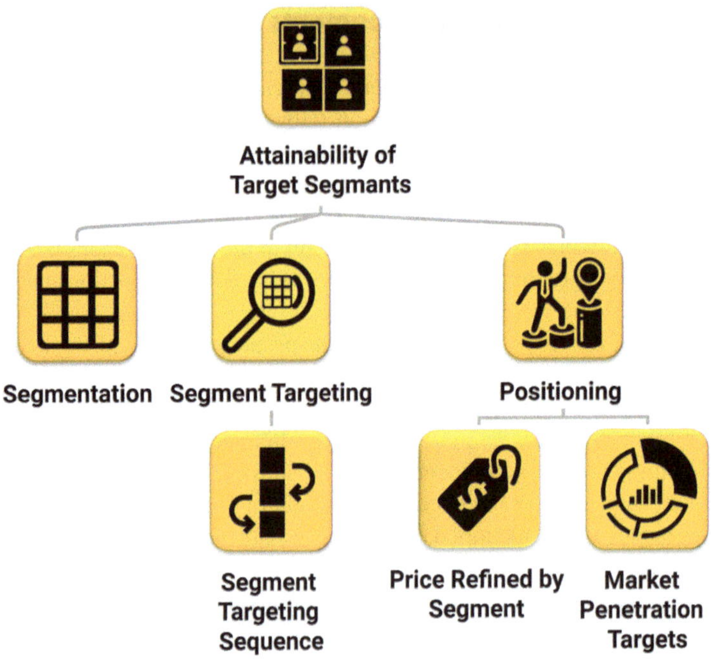

Fig. 7.45 The founding problem clusters of segmentation, targeting, and positioning in medical technology innovation

7.6.1.1 Geographical Criteria Relevant for Segmentation in Medical Innovation

There are a few typical criteria that drive most of the life sciences' industry geographical segmentation:

Current Spending in Healthcare or on a Specific Disease

The allocation of healthcare resources varies significantly between high-income and low-income countries. High-income countries typically spend around US$3000 per capita on healthcare, while low-income countries allocate as little as US$30 per capita (Raghupathi & Raghupathi, 2020). This disparity in spending has profound implications for health outcomes, as lower healthcare expenditure is associated with higher rates of morbidity and mortality, mainly from preventable diseases (Novignon et al., 2012). On the other hand, this disparity in spending has profound implications for companies seeking the profitable commercialization of their innovative medical technologies. One of the first market prioritization criteria employed by life sciences companies is precisely represented by the geography's level of healthcare spending, both as a global percentage of the GDP and as the spending budgets allocated to specific diseases. This spending level includes both the public and private healthcare delivery systems.

Country's Regulatory Environment

As mentioned in the previous chapter, the regulatory environments governing medical practices and products vary significantly across countries, influenced by factors such as economic status, healthcare infrastructure, and cultural attitudes toward health. The disparities in regulatory environments across regions can lead to inconsistencies in the availability and quality of medical products. This inconsistency affects not only patients' access to new treatments but also innovative health initiatives. For these reasons, the regulatory context and practices must be utilized to segment and prioritize the opportunities in different geographical areas.

Country's HTA Practices

HTA approaches vary significantly across different regions, influenced by local healthcare systems, economic conditions, and cultural values. For instance, in Europe, HTA processes are characterized by a diverse landscape in which methodologies and outcomes differ significantly between countries. While harmonizing efforts such as the EUnetHTA aim to foster collaboration and knowledge-sharing among European countries, the diversity in HTA frameworks and methodologies underscores the challenges in achieving harmonization across HTA practices globally. Medical innovations that rely heavily on their claimed health economic advantage, such as therapeutic and surgical add-ons, must carefully prioritize countries or regions where the HTA perspective is concurrently critical and approachable.

Country's Reimbursement Practices

Exactly similar to HTA, reimbursement practices in healthcare vary significantly across different regions, influenced by local policies, economic conditions, and healthcare system structures. Medical innovations that rely heavily on the existence or the achievability of sizable reimbursement prices must carefully select their geographical prioritization according to different regional reimbursement practices.

Regional Incidence and Prevalence

Understanding these regional variations in disease incidence and prevalence helps identify regions with more urgent medical needs, which are likely to allocate larger resources to specific public health interventions. Evidence shows that disease incidences and prevalence can vary across different regions due to various factors.

For instance, regional differences in industrialization and development can influence the epidemiology of inflammatory bowel disease, affecting newly industrialized and emerging regions in Africa, Asia, and Latin America differently than it impacts early industrialized regions such as North America, Europe, and Oceania (G. Kaplan et al. 2025; G. G. Kaplan and Windsor 2021). Similarly, Parkinson's disease has shown geographical and ethnic variations in the USA, with higher incidence in the upper Midwest and Northeast regions, potentially due to environmental factors associated with industrialization (Wright Willis et al., 2010).

Hodgkin lymphoma has also exhibited regional differences, with decreasing incidence in low sociodemographic index (SDI) regions and stable incidence in higher SDI regions globally (Zhou et al., 2019).

Spatial and temporal analyses have revealed regional variations in the distribution of disease outbreaks, such as lumpy skin disease in Uganda (Ochwo et al., 2018) and Lyme disease in the USA (Kugeler et al., 2015).

Spatial variability in disease incidence has also been observed for conditions such as campylobacteriosis in Manitoba, Canada, and hereditary anemias in Saudi Arabia (Green et al., 2006). These regional differences may be influenced by factors such as environmental characteristics, socioeconomic status, and genetic factors (Memish et al., 2011; Alharbi et al., 2021; Green et al., 2006).

In summary, disease incidences can vary significantly across different regions due to a combination of factors, including industrialization, environmental conditions, animal husbandry practices, genetic factors, and reporting biases.

Geographical segmentation typically involves multiple layers that draw the entrepreneur's attention from large regions to specific countries, then to provinces or districts, and finally to specific cities.

7.6.1.2 Finer Segmentation Criteria

Regional geographic segmentation always leads to a finer degree of segmentation. Within countries, there are several ways medical entrepreneurs can characterize smaller segments.

If, for instance, the medical venture plays in the space of bedside infusion systems, then groups of specific users or buyers can represent valid sub-segments. Hospitals, nursing homes, ambulatories, and home care organizations are good sub-segment examples. They purchase and utilize infusion system technologies in various settings and contexts, primarily for different patient types or at different stages of recovery or disease stages.

Within the hospital segment, it is helpful to distinguish between private and public hospitals, as they typically decide to acquire new technologies following different criteria. Moreover, within each hospital, it makes sense to segment further, distinguishing between the specific departments that often hold different unmet medical needs and also specific user requirements. The sequence of Tables 7.8, 7.9, and 7.10 provides a graphic rendering of these considerations and an example of a graphic tool that has proven to be very useful for market segmentation in medical technology:

Table 7.8 Segmentation across geographic regions

Product Market Matrix	
By Region↓	Segments
North America	
Europe	A
Asia Pacific	
Latin America	
Middle East and Africa	

Table 7.9 Segmentation across countries within the same region

Segmentation Matrix of "A"		Infusion Pumps			
		Large Volume Infusion Pumps			
By Country ↓	By End User →	Hospital	Nursing Home	Ambulatory	Home Care
Europe	Germany				
	France	B			
	Italy				
	Spain				
	UK				
	ROE				

Table 7.10 Segmentation across types of hospitals and hospital departments

Segmentation Matrix of "B"			Infusion Pumps	
			Large Volume Infusion Pumps	
			Hospital	
By Department/Usage ↓		By Type of Hospital >	Public Hospital	Private Hospital
Europe	France	Oncology		
		Gastroenterology		
		Pediatrics		
		Neonatology		
		Pain Management		
		ICU	C	

from the selection of a region (europe) to the selection of a country (France) and of an health care segment (Hospitals), down to the selection of the sub-segment of ICUs in Public Hospitals.

Segmentation in life sciences can reach a very high level of selectivity and, therefore, precise identification of targets. In addition to the criteria already mentioned for the life sciences industry, other typically implied segmentation criteria are:

Behavioral Segmentation

- Prescription/usage behavior of physicians (early adopters vs. conservative prescribers)
- Patient adherence to treatments (compliant vs. non-compliant)
- Technology adoption rate in hospitals (e.g., telemedicine, robotic surgery, AR smart glasses)
- Purchasing processes: centralized tenders vs. decentralized procurement

Clinical/Medical Segmentation

- Therapeutic area (oncology, cardiology, orthopedics, diabetes)

- Disease stage/severity (early vs. late stage cancer)
- Patient risk profile (high-risk vs. low-risk populations)
- Diagnostic markers (biomarker-based segmentation in precision medicine)

Psychographic Segmentation

- Attitudes toward innovation (physicians open to AI-driven diagnostics vs. skeptics)
- Patient lifestyle and values (preventive health seekers vs. reactive care seekers)
- Hospital philosophy (cost-minimizers vs. quality/max-innovation institutions)

Technological Segmentation

- Digital readiness (high-tech hospitals adopting tele-surgery vs. low-tech centers)
- EHR/EMR integration capability
- AI and genomic medicine adoption

Figure 7.46 provides an example of how details and multiple criteria can fine-segment, for instance, MS patients. Criteria include not only the exact type of MS disease but also the related severity, the recurrence of symptoms, and even the type of treatment already in use. Such a highly detailed segmentation can lead even to the

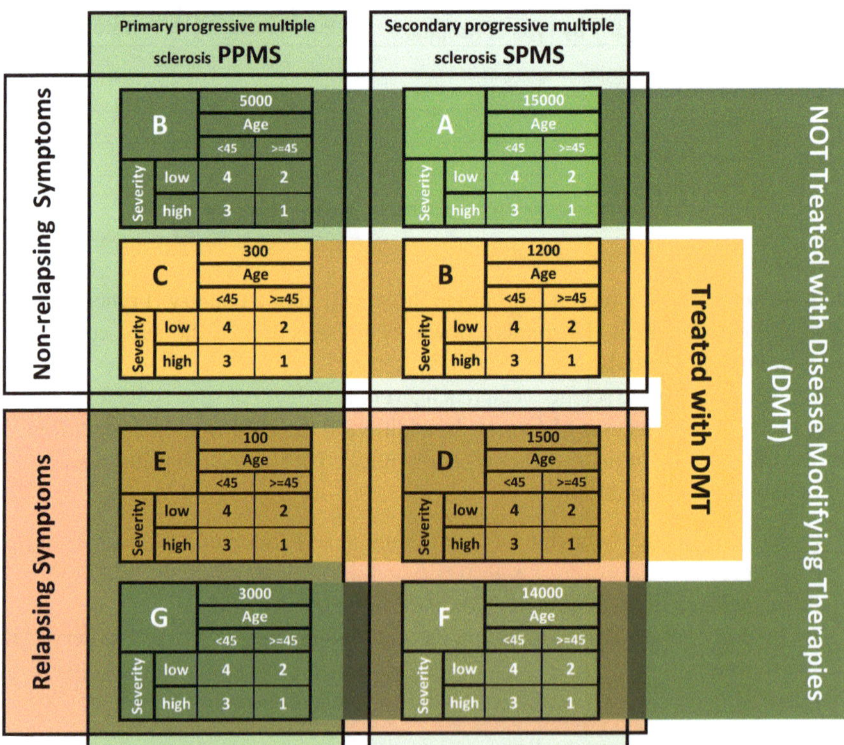

Fig. 7.46 An example of highly refined segmentation of MS patients

identification of the exact number of patients belonging to segments characterized by age and disease severity. Letters A to G help the decision-maker to prioritize and, therefore, to target only specific segments.

7.6.2 Targeting: Prioritizing Among Monetization Opportunities

Very often, segmentation is designed to identify homogeneous market segments concurrently and also to distinguish segments through their degree of attractiveness. Therefore, once the criteria to prioritize the most attractive segments are framed, the decision-maker must apply them to the segmentation maps to identify the target segments. These criteria typically fall in the areas already encountered by the venture, following the path of the K9 map:

Patient Population Characteristics:

- Epidemiology: Incidence, prevalence, and growth trends of the target disease
- Segmentation: By disease stage, genetic/biomarker status, comorbidities, or risk profiles
- Accessibility: How easy it is to identify and reach patients (e.g., rare vs. widespread conditions)

Unmet Medical Need:

- Segments characterized by serious patients' conditions with limited or no effective therapies gain faster adoption
- Disease burden, prevalence, mortality, quality-of-life impact, and gaps in the current standard of care

Clinical Evidence and Differentiation:

- How well the product demonstrates superior safety, efficacy, or usability compared to alternatives in a specific market segment.
- Differentiation can be in clinical outcomes, mode of action, ease of use, or patient adherence compared to the solutions adopted in a specific segment.

Reimbursement and Health Economics:

- Targeting segments where payer value is strongest (e.g., high unmet need + cost offsets like fewer hospitalizations)
- Targeting segments where payer willingness to pay is based on cost-effectiveness, budget impact, and health outcomes data (QALYs, ICERs) that fit with the venture's claims

Competitive Landscape:

- Segments characterized by few (or none) existing competing therapies or diagnostics
- Segments that can potentially provide first-mover advantages

Regulatory Pathways and Designations:

- Regions or segments with known, predictable, and financially approachable regulatory processes
- Regions or segments where the innovative solution could enjoy fast-track approvals can make smaller but high-value patient groups attractive targets

Prescriber and Care-Setting Dynamics:

- Segments with clear and reachable treatment decision-makers
- Targeting often starts with influential KOLs and high-volume centers before expanding to community practice

Other Adoption Barriers and Enablers:

- Barriers: Segments with higher training requirements, more complex integration with hospital workflows, or IT/EMR compatibility
- Enablers: Segments with existing diffusion of digital health support tools, companion diagnostics, or telemedicine integration.

Strategic Fit and Portfolio Priorities:

- Segments that allow synergies with existing commercial infrastructure, such as distribution channels and their respective portfolios
- Segments fit for an alignment with company strengths (e.g., oncology franchise, cardiology focus)

7.6.2.1 Targeting Using the Ansoff Matrix

Due to the discussed complexities of the life sciences industry, segmentation is often based on multiple, overlapping criteria that may lead to highly granular market partition maps. Among the several well-known maps available for visual market segmentation, we have found that the maps mixing market segments with product or service offerings are among the most suitable decision-making tools. For instance, the framework offered by the Ansoff matrix (Richardson & Evans, 2007; Hussain et al., 2013; Ecocibi, 2017) precisely achieves this goal. Ansoff developed a segmentation and targeting framework for ventures to map all their product-market segment options, called the product market matrix (PMM). For established businesses, possible growth opportunities are discovered by combining existing and new products or services with existing and new markets. The product-market options are then visualized through a matrix highlighting all the distinct strategic alternatives (Table 7.11).

Table 7.11 A basic version of the Ansoff's matrix

		PRODUCTS	
		EXISTING	NEW
MARKETS	EXISTING	Market penetration	Product development
	NEW	Market development	Diversification

Table 7.12 Segmentation across hospital types and hospital departments

Segmentation Matrix of Hosp ital Segment in France			Infusion Pumps	
			Hospital	
			Public Hospital	Private Hospital
Europe	France	Oncology		
		Gastroenterology		
		Pediatrics		
		Neonatology		
		Pain Management		
		ICU	C	

Table 7.13 An example of product market matrix in medical technology

Product Market Matrix		Infusion Pumps				
		Current Products		Products Under Development		
By ICU Type↓	By Type of Pump →	Large Volume Infusion Pumps	Syringe Infusion Pumps	Insulin Pumps	Enteral Feeding Pumps	Implantable Pumps
Cardiac ICU						
Isolation ICU						
Long-Term ICU		D				
Surgical & Trauma ICU						

One practical use of the PMM concept for new medical ventures (often focused on only one therapeutic or diagnostic product) consists of utilizing criteria such as specific therapeutic indications as an additional level of distinction within a single product group. Graphically, a one-product column can be divided into sub-columns related to different therapeutic or diagnostic indications. The market rows, instead, are often associated with different geographical areas and can be further subdivided into smaller geographical regions, which, in turn, can be subdivided into individual market segments, for instance, of specific patients with specific disease stages.

Managers can establish prioritization criteria for both products and markets, and in this way, the PMM becomes an excellent tool for simultaneously visualizing segmentation and targeting. The PMM does not need to be used as a strategic tool for long-term portfolio planning; instead, it can serve as a rational and efficient segmentation and prioritization instrument that supports the medical entrepreneur in identifying the most attractive patient segments for new therapies (or new diagnostics) and for each individual indication.

Tables 7.12 and 7.13 illustrate a typical example of a sequential, multi-layer segmentation of the market of infusion pumps leading to a PMM. Specifically, the segment "France, public hospitals" is initially sub-segmented by the different relevant hospital departments.

The ICU segment is further explored, revealing different types of ICU and then identifying specific, commercially viable products for each of the identified segments. This process leads to the formation of a PMM.

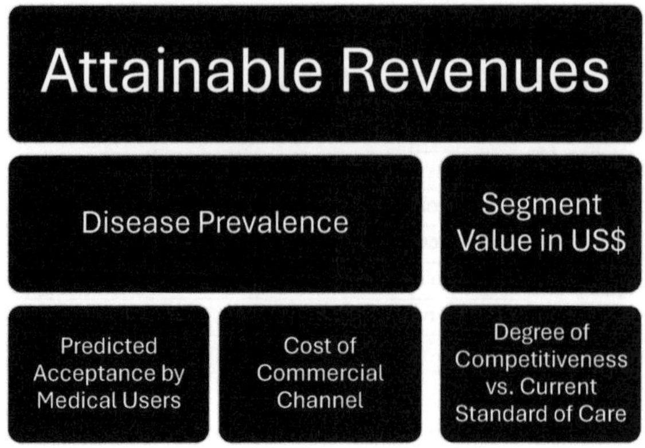

Fig. 7.47 An example of how a single segment PMM element can be visually organized to highlight the key criteria and the respective parameters driving its prioritization

Indeed, Table 7.13 represents a PMM in which the rows correspond to the identified five types of ICU, and the columns refer to the six primary product categories. In this example, large-volume infusion pumps and syringe pumps are the core products that have already been developed, while the others are still under development. In this instance, the venture's management could prioritize the commercialization of large-volume pumps in Long-Term ICU in French public hospitals, targeting the segment represented by the letter "D."

Following this practical, visual method, medical ventures use their set of prioritization criteria to point their attention to the most attractive segments. Each of the product market segments must be prioritized using clearly defined and measurable criteria. The selection of prioritization criteria requires critical managerial decisions. The management must elect prioritization criteria among relevant parameters such as segment size (population, incidence, and prevalence), segment value (current healthcare spending in certain disease prevention, diagnosis, and therapy), expected product acceptance, expected adoption rates, expected market penetration, expected pricing levels, expected reimbursement, overall competitive rivalry, presence of direct competition, and commercial viability.

Figure 7.47 provides an example of how a single segment (a single PMM or SM element) can be visually organized to highlight the key criteria and parameters driving its prioritization: attainable revenues, disease prevalence, segment value, predicted acceptance by medical users, cost of the commercial channels and degree of the new technology's competitiveness.

7.6.3 Positioning

Once the target segments are identified, decision-making must focus on positioning the new medical technology in each segment. Doing so is essential for determining (or predicting) the likelihood of commercial success as well as the speed and extent of market penetration.

A decision-making bias we frequently observe in medical ventures involves underestimating the positioning statement's critical role before and during the commercialization phases. In many of these cases, the positioning statement is considered a theoretical, or even cosmetic, marketing communication tool—a PowerPoint slide to be assigned to some consultant with a commercial background—at some point, usually too dangerously close to the actual commercial launch.

The positioning statement identifies the target customers, the product's specific features in the context of its category, its particular benefits, and its differentiation claims from the nearest competitive alternative. It is an important, rationally constructed statement that is critical to ensuring internal alignment and that, later, becomes the core messaging tool for the marketing campaign. The positioning statement is an evolution subset of a value proposition. It not only makes the value proposition ready for external marketing communications but also helps significantly in clarifying different competitive claims internally (Gospe, 2011).

The fundamental importance of a disciplined formulation of the positioning statement is "forcing" medical ventures' teams to answer basic questions. To determine the product positioning, entrepreneurs must consider questions such as: What criteria do clients (stakeholders) use to evaluate new medical technologies? How important is each of these criteria in the decision-to-adopt process? How does the product (compared to the competitive products) perform on these prioritized criteria? While buying, what decision-making processes do customers (stakeholders) use? (Ostasevičiute & Šliburyte, 2008).

More general approaches are based on providing the answers to different types of questions concerning the offered product or service (Kapferer, 2012). The following are some examples:

- For whom is the product or service designed? (segmentation)
- For which market is the product or service particularly suitable? (target market)
- What does the product or service promise? (key elements of the brand)
- What proves the value of the product or service? (evidence supporting the value proposition)

A more prescriptive and practical approach structures the positioning statement through the formation of sentences based on specific prepositions:

- **FOR** the *target customer*
- **WHO** *statement of need or opportunity*
- *Product name* **IS** *a product category*
- **THAT** *statement of benefit*
- **UNLIKE** *competing alternatives and product names, statement of primary differentiation*

Contrary to the most common practices of most early-stage medical ventures, the centrality of the decisions around positioning is well understood and broadly recognized in the corporate life sciences industry. In large multinational corporations, positioning is taken very seriously, and activities focused on preparing the positioning statement begin at least 12–18 months before product launch. These activities are sometimes even an integral part of the R&D objectives and are usually driven by a cross-functional team. This preparation is centered on a template that splits the preparation of the positioning statements into four areas and designates specific management roles in relation to each area (Vanderveer & Pines, 2007):

- The "target audience" identifies the customer segments addressed in the statement
- The "frame of reference" describes the aspirational placement of the product within the larger treatment or disease context
- The "benefits" statement describes the main clinical advantages for patients and physicians
- The "reason to believe" offers additional advantageous elements to substantiate the main benefits.

After reviewing some of the most common approaches, we suggest developing the positioning statement using a template that is broadly recognized in the life sciences industry and compatible with the reviewed approaches. It is typically framed around five phrases of a statement using the words "for," "…is the," "that," "because," "so that," and "unlike." Table 7.14 shows an example of a positioning statement for a hypothetical (unfortunately hypothetical) medical helmet that can support the treatment of neurological diseases such as MS.

7.6.3.1 Positioning Statements for Medical Technologies Require Proofs

One of the most relevant efforts required by the positioning statements is the need to substantiate each sentence with data and facts. Each of the above-described six summary sentences must be solidly supported by results such as those from the clinical and usability tests that have been implemented, ideally, peer-reviewed, and published in the mainstream literature. More precisely, every word of the positioning statement must be carefully weighed and supported by hard facts. Medical innovators should be prepared to spend significant time supporting virtually every word of the positioning statement.

For instance, the fifth part of the positioning statement states, "so that patients are, over time, relieved of worsening symptoms and grow hope for future improvements." A thorough statement reviewer should ask questions about keywords such as "over time," "patients," "relieved," and "hope."

Table 7.14 Example of positioning statement for an hypothetical treatment of multiple sclerosis

For	primary and secondary progressive MS patients suffering from worsening neuronal damage progression
product is	A new class of disease-modifying therapy (DMT) based on a novel, non-pharmaceutical mechanism of action
that	Reduces MS relapses and promotes disability regression
because	Of the results demonstrated compared to the placebo in the clinical trial, and according to real-world data for more than 50,000 patients globally
so that	The patients are, over time, relieved of worsening symptoms and grow hope for future improvements.
Unlike	All current DMTs that modulate the peak symptoms of relapses without affecting the disease progression.

- *"Over time"*
  - What time frame characterizes "over time?"
  - Which tests were implemented in different time frames?
- *"Patients"*
  - Which group of patients is targeted? At what disease stages are these patients?
  - What percentage of patients may enjoy these benefits?
  - Which tests prove the improvements in the target patients?
- *"Relieved"*
  - How is "relief" measured? Is it measured by using radiological evidence or assessing the actual patient's perceived psycho-physical relief?
  - Which tests support the expectation of relief?
  - What is the extent of the relief?
  - Is the relief lasting or temporary?
  - Which tests prove the duration of the relief?
- *"Hope"*
  - How was "hope" detected or measured?
  - Is this statement about "hope" supported only by internal logic or by actual tests?

7.6.3.2 Positioning Statements Are Key During Product Development

Nothing impedes a medical venture from using the positioning statement before developing a final testable prototype. On the contrary, working on a positioning statement before or during the development of new medical technology helps developers re-contextualize their chronic focus on clinical and technical features within the larger framework of patients, users, medical decision-makers, buyers, and competition. Drafting an ideal positioning statement during the earlier phases of product development helps anticipate market issues that may otherwise become known when it is already too late, when the medical technology is ready and, somehow, marketed. Moreover, early positioning statements represent an invaluable communication tool to explain to investors the potential value of the new medical technology. For instance, when communicating with non-specialized VCs, translating the details of the design and clinical trial results into insights about adoption rates, pricing ranges, and potential revenues can be challenging. A clear value proposition and a crisp positioning statement are ideal to bridge this communicative gap. The entrepreneur can say, "If my predictions about clinical outcomes and treatment cost are proven, then my value proposition and positioning statement will be ....".

I have often observed how medical technology inventors tend to make strong but unverified assumptions that link sought but untested clinical advantages to powerful and broad value propositions and positioning claims. Only in later venture phases, almost at the time of the market launch, do the optimistic entrepreneurs start seeking the actual clinical and competitive fact-based justifications necessary to support the predicted adoption and monetization.

7.6.3.3 Positioning and Pricing

Only after identifying the key target segments and the related positioning can medical entrepreneurs set more precise and segment-tailored pricing levels and policies.

Indeed, precise price-setting can be rationally performed only when dealing with known segments, each one characterized by a significant degree of uniformity. As already mentioned, this uniformity encompasses elements such as: patients with similar unmet medical needs, caregivers with similar unmet needs, buyers with similar buying-decision pathways, and payers with similar needs and ability to pay, operating in a homogeneous regulatory framework. When medical products are not subject to established reimbursement by public or private intermediaries, each medical technology player enjoys the choice among a wide range of prices to adopt to maximize penetration and profitability while ensuring coherence with the product positioning in the competitive landscape.

Section 7.3 uses the definition of price ceiling (PCC) to set an essential point of reference for price-setting decisions. PPC is the price level at which the manufacturer sets to zero the net customer's utility. Therefore, the predicted price level (PPL) must be smaller than the PCC to ensure that a positive, residual customer utility (PCU) can be delivered. The PPC for the new medical technology can be calculated as follows:

$$\text{PPC} = \sum_{i=1}^{n} \text{VF}_i$$

Here, a technology with "*n*" features contributes with "*n*" values (VF_i) corresponding to each feature.

Maximum levels of PPL (*max*PPL) are, therefore, limited by the need to reward buyers with an appreciable and competitive economic utility.

$$\text{PPC} - \text{PPL} = \text{PCU} > 0$$

Minimum levels of PPLs (*min*PP) are, instead, limited by the minimum profitability requirements of the venture selling a product at a predicted minimum profitability level (PCL).

$$\text{PPL} - \text{PCL} = \text{Minimum Profit} > 0$$

Only within the range [minPPL – maxPPL] can the entrepreneur identify the price level that, in each specific segment, reflects the positioning strategy of the medical product or service (Fig. 7.48).

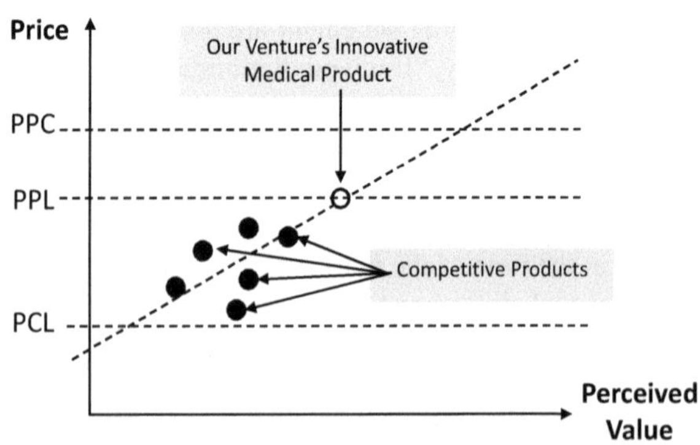

Fig. 7.48 A qualitative representation of the statistical relationship between perceived value and price: it is an helpful tool to establish prices of new products when using PCL,PPL, PPC and competitive prices as references

7.6.4 Key Assumptions and Decision-Related Questions (Table 7.15)

Table 7.15 Attainability of target segments: core assumptions and decisions affecting the venture's value

Attainability of target segments	
Sub-cluster	Core assumptions and decisions affecting the venture's value
Segmentation	What are the geographical segmentation criteria?
	What are the patient segmentation criteria?
	What are the buyers' segmentation criteria?
	What are the users' segmentation criteria?
	What are the reimbursement criteria?
	What are the HTA-related criteria?
	What other stakeholders are relevant to the segmentation?
	What are the precise disease-based market segmentation criteria?
	What are the regulation-based market segmentation criteria?
	Are the selected segmentation criteria translatable into measurable variables?
Targeting	What is the PMM (or what are the PMMs) of reference?
	What are the segment prioritization criteria?
	What are the geographical regions of interest?
	Are the prioritization criteria translatable into measurable variables?
	What is the sequence of regional market entrances?
	What are the selected key target segments? How can these choices be rationally supported?
	Will countries be targeted concurrently or individually, following a specific sequence? What precise sequence? Can the identified sequence be rationally supported?
	Within each country, will segments be targeted concurrently or individually, following a specific sequence? What precise sequence? Can the identified sequence be rationally supported?
Positioning	What are the positioning statements for each of the top-ranked target segments?
	Which clinical trials, usability tests, and other experiments support each promise contained in the positioning statement?
	Which additional tests, trials, and experiments must be performed to support the entire positioning statement?
	How will the medical product or service be positioned in each segment compared to the current standard of care?
	How will the medical product or service be positioned in each segment compared to that of the competition?
	What field tests have been implemented to test the attractiveness of the value proposition and of the positioning statement?
Pricing	What prices will support the sought positioning in each of the targeted segments?
	What is the assessment of the respective willingness to pay the set prices in each of the target segments?
	Which field tests can be utilized to support price decisions in each of the target segments? Which tests have already been implemented?
	Does end-buyer pricing take into account actual shipping and import duties for target segments identified in foreign countries?
	Does end-buyer pricing take into account the actual cost of the commercial channels in all target segments identified in both domestic and foreign countries?
	What pricing policies will mitigate the risk related to exchange rate fluctuations in those target segments identified in foreign markets?
	Can tailored pricing policies help mitigate the collection risk in those target segments identified in markets?

7.7 Viability of Monetization and Commercialization

Once the target market segments are identified and the product's positioning and pricing are set, the medical entrepreneur disposes of the critical information necessary to decide how to generate revenues. As summarized in Fig. 7.49, this section focuses precisely on the two key areas of decision-making in relation to revenue generation: monetization and commercial model. The area about the commercial model is further subdivided into two categories of decision problems: the first related to creating demand, and the second, maximizing revenues.

7.7.1 Monetization Model

As mentioned in Sect. 7.4, a monetization model explains the mechanisms through which the venture intends to generate revenues. It articulates how a business converts its value proposition into financial income, which is crucial for long-term sustainability and competitiveness (van Angeren et al., 2022). It identifies the precise payer type as well as the type of contractual agreement that justifies all sales charges.

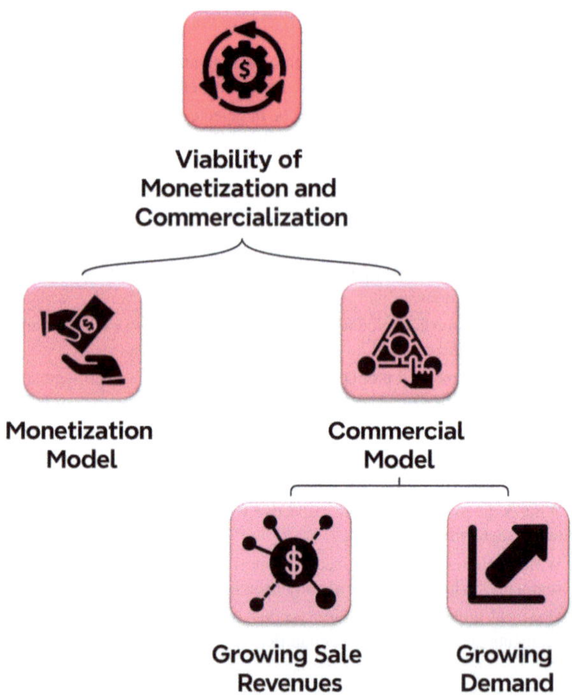

Fig. 7.49 The founding problem clusters relevant to the viability of monetization and commercialization

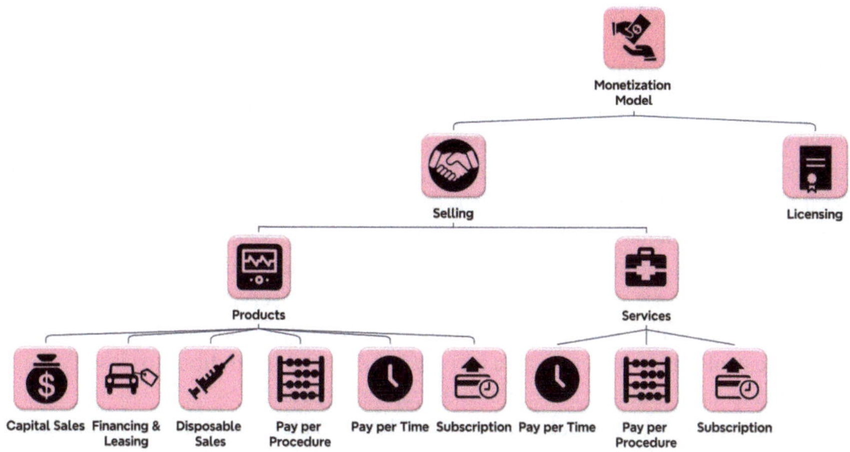

Fig. 7.50 The key decision problem clusters relevant to the monetization of medical technology innovation

Figure 7.50 summarizes the monetization models that are most used in medical technology. Licensing is a revenue-generating option popular in the medical technology sector. However, it is more commonly used by mature organizations rather than by young ventures that must still prove the ability to generate revenues. The licensing model is most used between manufacturers in the medical technology industry, more than between a manufacturer and a healthcare provider. Therefore, this chapter focuses on other, more popular monetization options among medical startups.

7.7.1.1 Sale or Capital Sale

The traditional monetization model for the sale of medical products involves a change of property title in exchange for one payment. This model (sale) remains the reference for relatively cheap products that do not require large cash transactions or financial amortization. When the marketed product consists of large and expensive equipment, this type of sale transaction is often called a "capital sale." Capital sales are typically characterized by high purchasing barriers, as they require upfront, large cash disbursement. Consequently, in the last 20 years, other monetization models, which are less cash-intensive, have become the most common alternatives to the capital sales of medical technologies.

7.7.1.2 Financing and Leasing

Leasing is currently the most common model, allowing customers (typically hospitals) to acquire or use a piece of large, expensive equipment without incurring large cash payments.

Due to constrained budgets, leasing represents a more economically viable financing method, allowing healthcare providers to utilize modern equipment without the associated upfront costs (Gamova et al., 2020). Leasing strategies are increasingly adopted in healthcare for high-value equipment, optimizing budget allocation (Jufriyanto et al., 2018).

Leases can be broadly categorized as operating leases (short-term, with no ownership) and finance leases (long-term, with potential ownership), with a colorful range of variations within each category. The flexibility provided by leasing options further enhances operational adaptability, enabling hospitals to align their equipment resources with fluctuating patient demands and technological advancements. Leasing arrangements can foster and include ongoing support and maintenance, which are critical for maintaining the operational efficiency of medical technology. Companies that lease medical equipment often include service and maintenance in their contracts, effectively reducing healthcare facilities' total cost of ownership. Leasing can also help lower taxable income by allowing institutions to classify lease payments as operational costs rather than capital expenditures.

It is particularly beneficial in countries where financial regulations favor operational spending to incentivize healthcare service delivery (Jazzar, 2023). Additionally, leasing can drive improvements in financial performance, particularly in emerging economies, where access to capital for healthcare innovations remains limited (Malkin & von Oldenburg Beer, 2013).

Several viability issues must be considered while planning to utilize these models. The first issue concerns the capabilities required to perform leasing agreements. Organizations must maintain a robust understanding of the leasing terms, limits, and obligations surrounding current devices.

Another critical issue is the ability to replicate a leasing model abroad. Young medical ventures may opt to rely on international distributors for international commercialization. Designing a model in which distribution companies lease from the manufacturer and sublease to the final customer is highly complicated. International distributors in the medical industry are usually more inclined to apply their own core business model, which is to buy and resell. Some distributors may be able to buy from the manufacturer and lease to the final customer. In this case, the manufacturer's monetization model still falls under capital sale, while most of the mentioned advantages of leasing agreements are enjoyed only by distributors and their customers.

7.7.1.3 Disposable Sale

Another popular model is employed for large equipment that requires the use of disposable sets, which are normally replaced after each procedure on a patient. Apheresis machines and extracorporeal circulation machines for open-heart surgeries represent typical examples of this equipment. Indeed, their users must replace disposable bloodlines each time a new patient is connected to the system.

Companies selling this type of product have historically moved toward models that make the equipment available in exchange for a relatively small lump sum (sometimes for free) but require a commitment to purchase a minimum amount of disposables per period at a surcharged price. Historically, this selling process has moved from capital sales to a reduction of the equipment's price in exchange for higher disposable prices, and finally to the securing of the disposable sales and

Fig. 7.51 The trend in capital sales vs. disposable sales and leasing in medical technology (from the 80 s to the current times)

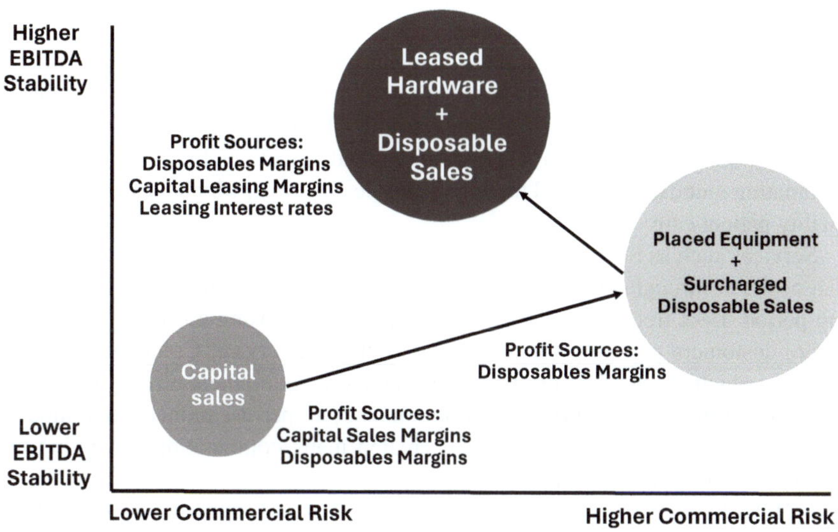

Fig. 7.52 The relationship between the monetization model, the degree of stability of the EBITDA and the commercial risk

leasing of the equipment (to more aggressively mitigate picks of cash disbursement from customers) while taking advantage of the additional revenues generated by leasing services and interest-based fees (Fig. 7.51).

Figure 7.52 qualitatively shows how the transition from capital sales to the sale of free-placed equipment with disposables helps stabilize revenues and EBITDA while increasing the commercial risk associated with attaining the minimum number of disposables sold. Indeed, the revenue line is not attained by the sale of large equipment but is built by the recurring purchase of disposables. The next transition toward leased hardware and sold disposables further stabilizes the company's pre-tax profit while reducing the risk of not selling an adequate number of disposables.

Indeed, profits cannot be hit by the significant cost of placing large equipment for free.

When a large equipment does not require disposables to operate, the most trendy financing alternatives to capital sales and leasing are based on paying per procedure (also known as "pay per use") or paying per time. These approaches are typically derived from the commitment of a minimum number of purchased procedures or a minimum amount of purchased time during each period.

In pay-per-use cases, when the manufacturing cost of commercialized medical equipment is relevant, the seller can mitigate risks related to low equipment usage by asking for initial lump sums. Subscriptions are also used, but they are most frequently used for non-expensive equipment that requires regular access to remote online services such as remote AI platforms. In most cases, the subscription fee represents a payment for both the remote cloud service and the concurrent use of the equipment.

7.7.1.4 Subscription-Based Sales

We now focus on the monetization of medical services. Relevant examples of services are those provided by AI-based diagnostic supports offered by companies such as Arterys. Its FDA-cleared, deep learning AI-powered platform allows faster radiology image examination and reduces missed detections by up to 70% while automating mundane tasks and freeing up physicians' time, which they can invest in quality patient care.

Services such as the one provided by Arterys can be charged per procedure (per diagnosis) or through a periodic subscription that includes all usage during the billing period. Less frequently, these services are charged using a time-based model, where customers pay only for the time they use the service. Companies such as Arterys, which mainly rely on the data generated by equipment already owned by customers, usually adopt a pure subscription or a pay-per-use monetization model.

The medical technology market has created many combinations involving the above-mentioned monetization models.

WHOOP, Inc. is a digital company (a former startup) in the fitness technology space, which has been evolving its business model in the competitive wearable health tracker industry. Although it is not yet fully established in the medical devices industry, it represents a good example of the monetization trend in medical technology. After launching as a wearable device company, it evolved its monetization model from selling a device to providing software as a service (SaaS). In early 2020, WHOOP launched research initiatives to track user well-being amid the COVID-19-induced social-distancing conditions imposed globally (Qermane & Mancha, 2021).

WHOOP currently sells subscriptions, providing customers with one device in exchange for a commitment to pay the first 12 months of the subscription. By continuing to pay the subscription, customers receive the AI-based analytical power of Whoop's platform, together with all the upcoming band firmware, new band versions, and upgrades developed in their cloud-based AI analytics.

7.7.1.5 Pay Per Use

The pay-per-use (PPU) model in medical technology is a pricing strategy designed to maximize accessibility and adoption. This model enables healthcare institutions to utilize medical technologies as needed, eliminating the need to confront upfront capital costs and aligning expenses with actual usage. Numerous studies have highlighted the benefits and challenges of implementing this model in healthcare settings.

By utilizing this payment method and by leveraging demand forecasting, hospitals can optimize their operations while reducing the financial burdens linked to purchasing (Yu et al., 2024). Additionally, this model aligns well with value-based healthcare principles, as it allows hospitals to pay solely for the actual utilization of medical devices, thereby directly linking costs to care outcomes (Rahmani et al., 2021).

Similar to disposable sales, PPU models bear the seller's risk of low usage, which can lead to profitability issues for the manufacturer. For this reason, minimum consumption clauses or initial lump-sum payments are typical of PPU agreements.

7.7.1.6 Closing Remarks

Many medical technology startups overlook the decisions concerning the monetization model, with the frequent consequence of postponing it for too long (Ramanujam & Tacke, 2016). In the absence of other implemented market tests, testing the viability of the monetization model is delayed until the product launch. In many cases, it also entails postponing the market test of pricing until the product launch.

In other cases, medical startups, attracted by the most financially attractive options, take on the risk of making early, untested assumptions and decisions about their monetization model by forgoing feasibility testing across international distribution channels or attractiveness testing for both intermediaries and end-customers. For instance, medical startups commercializing devices that require the real-time use of remote cloud platforms or remote AI-based analytics may elect the PPU model early without considering the possible barriers from international commercial channels.

As already mentioned, it is essential to underline that the typical international distributors of medical devices are well acquainted with the processes of buying, storing, promoting, and reselling. Therefore, they may be unable or reluctant to adopt a model entirely based on subscriptions.

Among other difficulties, these models have additional accounting and margin-split complexities when applied to an international distributor.

Alternatively, the manufacturer may sell the devices to the distributor, who may later employ models such as rental, leasing, or subscriptions. However, in this last case, medical ventures would still see revenue lines built on standard selling-buying transactions, experiencing cashflow bumps and leaving all the mentioned financial advantages to distributors and, somehow, to their end-customers. With some degree of simplification, while focusing on the topic of monetization models, the medical entrepreneur must answer the following questions:

- What monetization model best assures smooth penetration and long-term profitability?
- What payment model most reflects the value delivered to customers?
- What is the status quo of the customer's purchase standards?
  - Are target customers locked in by the practice of public tenders?
  - Can customers administratively manage to
    pay subscriptions?
    pay per use?
    pay through leasing agreements?
- Is the elected monetization model suitable for international distribution channels?
  - Why and to what extent?
  - Which market tests are necessary to prove the model's suitability?
- Is the international distribution channel open to mixed approaches that, in addition to standard purchases, may include subscriptions or payments per use?
- What impact will the elected monetization model have on revenue growth and predictability?
- What reflection will the elected monetization model have on profitability and its stability?

7.7.2 Commercial Model

Here, the term "commercialization" specifically refers to the planning and execution of activities intended to maximize sales revenues and, concurrently, grow the demand for new medical technologies.

The election of a commercialization model entails all the choices related to the sales process, the marketing process, the sales organization, and all the commercial activities implemented using internal or external company resources. Internal resources include, for instance, a direct sales force, product specialists, and service specialists. Outside resources may consist of, for example, business partners, wholesalers, distributors, and sales agents. Similar to established companies, medical startups must attempt to mix internal and external resources to achieve their sales goals while reducing the cost of customer relationships and enhancing customer impact (Zoltners et al., 2004).

In medical innovation, as in other technology-rich industries, it is helpful to separate the commercial effort into two components: the first focused on growing sales and the second focused on growing demand (Fig. 7.53). In the technology market, sales effectiveness and demand generation are distinct efforts. Sales effectiveness focuses on improving the efficiency and productivity of the sales process, while demand generation is centered on creating awareness and interest, which can lead to enhanced demand and customer acquisition.

7.7.2.1 Growing Sales Revenues

Company-owned (direct) channels typically include direct sales forces, telechannels, commercial websites, and company-owned stores. The category of

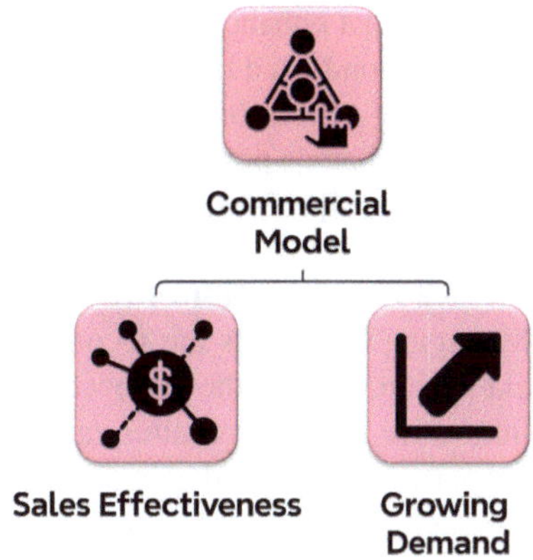

Fig. 7.53 The two pillars of a commercial model: sales effectiveness and demand generation (growing demand)

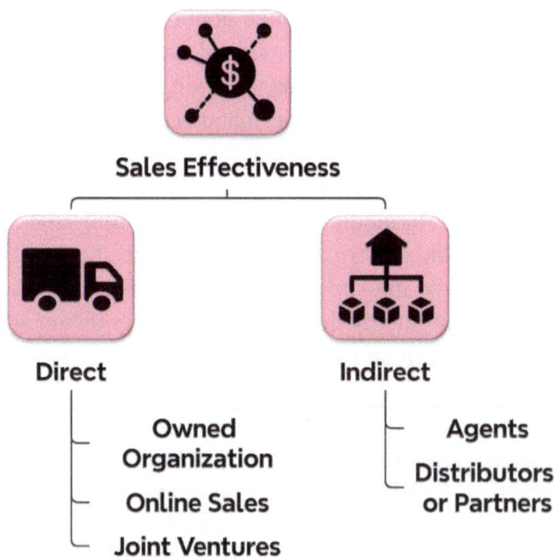

Fig. 7.54 The two most basic global commercialization models

intermediaries involves organizations or individuals that are not part of the company, such as agents, brokers, distributors, integrators, and Internet intermediaries (Fig. 7.54).

The decision about the type of channels is made by weighing several factors, such as the current understanding of customers' decision-making process, the specific competencies of alternative commercial channels, the need for control over the selected channel, the maximum acceptable price for end-buyers, and the economic affordability of multiple layers of intermediaries.

Each channel comes with a different cost profile (fixed and/or variable) that must be reflected in the venture's total profitability and, thus, ultimately in the prediction of the venture's value.

Sales excellence can be defined as the systematic approach to achieving superior performance in sales through a combination of effective strategies, behaviors, and practices that enhance individual and organizational outcomes. These activities are typically conducted by sales reps or key account managers when the call points are the buying offices of large institutions, and when it makes sense to leverage the entire company's portfolio. In some cases, primarily those involving public institutions, tender specialists must navigate the often-complex bidding frameworks.

7.7.2.2 Demand Generation

Managing demand is the second important pillar of medical technology commercialization. Neither the presence of early adopters nor the award of public tenders to innovative medical ventures can guarantee sustainable or growing revenues unless supported by actual demand. The issue of demand generation includes both the adoption speed and the degree of usage or consumption by adopters.

Scientific communication, through publications, congresses, and the endorsement of scientific associations, can heavily aid the adoption of medical technology. Moreover, with the ultimate goal of supporting the quality, safety, and efficacy of healthcare technologies, medical associations often design or endorse existing clinical protocols or guidelines involving the use of specific medical devices. To achieve this type of endorsement, manufacturers must conduct all scientific and technical activities necessary to prove the adequacy of the marketed devices or solutions.

In some cases, medical associations may endorse technical characteristics. For instance, 16 medical societies, including the Massachusetts Medical Society (MMS), the American Society of Anesthesiologists (ASA), and the Society for Technology in Anesthesia (STA), have endorsed the value of medical device interoperability, believing it leads to significant advances in patient safety. Beyond adoption, demand-focused organizations are concerned with the long-term usage or consumption rates of adopters. Because they know that not all adopters will meet the usage required to support growing profits and profitability when adopting monetization models based on actual usage (i.e., PPU or disposable sales). Many healthcare delivery institutions have stored medical equipment they once purchased but then rapidly discarded. These situations are particularly harmful to medical ventures that have adopted a monetization model based on subscriptions, pay–per–use, or the recurring sale of disposables. Once a customer decides to stop using a specific technology, they stop buying disposables, performing and paying for procedures, and are likely to stop paying subscriptions after the minimum committed period. Many healthcare providers buy technologies based on their initial evaluations. Still, sometimes, they dismiss them after a period of actual usage and practical experience. Day-to-day comparisons with existing alternatives and issues may only be detected after substantial usage and practice. For these reasons, manufacturers must ensure adequate customer support, which is often delivered through highly specialized product training and highly responsive technical support. This is why established medical technology companies design and deploy demand-focused resources to influence the actual usage of newly acquired medical technologies through user training, education, field support, generation of additional evidence of clinical superiority, or identification of new indications or applications for the same technology (Fig. 7.55).

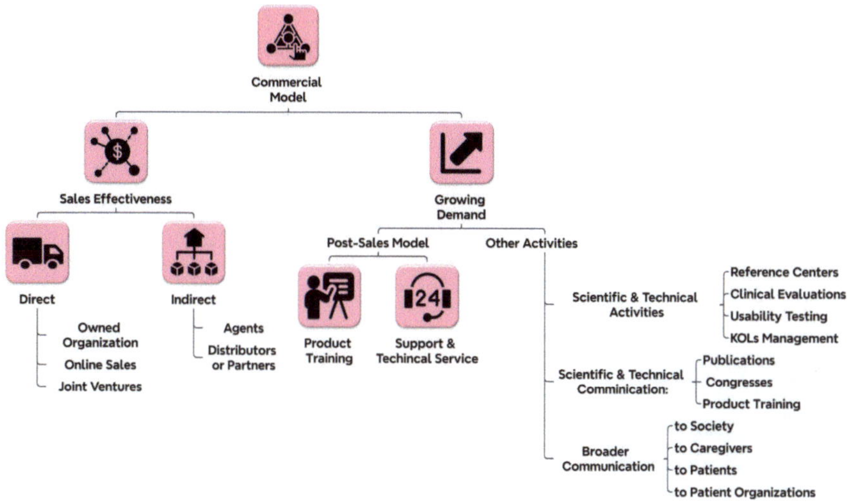

Fig. 7.55 The key decision-related problems relevant to predicting and influencing a growing demand

7.7.2.3 Closing Remarks

Figure 7.56 provides a simplified example of a typical, direct commercial organization matrix for a medical device company. This framework is a valuable tool for deciding where and how to focus sales effectiveness and demand generation activities by clarifying their objectives and the required resources. It facilitates a visual, graphic approach to allocating commercial resources across product lines (or indications) depending on their priority (H, M or L). Creating and filling this type of map requires the medical entrepreneur to make and test several assumptions about the adoption process and decide the organization's specific objectives in demand generation and sales effectiveness. This exercise helps clarify the products for which the commercial resources must be allocated and prioritized and the competencies that are required for each commercial role. Indeed, medical ventures must decide relatively soon what profiles may best support the identified commercial goals and tasks, as the required profiles are not always easy to find in the job market and are not the cheapest headcounts to attract and retain.

In summary, there are several decisions and assumptions that must be addressed in the area of commercialization:

- Which stakeholders will be involved in the commercialization process?
- What type of interactions will be established with each of these stakeholders?
- What are the main goals of each relationship?
- What are the activities and the roles that will support sales effectiveness?
- What are the activities and the roles that will support demand growth?
- Which profiles best ensure successful implementation?
- Does the venture currently employ these profiles as team members? If bot, when and where will the venture hire these profiles?
- Are there commercial intermediaries or distribution companies that offer these critical capabilities or profiles?

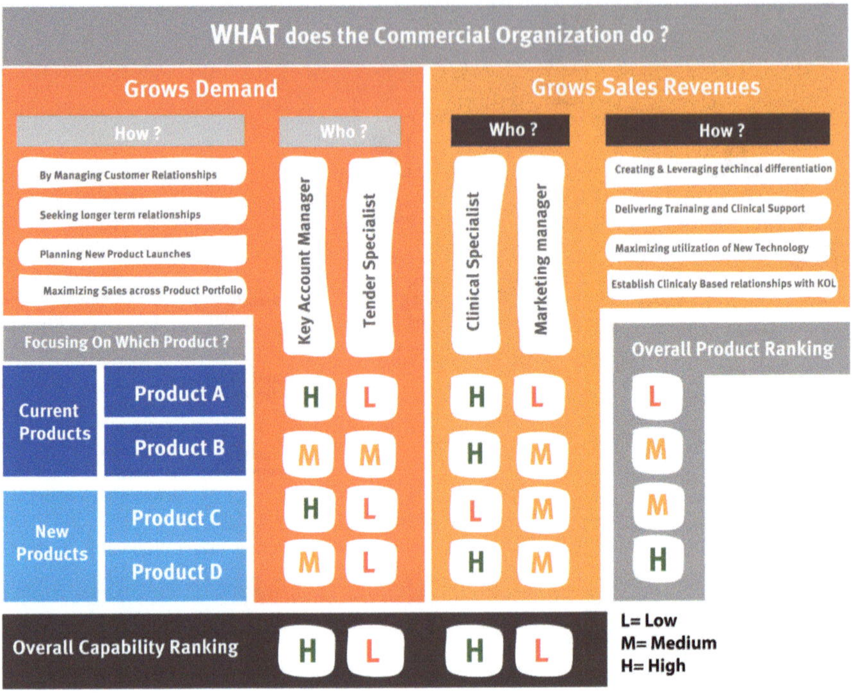

Fig. 7.56 An example of a typical, direct commercial organization matrix for a medical device company

7.7.3 Cost of Commercialization

In the venture's continuous efforts to sharpen the prediction of future profits it is vital to include the prediction of the cost structure related to the designed commercial model. Without entering the details of the cost components of commercialization, it is worth recalling that global commercial channels, in most cases, bear a significant cost, whether the resources are internal (such as employees) or external (such as distributors and agents). A frequent decision bias of medical technology startups is related to planning the use of a globally distributed chain of intermediaries without precisely and realistically accounting for the margins required by each of them and the consequences for the end-buyer price and the venture's profitability.

These decision biases often cause the late discovery of hard facts, such as the economic challenges in building an international commercial channel that concurrently delivers satisfactory margins, financially incentivize each of the intermediaries and maintain an accessible end-buyer price. When the decision to build a chain of international intermediaries is tested too late in the venture's development process, it often results in painful wake-up calls that undermine the viability of the financial plan (Ramanujam & Tacke, 2016). An early prediction of the cost to commercialize a medical technology across countries is due very early in the venture life: it must begin when the pricing-level decisions is made, normally, in the third cluster of the K9 map. These simulations should not be limited to accounting for the margins required by the different layers of intermediaries but must include all other expenses, such as the shipping costs and the import taxes of each of the selected countries. The "experiment" of engaging in early contractual conversations with potential distributors provides excellent insights on this subject.

In summary, the medical entrepreneur, confronted with the issue of the cost of commercial channels, must answer questions such as the following:

- What is the cost of the national and international channels?
- What are the costs related to transportation and taxes?
- What are the costs of agents' commissions and distributors' margins?
- Are the margins economically attractive to intermediaries?
- Are the costs of the international channels compatible with the planned price for end-buyers?
- Are the costs of international channels compatible with the planned venture's profitability?
- Which channel choice represents the best compromise between cost and effectiveness?

7.7.4 Suitability of the New Medical Product to International Commercial Intermediaries

Most medical startups test their monetization and commercial model in their home countries. As mention, unless they can operate through directly owned foreign legal entities, they often rely on international intermediaries, mainly distribution companies, as local agents do not offer sufficient logistical, regulatory, and promotional support. We have observed that a frequent bias of medical startups consists of making untested assumptions about international distributors' actual interest and availability in allocating meaningful resources to commercialize a specific new medical technology. As discussed in Chap. 4, this bias is often rooted in the entrepreneur's emotional attachment to the invention (Scheier & Carver, 2018). The assumption that medical technology distribution companies are sitting in their respective countries waiting for their phones to ring to finally fall in love with our mecial invention is as common as detrimental.

When such an optimistic set of assumptions remains untested until market entry, it may generate disappointing surprises. For instance, medical ventures may realize the following:

- Most successful medical distributors already commercialize a large, profitable, and well-thought-out portfolio of medical technologies.
- Distributors carefully select any new addition to their current portfolio by prioritizing the following:
  - Products that fit the portfolio
  - Products that are likely to exert an early, positive, and significant impact on their revenue lines

- Products that require a limited commercial and promotional effort compared to their potential future revenues and the rest of their portfolio
- Products from well-known MedTech or BioTech companies that carry branding advantages, quality assurance, and supply continuity
- Products that do not directly compete with the ones that are already distributed
- Products that represent a meaningful complement to the existing portfolio
- Products for which the knowledge of existing and known call points can be leveraged

The issue concerning the distributor's motivation to adopt a new medical technology from a young venture has significant implications on the venture's ability to influence the dealer's actual efforts and compliance with the signed agreements. For instance, a distributor not truly interested can certainly (and sometimes lightly) decide to sign an exclusive country-wide commercial agreement based on the purchase of minimum quantities. On the other hand, this company recognizes that its worst-case scenario is failing to meet the agreed minimum quantities, and its highest risk is losing exclusivity on a product that is not a commercial priority.

While a medical entrepreneur might feel the urgency of showing investors a newly signed distribution agreement for a large and attractive country, when the intermediary's motivation in investing resources in commercializing a new medical technology is not genuine, these agreements might result in months or years of commercial paralysis in the concerned regions.

In general, any credible proof of intermediaries' genuine interest in the acquisition and commercial development of a new medical technology represents the strongest predictive signal for a sound and healthy collaboration with the innovating venture, that respects commercial agreements.

For this reason, a good initiative that helps test the most basic assumptions about external distribution channels consists of early surveys about the existence of intermediary organizations that, for self-evident financial, strategic, or portfolio reasons, may be genuinely interested in investing time and resources in the commercialization of our new technology.

In summary, there are several additional questions concerning the intermediaries' motivation that the entrepreneur must consider well in advance of market entry:

- Will the venture commercialize new medical technology as a standalone product or as part of a more extensive product portfolio?
- Suppose the new technology won't be offered as a part of a compelling portfolio. Are there international intermediaries that, from a portfolio perspective, would benefit from the addition of the venture's new medical technology?
- Are these benefits potentially comparable to the time and resources required by these intermediaries to create the demand for new technology, launch, and commercialize it?
- Which early field tests can provide valuable information about the product attractiveness for international distribution channels?

7.7.5 The Complete Map (Fig. 7.57)

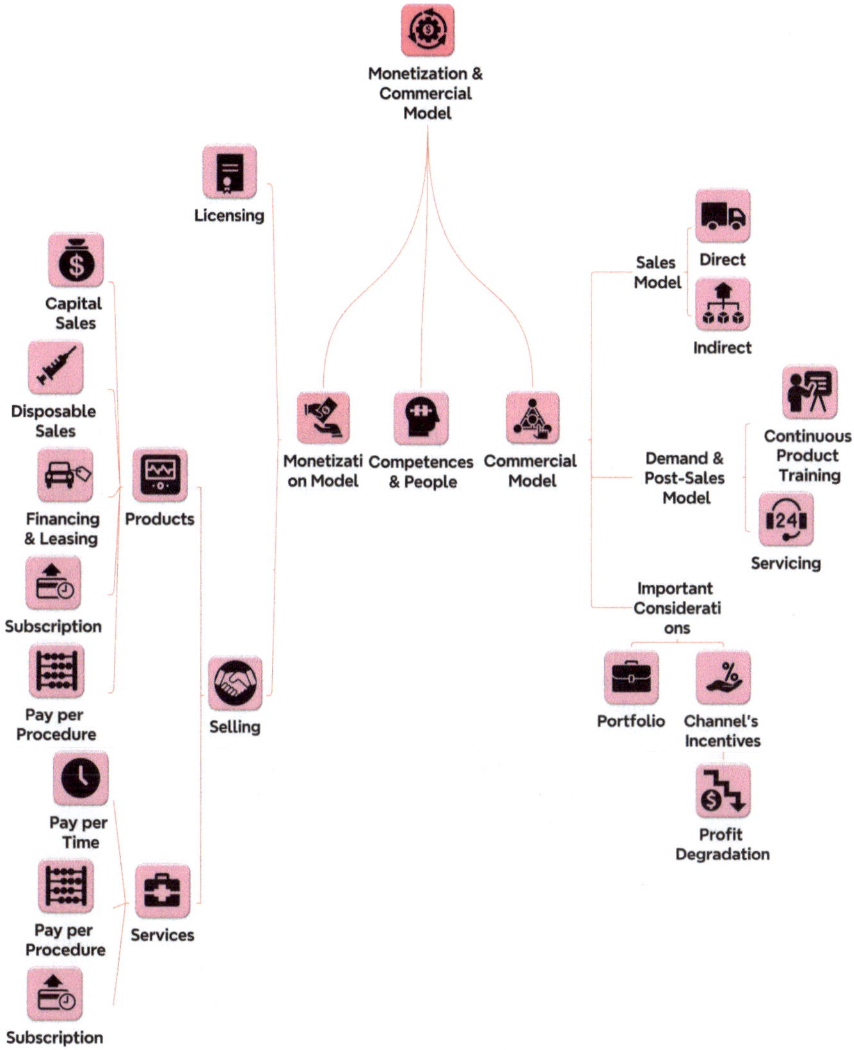

Fig. 7.57 The complete map of decision-related problem clusters relevant to the viability of monetization and commercial model

7.7.6 Key Assumptions and Decision-Related Questions (Table 7.16)

Table 7.16 Viability of monetization and commercialization: core assumptions and decisions

Viability of monetization and commercialization	
Sub-cluster	Core assumptions and decisions to be made
Monetization	What is the most attractive and profitable revenue-generation model?
	Will revenues come from selling products, services, or both?
	Will subscriptions be part of the revenues? To what extent?
	Will revenues come from licensing?
	Will we sell directly to the patient? Physicians, hospitals, or governments?
	Will intermediaries be used? Which types and where?
	How much will national and international intermediaries cost?
	Which monetization model best assures smooth penetration and long-term profitability?
	Which payment model most reflects the value delivered to customers?
	What is the status quo of customer purchases? Are target customers locked in by the practice of public tenders? Can they administratively manage to pay for subscriptions? Can they sign long-term rental agreements?
	Is the elected monetization model suitable for international distribution channels? Why and to what extent? Which tests are necessary to test this assumption?
	Is the international distribution channel actually open to combined billing approaches that, on top of standard purchases, may include subscriptions or payments per use?
	What reflection will the elected monetization model have on the growth, sustainability, and predictability of revenues?
	What reflection will the elected monetization model have on profitability and cash inflow stability?
	Which monetization model best assures smooth penetration and long-term profitability?
	Which payment model most reflects the value delivered to customers?
	What is the status quo of customers' purchase habits and procedures? Are target customers locked in by the practice of public tenders? Can they manage to pay for subscriptions administratively? Can they sign long-term rental agreements?
	Is the elected monetization model suitable for international distribution channels? Why and to what extent? Which tests are necessary to test this assumption?
	Is the international distribution channel open to combined billing approaches that, on top of standard purchases, may include subscriptions or payments per use?
	Are there legal obligations and constraints that require a clear transfer of title (and thus a straight sale transaction) to the international reseller?

(continued)

Table 7.16 (continued)

Viability of monetization and commercialization	
Sub-cluster	Core assumptions and decisions to be made
Commercialization	Which stakeholders will be involved in the commercialization process?
	Will the channel's profitability be attractive for intermediaries?
	What are the main goals of each of the commercial relationships with the intermediaries?
	What type of agreements will be sought with these stakeholders?
	What are the activities and roles that will support sales effectiveness?
	What are the activities and the roles that will support demand growth?
	What is the cost of the national and international channels?
	What are the costs related to transportation and taxes?
	What are the costs of agents' commissions and distributors' margins?
	Are the margins economically attractive to intermediaries?
	Are the costs of the international channels compatible with the planned price for end-buyers?
	Are the costs of international channels compatible with the planned venture's profitability?
	What channel choice represents the best compromise between cost and effectiveness?
	Will the new medical technology be commercialized as a standalone product or as part of a more extensive product portfolio?
	Are there international intermediaries that, from a portfolio perspective, would benefit from the addition of the venture's new medical technology? Are these benefits potentially comparable to the time and resources required by these intermediaries to create the demand for new technology, launch, and commercialize it?
	If only one product is commercialized, will the related revenues be attractive to the intermediaries?
	To what extent is the monetization model suitable for international medical technology distribution companies?
	Which early field tests can provide valuable information about the product attractiveness for international distribution channels?
Competencies and people	What competencies are necessary to support the monetization and commercial models?
	What cost is linked to the acquisition of the core competencies?
	Which profiles best ensure successful relationship management?
	Does the venture currently employ these profiles as team members?
	If not, when and where is it critical to start hiring these profiles?
	Are there intermediaries or distribution companies that offer these profiles?

7.8 Financial Viability

The financial viability of a medical startup is evaluated based on predictions of cash flow and the available cash reserves. Financial viability must be predicted and analyzed in three main phases: the pre-market phase, the post-market pre-breakeven phase, and the post-breakeven phase. Figure 7.58 presents an example of a cash reserves chart for a medical venture. It clearly shows how important it is to predict

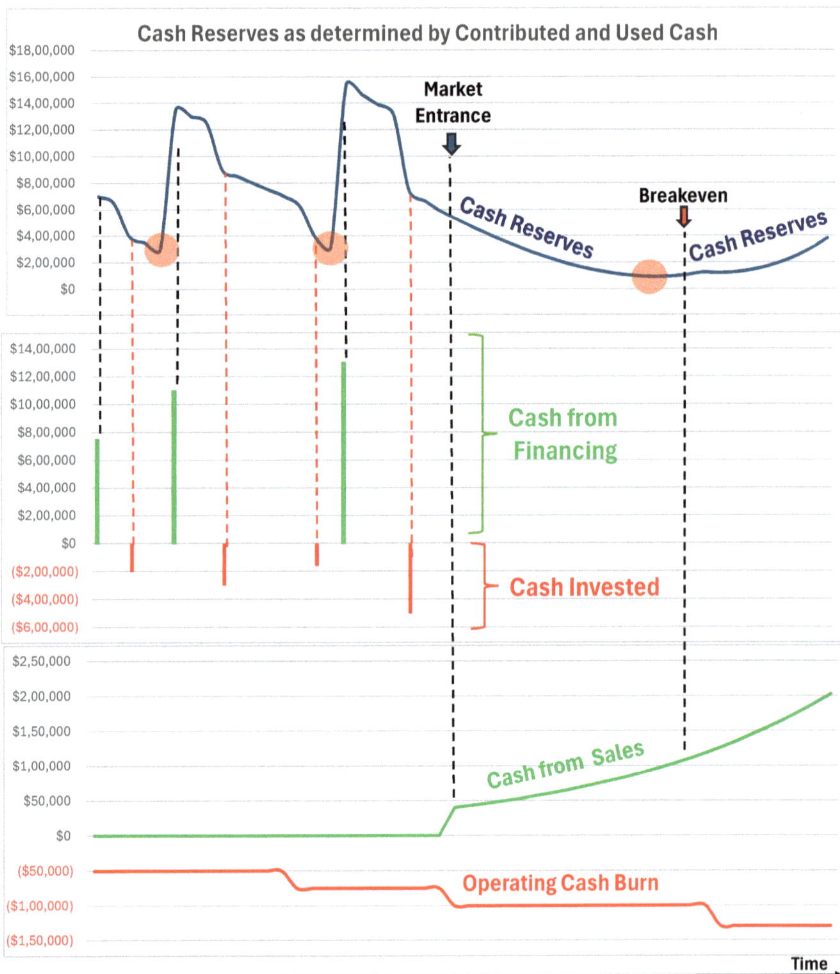

Fig. 7.58 An example of a cash reserves chart for a medical venture. It shows how important it is to predict the time of occurrence of cash reserve bottoms (periods of potential cash distress, highlighted by the bubbles), their depth, and their durations

the time of occurrence of cash reserve bottoms (periods of potential cash distress, highlighted by the red bubbles), their depth, and their durations. Before the market entry, medical ventures are particularly exposed to the risk of unpredicted liquidity depletion due to

- Significant investments in product development
- Large investments in clinical trials (often characterized by uncertain duration)
- Uncertainty over the regulatory and registration processes across the world and
- Unexpected withdrawal of previously committed investors

Before reaching the market launch milestone, cash reserve bottoms can be estimated by diligently forecasting operating expenses and investments, as well as predicting investor contributions and their timings. Despite these forecasting difficulties, in this phase, investments and operating expenses are far less unpredictable and more controllable than those required by post-market phases. After the market entry, before breakeven, medical ventures are heavily exposed to additional risk of unpredicted liquidity depletion due to

- Potential new investments in post-market clinical trials
- Investments in the direct commercial organization
- Difficulty in predicting the adoption speed
- Challenges related to predicting actual average selling prices and
- Issues pertaining to commercial efficacy and effectiveness

In summary, before the breakeven point, the leading financial goal of the medical venture is to avoid liquidity depletion to the point of insolvency. Like a deep-water scuba diver does with oxygen reserves, startups must avoid depleting their cash reserves for an extended period. The only exception to this rule is represented by very early-stage, informally organized medical ventures that can pause or minimize their cash burn based on personal financial sacrifices.

After breakeven, the core areas of financial concern include ensuring the venture's working capital and delivering the promised cash results and return on investment. When the predicted cash flow from sales starts to influence total cash flow significantly, forecasting efforts become even more vulnerable, as the uncertainty of sales volumes and growth rates represents more impactful sources of uncertainty.

For this reason, revenue growth and profitability must align sufficiently to support the delivery of the promised stream of profits. So much attention must be dedicated to cash flow forecasts because it is crucial to predict the possible exhaustion of cash flow reserves and the consequent business continuity or potential insolvency during the "cash valleys" of medical startups.

A very concise representation of the critical elements determining the startup's future cash flow is presented in Fig. 7.59: the entrepreneur must make assumptions about the cash required for investments, the cash needed for operating expenses, the cash expected from commercial revenues, and the cash received from investors.

Except for cash from financing, which embeds significant elements invisible to the entrepreneur, most of the prediction accuracy related to the issue of financial viability is supported by the quality of decisions and assumptions made in the

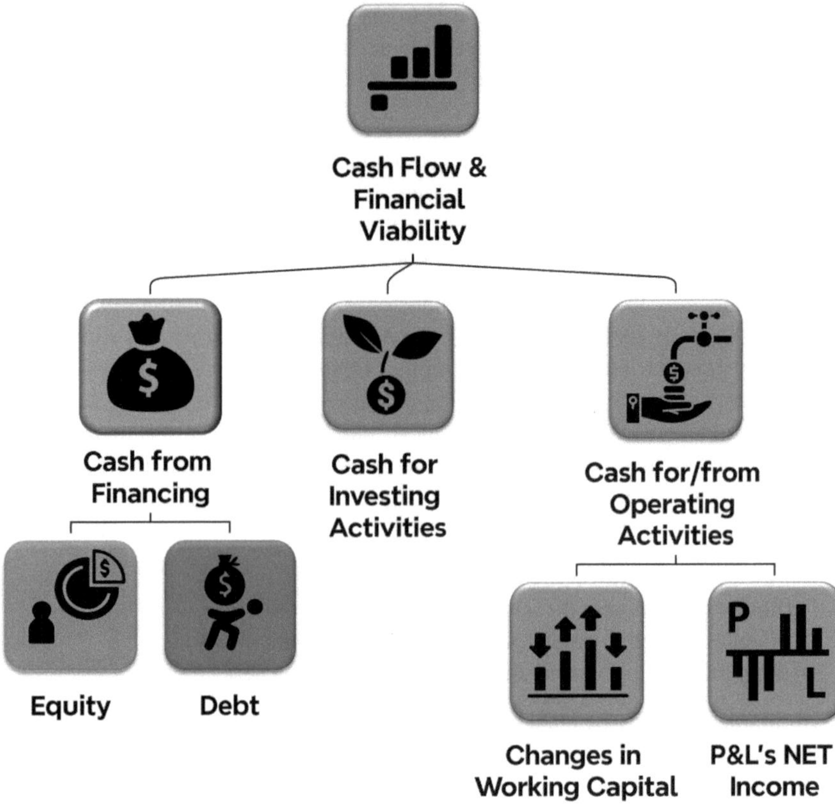

Fig. 7.59 The three pillars of cash flow and financial viability

previous seven clusters of the K9 map. For instance, the cash for investment activities is directly determined by the investments planned for product development, clinical trials, regulatory activities, and IP protection investments. The pre-market cash burns heavily depend on the talent acquisition and retention choices that typically account for most of the operating expenses. And, as already mentioned, considerably more difficult is the prediction of the post-market cash from operations, which, in turn, depends on the uncertain P&L performances. The changes in working capital also represent a predictive challenge, but are strongly tied to the assumptions already made when predicting operational (P&L) performances.

7.8.1 Predicting Revenues

The challenging estimation of post-market revenues involves the prediction of market penetration and of the average selling prices concurrently (Fig. 7.60).

While predicting adoption is the core concern, revenues are derived in a second step, applying to adopting customers the business model and the pricing defined in

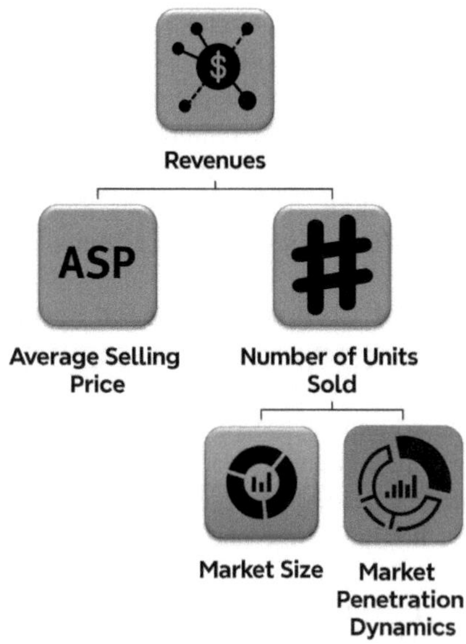

Fig. 7.60 The basic drivers of revenues and of their prediction

previous clusters, such as the business and commercial model, value proposition attractiveness, and customers' willingness to pay. As discussed in Sect. 7.7, in medical technology, revenues originate following one or a mix of a few classical models: capital sales, leasing or renting agreements, disposable sales, per-procedure charges, and subscription charges.

Product adoption is referred to by various names, including adoption curve, uptake model, product diffusion, and dissemination curve. All represent the time required for a product to be tested and used in the market (Cook, 2006). As for all other predictive efforts, entrepreneurs can more confidently predict the adoption curve by using a scientific approach to scenario decisions.

Unfortunately, a too common method to predict future market share consists of simply entering the number of units to be sold each year with the help of some best-guessed growth rates. This "formula-pulling" approach has been demonstrated to be as weak as it is fast, mainly because it does not require the forecaster to cross-check the revenue numbers with other known reference points, nor to be intimately confident about all the explicit and implicit underlying business assumptions.

Moreover, startups often assign these forecasting activities to their financial managers or analysts, who, since they are considered the venture's "model experts" and "know how to crunch the numbers," are expected to have oracular skills and end up being assigned unrealistically complex predictive assignments. In reality, they commonly lack exposure to the harsh reality of the adoption and commercialization of medical products. These analyses primarily rely on the marketing and competitive information usually provided by the CEO or by a commercial head. Even this core market information, in my experience, often suffers from predictive biases due to entrepreneurial optimism (Hmieleski & Baron, 2017) and a lack of assumption testing (Camuffo et al., 2024a, b).

Even when the marketing information provided to the analysts is realistic, the forecaster may still be a victim of the typical, silent, unofficial organizational expectations that want people to "be positive" and compile numbers that are somehow "expected" by the venture's majority shareholders or leaders. These quiet, implicit requests to "believe in the product" or "believe in the organization" must be taken into serious consideration as they increase the probability of predictive biases (Ehrman & Shugan, 1995; Markovitch et al., 2015; Cook, 2006).

7.8.1.1 Competition-Based Forecasting

More rigorous approaches mitigate these sources of predictive bias by using external reference points to guide the forecaster's judgment. For instance, a competition-based approach uses the relative shares of all competing products (or services) as external boundaries to the penetration speed of the new medical technology. For each period, the forecaster enters the projected share for all the competing products in a competitive matrix, as graphically represented in the example provided in Fig. 7.61.

This method heavily relies on the forecaster's ability to identify all the competitive dimensions and translate their relative strengths into market shares. A predictive effort that focuses on head-to-head value proposition comparisons as the main predictor of future shares. The positive aspects of this approach are mainly two:

- It takes active advantage of all the assumptions made in the precedent clusters, such as those related to value proposition attractiveness and competitiveness.
- It adds accuracy because it is based on precise answers to basic questions about competitiveness, such as:

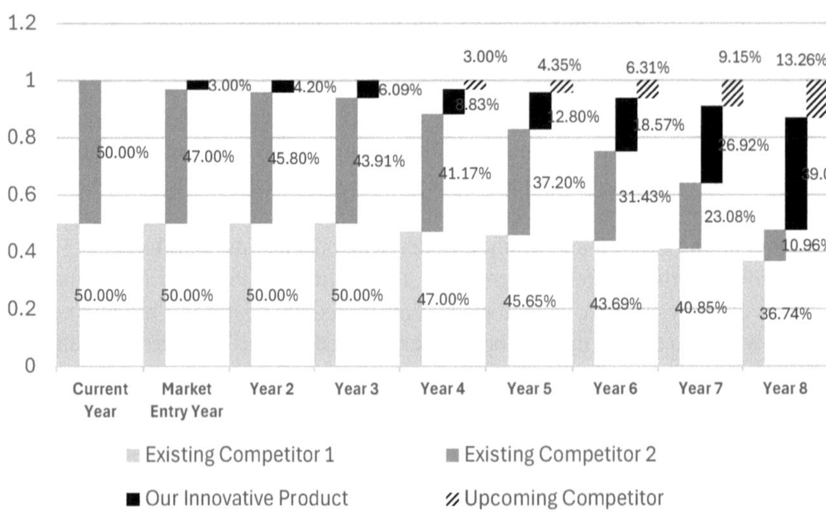

Fig. 7.61 A simplified predicted penetration dynamic for an innovative medical product in a stagnant market context

- Is the venture focusing on gaining shares from a specific competing product, or will it gain shares from all the competitors?
- Which competitors will lose share? Why? How much?
- Is the total market growing? If yes, which competitor is best positioned to exploit the future market's growth?

Providing forecasters with specific, fact-based answers to such questions enhances their ability to assess the validity of many underlying hypotheses.

Figure 7.61 represents a simplified predicted penetration dynamic for an innovative medical product in a stagnant market context, and, for each competitor, it shows the predicted share-related consequences of the new market entrant. For instance, existing competitor 2 is the market player that will suffer the most due to the new product's market entry. The forecaster must be able to diligently support this strong assumption by explaining what specific competitive strength of the venture's technology can support such a loss of share by an established market player. The forecaster may decide to test this assumption by interviewing buyers (or other decision-makers) about their explicit reasoning process that leads to the choice between the venture's product and the product commercialized by an existing competitor. This methodology breaks predictive problems into minor, less complex, connected problems that better support testing activities, representing more suitable targets for the application of a scientific approach. Hypotheses must be all tested and sharpened through disenchanted, cynical cross-functional discussions within the organization. Still, the output needs to be handled with care and should not be treated as "the" forecast but as one of the multiple refined signals supporting the ultimate, final penetration prediction.

7.8.1.2 The "Peak Share and Time to Peak" Forecasting Method

In many cases, advanced medical innovations have no competitors except, in some cases, the current standard of care. Therefore, using competing products as forecast references or determining growth constraints is not viable. In these contexts, "peak share and time to peak" is the most common method of predicting the dynamics of new product adoption. This method also breaks the complex forecasting problem into smaller, more approachable, and ultimately more testable assumptions.

Indeed, the forecaster must focus on at least two different parameters: the peak share (or pick adoption rate) of the product being forecasted and the amount of time required to attain this peak share. This is how in Fig. 7.62 the lower curve, which represents the frequency of adoption (number of new adopters per period), influences both the shape and the maximum value of the S-curve, representing the total number of accumulated users. To estimate the peak share, the forecaster will focus on the product price and value proposition features that support a target maximum penetration target. The slope of the S-Curve (or the time to peak) will be estimated by examining the planned commercial processes and resources that significantly impact the efficacy and duration of the penetration efforts.

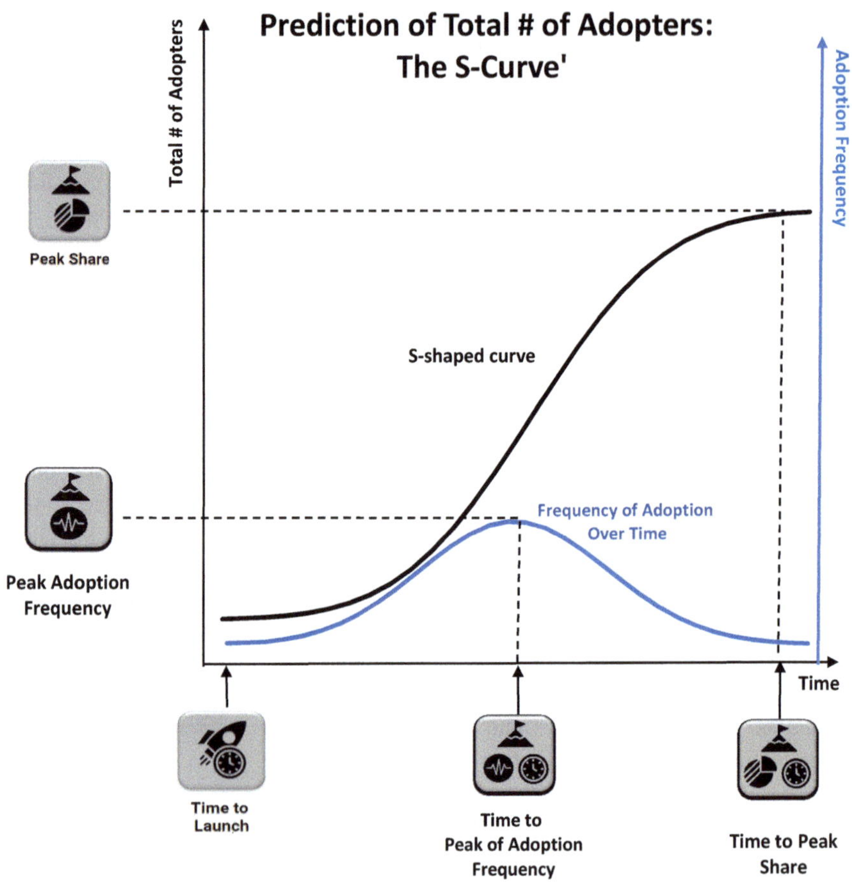

Fig. 7.62 Example of an S-shaped curve and of the corresponding frequency of adoption curve

7.8.1.3 Predicting Peak Shares Using Product Attributes

While the peak-share and peak-time method helps predict adoption in markets without active competition, it can also be used when a market holds active competition. The most reliable methods for predicting future market shares are based on the weighted quantification of the competitive performances of the new product's attributes against those of each competing product.

In this context, the forecaster must construct a list of product features that are relevant to customers, assign respective weights indicating the importance of each feature to customers, and mathematically construct weighted averages that constitute a quantitative measure of estimated product attractiveness. This method assumes that when an innovative medical product achieves its peak share, the distribution of market shares across competitors will retrace the distribution of relative product attractiveness, as illustrated in Table 7.17.

Table 7.17 An example where three product features have been used as determinants of product competitiveness, supporting the predicted share

		Therapeutic Efficacy	Treatment Safety	Ease of Use	Total Score	Share
	Weigh	50%	30%	20%		
Scores	Our Product	9	7	4		
Scores	Product 1	4	5	6		
Scores	Product 2	6	7	3		
Scores	Product 3	4	3	9		
Weighted Scores	Our Product	4.5	2.1	0.8	7.4	**33%**
Weighted Scores	Product 1	2	1.5	1.2	4.7	**21%**
Weighted Scores	Product 2	3	2.1	0.6	5.7	**25%**
Weighted Scores	Product 3	2	0.9	1.8	4.7	**21%**
					22.5	**100%**

The strength of this method relies on its rational approach and on the transparency of the assumptions and, therefore, on the enhanced defensibility. Nobody can impose "more optimistic" market shares without motivating changes in individual performance ratings. On the other hand, the forecaster adopting this method must pay close attention to new sources of potential bias, such as those related to:

- The determination of the list of key comparative features
- The assignment of individual weight to each feature and
- The quantitative performance scoring process

7.8.1.4 Predicting Peak Shares Using Product and Marketing Attributes

The distribution of shares may not purely depend on product features but also on competitive marketing pressure and commercial reach. Therefore, forecasters may add these non-product criteria to the prediction of shares.

When identifying the most effective criteria, three factors were found both in our field experience and in the literature to represent the major explanatory variables of the adoption of medical devices: (1) the subjective expected value of the device, (2) information and learning, and (3) the innovativeness of the adopting unit (Roback et al., 2007).

Using this framework, a forecaster can add parameters related, for instance, to the issue of "information and learning" as it refers, not only to the strength of the marketing arguments about product superiority but also to the venture's ability to consistently communicate them globally. It can be assessed, for instance, using measurable parameters such as the number of published papers that confirm the superiority of a particular therapy or diagnosis and the financial investment in marketing organizations and tools. Table 7.18 provides an example of such an approach.

In summary, in the attempt to take into account most of the competitive divers, the forecaster may use a wide range of product—and non-product-related—parameters to estimate the company's peak share.

Table 7.18 An example where three product features and two non-product attributes have been used as determinants of product competitiveness, supporting the predicted share gain

		Therapeutic Efficacy	Treatment Safety	Ease of Use	Marketing Investments	Scientific Publications	Total Score	Share
	Weigh	40%	30%	10%	10%	10%		
Scores	Our Product	9	7	4	4	4		
	Product 1	4	5	6	6	6		
	Product 2	6	7	3	10	10		
	Product 3	4	3	9	9	9		
Weighted Scores	Our Product	3.6	2.1	0.4	0.4	0.4	**6.9**	**29%**
	Product 1	1.6	1.5	0.6	0.6	0.6	**4.9**	**21%**
	Product 2	2.4	2.1	0.3	1	1	**6.8**	**29%**
	Product 3	1.6	0.9	0.9	0.9	0.9	**5.2**	**22%**
							23.8	**100%**

7.8.1.5 Predicting Time to Peak

While predicting adoption, our field observations suggested that, with a few limitations and exceptions, adoption in medical technology behaves in a somewhat similar manner as that in the pharma world:

- High-quality brands achieve peak sales earlier, with higher peak sales levels
- Early entrants achieve peak sales later and have higher peak sales levels (Fischer et al., 2010)

Young medical startups are not high-quality brands and that—by the definition of innovators—they plan to be early entrants. Therefore, unless a venture sells a long-awaited lifesaver for a high-prevalence disease, entrepreneurs must assume longer times to peak than what their therapeutic or diagnostic superiority alone may suggest.

We have observed that early-stage medical startups frequently predict the time to peak and peak share by focusing solely on product features, without considering the challenges of being the first entrants and the disparities in commercial spending between their constrained budgets and those of established global companies. This type of bias is critical because, by optimistically placing the milestone of time to peak share, startups predict unrealistically early breakeven milestones, increasing the post-market risk of running out of cash.

The forecaster must consider many dimensions. For instance, the size and geographical distribution of commercial organizations play an essential role in the speed of market penetration. The choice of commercial channels and organization has substantial implications for predicting how long it will take to achieve the target market share.

What are the major drivers of time to peak share for a single, innovative medical product? Suppose we exclude the complex contributions of all the involved decision-making stakeholders and the potentially heavy effects of public or private reimbursement. In that case, the major drivers of time to peak share are the commercial effectiveness and the transactional cash burden for buyers (as it is in most industries). In other cases, when appropriate, the government or private insurance may award reimbursement; when this happens, the time to peak share is significantly shorter. On the other hand, obtaining reimbursement is neither easy nor predictable; thus, it is typically wiser to work concurrently on two predictive models: with reimbursement and without reimbursement.

The Static Friction in Medical Technology Adoption

In the medical technology space, it is rare to observe an adoption prediction being actually reflected in reality: the first significant signs of actual adoption often come much later than predicted. This type of biased assumption resembles one that can be made about Newton's second law of motion, expressed as F = ma. Since this law states that the acceleration of an object is proportional to the force acting on it, an incautious forecaster could expect to be able to move a parked train carriage with a simple hand push. The reason for the forecaster's disappointment can be found in a naïve or superficial adoption of the law of motion (F = ma) without consideration for the static friction, which counters and nullifies the pusher's force till it exceeds a certain threshold. Only once (and if) the pushing force exceeds this threshold, the carriage starts actually to move, and this usually happens later than initially expected.

The prediction of future adoption of medical technology commercialized by small and resource-constrained startups often suffers from the same type of bias described in predicting the ability of a man to move a train carriage seamlessly. Several factors play roles similar to that of the static friction, for instance: a slower than expected approach to hospital purchasing groups, a later than anticipated onboarding of key commercial personnel, a minor technical issue that was overlooked in the pre-market tests, an unexpected increase in the initial manufacturing leadtime, or a longer than expected sales funnel processing of commercial leads, and many other possible issues playing individually or jointly. For this reason, when formulating an adoption prediction curve, it is essential to cynically and transparently explore and weigh all known potential barriers to timely adoption success. It is therefore necessary to review and test all the explicit and implicit assumptions supporting the predicted adoption. In the absence of such a disciplined approach, at a minimum, forecasters should consider adding contingency lag times to represent unexpected, unplanned, or unknown causes of delays in the actual beginning of measurable adoption.

7.8.1.6 Predicting Adoption Rates of Individual Consumers

Consumers' decision to adopt wearable healthcare technology is affected by several factors. Fitness device users, for instance, care a lot about hedonistic motivation, functional congruence, social influence, perceived privacy risk, and the perceived risks associated with the consequences of unhealthy behaviors. According to a compelling empirical study on wearable technology acceptance (Gao et al., 2015), while medical device users consider the aforementioned dimensions, they pay more attention to the following:

Perceived performance expectancy is defined as the degree to which adopting technology will afford effectiveness to users in the performance of certain activities (Venkatesh et al., 2012). In the context of wearable healthcare devices, effectiveness can be seen as the degree to which the device can support users in monitoring daily physical conditions, making personal healthcare plans, reducing health-related threats, and so on (Gao et al., 2015).

Self-efficacy is the efficacy of self-monitoring and self-managing physical conditions, indicating the extent to which users are unable to effectively self-monitor and self-manage their own physical conditions.

Effort expectancy is widely known as the degree of ease in the consumer's use of (medical) technology (Venkatesh et al., 2012).

Perceived vulnerability refers to the potential exposure to a health threat, while **perceived severity** represents the extent of the threat from unhealthy behaviors (Ronald, 1975).

When consumers believe that adopting wearable healthcare devices can enable them to increase healthcare effectiveness, they are more likely to adopt the technology (Gao et al., 2015).

The context changes completely when, instead of fitness motivation, a medical prescription or recommendation is the primary basis for an individual's decision to adopt a new medical technology.

Since the mid-70s, when blood pressure monitors were commercially made available, physicians have started to learn the vital role of patients' compliance (often referred to as adherence) every time the use of a new medical technology is entirely under the patient's control. Despite an initial adoption decision, most patients later decrease or even completely abandon the use of both diagnostic and therapeutic devices for various reasons. Independently, by compliance, at some point, a patient might decide to stop using a device: Persistence refers precisely to the length of use of a device by an individual patient.

Even in the pharma space, where patients are typically simply asked to swallow a pill once or twice a day, adherence is a huge issue. Pharmacological medication adherence can have a more direct impact on patient outcomes than the specific treatment's efficacy; it is estimated that adherence to chronic pharmacological medications is about 50% (Patel et al. 2025). One can easily see how heavily the problem of compliance affects the medical technology space, where patients are asked to unbox a not necessarily simple device, learn how to use it, regularly turn it on, use it, and put it away.

Even when patients initially decide to adopt a device, if they end up not using it regularly, or even dismissing it, the financial damage to the innovating venture won't be long in coming. Indeed, if the monetization model is based on pay-per-use, the revenue damage is apparent; if it is based on rental, then automatic renewals will soon be cancelled. Even in the case of patients purchasing the device, there will be damage because, sooner rather than later, the prescribing physicians will realize that that particular device performs poorly in terms of compliance.

Compliance and persistence with the regular use of medical devices are tricky topics. I still recall the words of a well-known physiatrist from Florida who, referring to the use of physical rehabilitation devices for home use: "… most of my patients prefer to drive from home to the clinic, do the exercises here, pay, and drive back home than doing the same exercise using a virtually free medical device at home."

One of the key elements to consider when predicting users' compliance is the usability of the medical device. An optimal user experience is needed to ensure that older patients with chronic conditions engage effectively with these technologies. Usability represents, therefore, another key predictor of both adoption and adoption's sustainability.

The role of the home environment and social support also plays an essential part. For instance, the presence of an active, technology-savvy caregiver might positively and radically influence compliance with diagnostic or therapeutic procedures.

In conclusion, patient compliance is a critical issue for the successful adoption of medical technology for home use. The interplay between technology design, user education, and environmental support must be emphasized to ensure that such interventions are effective in enhancing medication adherence.

It is also worth observing that the more a patient feels or is aware of the physical benefits that compliance might deliver, the higher the likelihood of good and sustained compliance.

When predicting adoption, medical entrepreneurs must therefore use segmentation criteria focused on identifying patient groups capable of sustaining high degrees of compliance with the use of medical technology.

For instance, if, from the perspective of an elderly home patient, a medical technology cannot appreciably delay or smooth the symptoms of a chronic neurodegenerative disease, then it cannot rely on substantial patient compliance unless it is delivered by a very easy-to-use device that truly minimizes both patients' effort and the requirement for strong discipline.

7.8.1.7 Predicting Adoption in Hospitals

The buying decision in the context of a hospital is an outcome of the influence of organizational factors, organizational processes, individual characteristics, group factors, technological factors, and the external environment, along with their mutual interactions (Ghodeswar, 2008).

Medical technology entrepreneurs must understand in depth the direct effect and interplay of these factors and assess their role in influencing the hospital management's propensity to purchase certain innovative medical technologies. Contrary to dangerously common practices, the venture's management must use these mechanisms as precious guiding inputs throughout the different phases of the POC and the design of both the monetization and the commercial models at the earlier stages of technology development, instead of waiting till actual commercialization.

Following Ghodeswar's framework, organizational factors for hospitals are characteristics such as the size, the teaching status, the degree of specialization, and the reputation.

While technology adoption occurs across primary, secondary, and tertiary levels of care, it is particularly strong in tertiary hospitals. These institutions, with their highly specialized services, research capacity, and role as referral centers, are naturally positioned to become early adopters of new medical technologies.

Hospital's organizational processes and procedures refer, for instance, to the decision-making procedures, the level of decision-making centralization, the internal climate toward innovation and risk tolerance, which shape how receptive a hospital is to change its processes and procedures. Hospitals that foster entrepreneurial or learning-oriented cultures are understandably more likely to integrate new technologies into their practice.

Individuals inside the institution also strongly influence the decisions about new technology adoption. Physicians, administrators, and leaders bring their competence, professional experience, and personal orientation toward innovation into the decision. Those who have been more recently trained and those playing a pivotal role in the scientific community are often exposed to the latest medical practices and therefore become advocates for adoption.

Group factors are relevant when Hospitals rely on buying groups where different stakeholders—buyers, users, influencers, and gatekeepers—bring their perspectives. The way these groups communicate, negotiate priorities, and balance procedural concerns with clinical needs is central to how decisions are ultimately made.

The technology itself is another critical factor. Hospitals evaluate whether a new technology is compatible with existing technological platforms, whether it brings clear advantages over the current practices, and whether it represents just a marginal improvement or a radical shift. Radical innovations are more likely to demand more training, more changes in the established processes, and create more uncertainty. These factors may slow down the technology adoption processes, while incremental changes may be more readily embraced.

Finally, the external environment exerts significant pressure. The positioning of the hospital in the competitive framework, compliance with regional or federal regulations, and affiliation with a hospital network all push hospitals to respond to what their peers are adopting. Institutions are constantly weighing how new technologies affect their reputation, their ability to attract patients, and their compliance with evolving policy frameworks.

7.8.1.8 Other Observations

The "elephant in the room" in the adoption predictive efforts is often represented by the perspective of receiving reimbursement from the government or from private insurers.

A key contextual element that poses a challenge to reimbursement is based on the widespread assumption that medical innovation is responsible for constant increases in the cost of healthcare services. In the USA, for more than 30 years, medical innovators have fought against an understandable but sensibly biased conviction that "the nation's commitment to health cost control will potentially conflict with its deeper commitment to innovation in medical technology—drugs, medical devices, diagnostic, and therapeutic procedures" [..]" (R. A. Rettig, 2017).

The core reimbursement questions ("if," "how much," and "when") are derived from the recurrent interactions of multiple stakeholders, often driven by competing priorities and goals. The cluster of status quo fit dealt with in Section 7.5. Provides additional guidance in navigating this complexity and in predicting the aggregated behavior of the involved stakeholders and the most likely purchasing orientations.

7.8.2 Using Previous Clusters to Build P&L Projections

This section uses an example to explain how revenues can be predicted based on assumptions and decisions made in the previous eight clusters of the K9 map. Figure 7.63 shows how information that is already refined and tested in the earlier eight clusters can be used to predict the actual revenues. This chart refers to a generic disposable device routinely used by patients (for instance, a blood glucose sticker for routine glucose-level measurements). The chart's color-coding helps visually and quickly identify the K9 map's cluster from which each predictive signal is derived.

The chart clearly shows how predicting disposable demand requires the results of predictions that have already been performed in upstream clusters of the K9 Map. In this example, revenues can be calculated as a function of the demand for disposables

Fig. 7.63 This map illustrates the importance of utilizing information (signals) derived from assumptions and decisions made in previous clusters to more rationally predict future revenues

and of the average selling price (ASP), which is a key parameter that has already been analyzed and predicted in previous clusters, such as those concerned with competitiveness (competitiveness), positioning statement (value proposition attractiveness), and fine-tuned positioning in target segments (attainability of target segments). The chart also suggests that once the size of the population of users is estimated, the forecaster must account for several usage determinants, such as compliance (how frequently it will actually be used compared to the prescription) and persistence (how long it will be used compared to the prescription). Not all patients using the new device will use it as regularly as prescribed (compliance), or they may continue to use it indefinitely (persistence). Compliance and persistence strongly depend on a long list of other parameters that have been examined in previous clusters. For instance, actual usage is often affected by the usability of the device, its reliability, its safety, and the venture's discipline in delivering regular training to patients on the proper use. Usage is also dependent on the physician's precise prescription and instructions provided to patients, as well as on the patient's and the caregiver's attitudes toward the given technology. This example helps us understand the importance of collecting, refining, and testing information in an orderly manner across the decisions and assumptions required by the clusters of the K9 map.

Figure 7.64 shows how to generalize the idea of linking the main assumptions from all other clusters in the K9 map to the prediction of key determinants of P&L

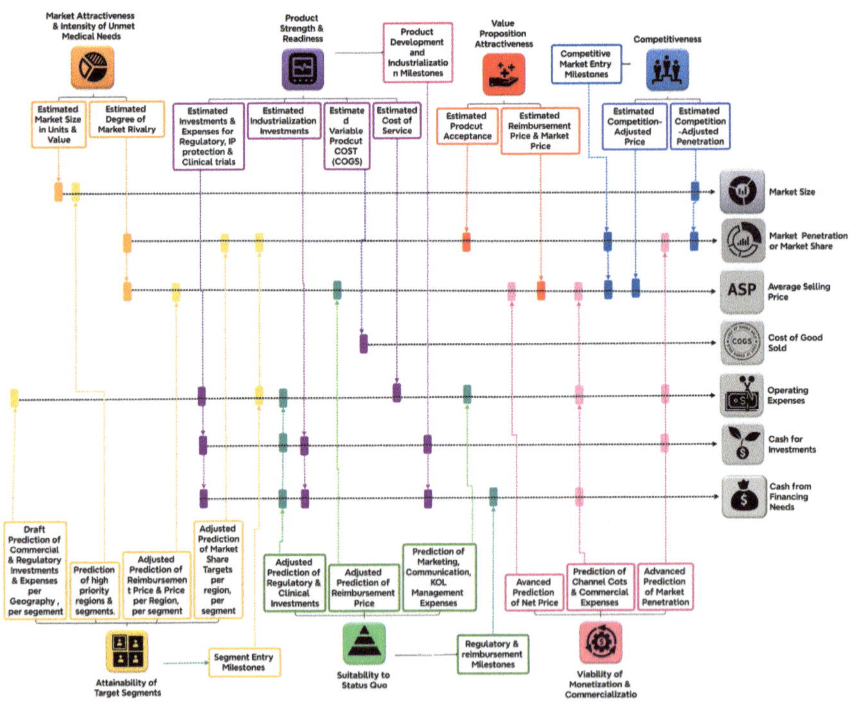

Fig. 7.64 This chart shows how to generalize the idea of linking the main assumptions and decisions coming from all clusters to support P&L and cash flow predictions

and cash flow. It explicitly highlights some essential relationships between the assumptions made while focusing on other clusters that affect key cash flow determinants. It shows which clusters best contribute to predicting which of the seven core financial inputs: cash from investments, ASP, market size, market penetration, cost of goods sold (COGS), cash from financing, and operating expenses.

For instance, the cluster related to the viability of monetization and commercialization offers fundamental inputs to the P&L prediction—for example, the predicted cost of the commercialization channels (which affects the post-market operating expenses) and a more precise estimation of the future average selling price (which affects the actual profitability). The cluster related to competitiveness offers obvious insights into the prediction of future market shares. Additionally, the cluster related to value proposition attractiveness provides relevant information on the sustainability of both market share and pricing. The cluster concerned with the attainability of target segments (yellow) offers additional predictive signals to the expectation of share and pricing in the context of specific segments and positioning.

In short, visual aids like the one represented in Fig. 7.64 help medical entrepreneurs to quickly tie any change in business assumptions and/or in business decisions to their impact on the financial value of a medical technology innovation. The goal of this approach is not to encourage the construction of an ultra-complex startup value model, potentially managed by an AI platform. The goal is to encourage medical entrepreneurs to rationally weigh the importance and value of each business assumption and decision, in light of their reflection on the value of their medical inventions. This posture should naturally push medical entrepreneurs to actively seek, organize, and analyze each piece of information coming from the market, from stakeholders such as users, sales reps, KOLs, and patients. This effort should also encourage entrepreneurs to use the most rational, refined, and proven approaches to process this information (or signals) when making assumptions and, therefore, making business decisions—approaches such as the scientific one.

7.8.3 Statistical Nature of the Financial Model

The disciplined use of the scientific approach for decision-making, which is applied throughout the recursive use of the K9 map, naturally provides high-quality inputs to the valuation methods based on predicting future revenues and cash flow. Figure 7.65 simply highlights how the prediction of P&L determinants, such as sold volumes and price, is central to this effort. When a scientific approach is applied, each of the relevant P&L predictive inputs is, in turn, provided by the related decisions and assumptions made over time. When these inputs have been obtained using a scientific approach, they most probably do not consist of single expected values but, instead, of their statistical distributions holding respective mean values (Fig. 7.66).

By employing simple statistical packages for the Monte Carlo analysis, these statistical inputs allow the prediction of the distributions of NPV, IRR, payback period, and their mean values. The study of these distributions will enable investors and entrepreneurs to assess the investment's fitness to their risk aversion or propensity. From an entrepreneur's point of view, it is an unmissable occasion to evaluate

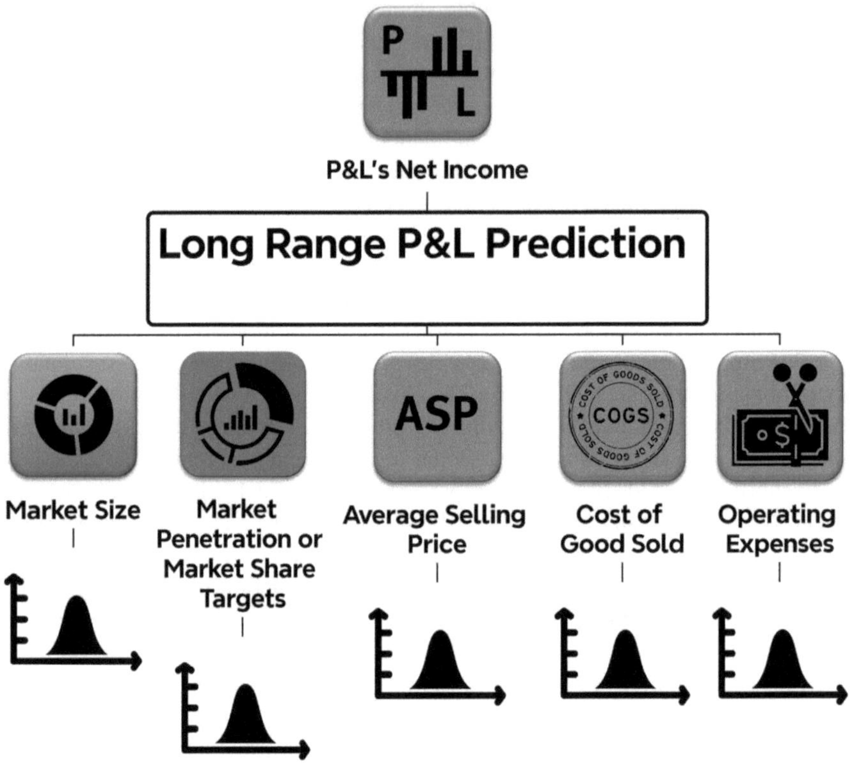

Fig. 7.65 If the main variables driving the prediction of the P&L are represented by statistical distributions, then EBITDA and each other line of the P&L will be represented by a statistical distribution

whether their innovative ideas can support return levels adequate to the high-risk profile of their medical ventures. Investors will not be confronted with a predicted (Internal Rate of Return) IRR but with a distribution of expected values of IRRs centered on the expected mean value. The investor will, hypothetically, read that there is a 70% probability of achieving a 50% or higher IRR or a 30% probability of achieving a 65% of higher IRR.

Rational investors choose projects based not only on returns but also on risks. The best projects tend to be those with the best returns, subject to some known and specified risks (Mun, 2004).

Investors are, therefore, likely to make decisions by referring to their respective risk-return profiles, which can be summarized in an indifference curve. Figure 7.67 represents an example of an indifference curve, which plots various combinations of risk-return pairs that an investor will accept to maintain a certain level of utility. If the combinations of risk and returns on a curve provide the same level of utility, the investor will be indifferent to choosing one. For a medical technology startup, it is essential to have both an estimation of the risk-return connected with its innovation and of the risk-return goals of all the involved potential investors.

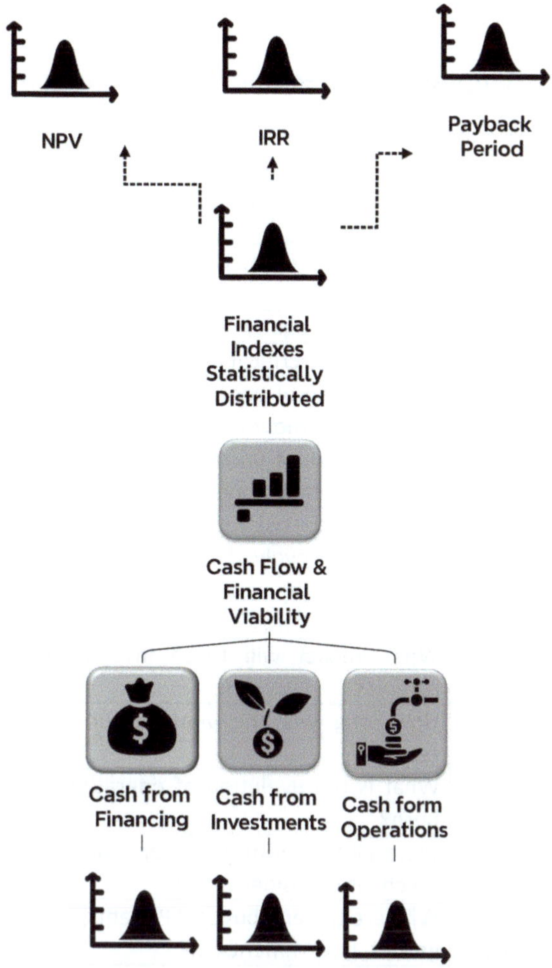

Fig. 7.66 If the main variables driving the prediction of the cash flow are represented by statistical distributions, then also NPVs, IRRs, and Payback Periods can be considered and approximated as statistical distributions

Fig. 7.67 An example of Risk-Return indifference curve

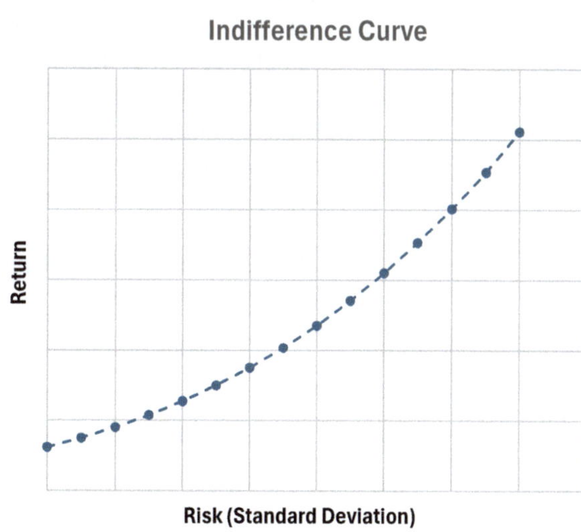

7.8.4 Key Assumptions and Decision-Related Questions (Table 7.19)

Table 7.19 Financial viability: core assumptions and decisions

Financial Viability	
Cluster	**Core Assumptions and Decisions to Be Made**
Cash from financing	How much cash will the startup need to reach breakeven?
	What mix of equity vs. debt financing is sought?
	What are the critical financing milestones?
	What degree of uncertainty characterizes the expected investment amounts?
	What degree of uncertainty characterizes the expected investment milestones?
	Is the financial risk-return compatible with the risk-return goals of the involved investors?
Investments	What are the amounts and timing of significant investments?
	What impact will these investments have on cash reserves?
	How long will the venture run with depleted cash reserves?
Operating expenses	What is the evolution of operating expenses and cash burn?
	Were all the K9 map clusters concerned with expenses taken into consideration?
Adoption rates and market penetration	Who are the end buyers? Patients, Hospitals, purchasing groups, governments.....?
	What is the decision-making process of the end-buyer?
	How well-suited is the process for adopting our new technology?
	If the end-buyer and user are individual patients, does the device's usability profile support sustained compliance? What usability tests have been successfully passed?
	If the end-buyer is a hospital, have all of the decision-making influencing forces been taken into account? What tests have been performed to support the related assumptions?
	What is the expected peak penetration rate (peak share) in each target segment?

(continued)

Table 7.19 (continued)

	What is the expected peak penetration rate in the target regions?
	What are the penetration dynamics in each of the target segments?
	How much of the product growth will come from total market growth?
	How much of the product growth will come from share gain?
	Which competitor will lose the most of its share due to the entry of the new innovative product?
	Which competitors will be impacted less? And why?
	How can the assumptions about competitive shrinking shares be tested?
	What is the impact of the identified future market entrants, and how will they affect the dynamic of market share redistribution?
	How does the HTA proposal perform against other competitors' HTA proposals? What assumption can be made? How can it be tested?
Revenues	What is the Average Selling Price (ASP) evolution?
	What is the evolution of the volume of sold units?
Gross margin	What is the forecasted gross margin? Which elements impact gross margin the most?
	What is the evolution and elasticity of the COGS?
Cash Valleys	When are the forecasted periods of low-cash reserves?
	How long are these low-cash reserves periods predicted to last?
	What contingency plans are in place in case of prolonged periods of low cash reserves?

7.8.5 The Complete Map (Fig. 7.68)

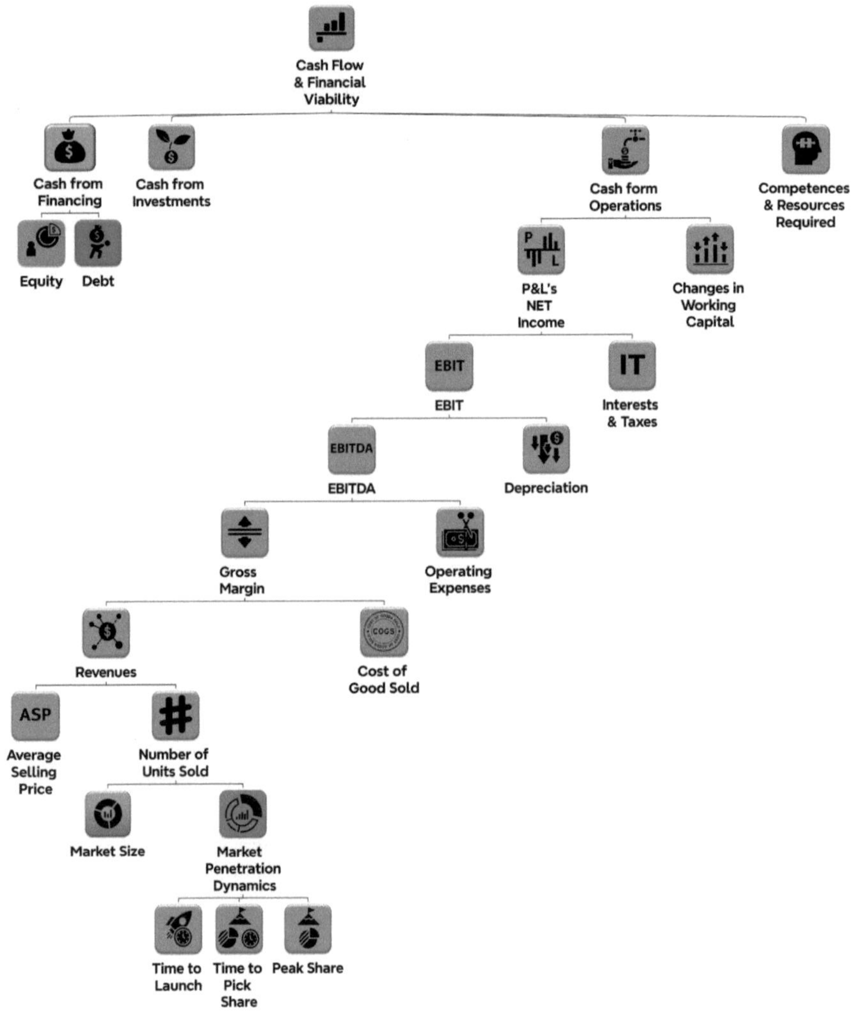

Fig. 7.68 The complete set of problem clusters impacting the prediction of financial viability

7.9 Valuation, Investment Return, and Risk

Nicola Redi
nicola@obloo.vc
Obloo Capital, Milan, Italy
MIB Trieste School of Management, Trieste, Italy

Value is broadly defined as how much something is worth in money or other goods for which it can be exchanged. When it comes to financial assets or liabilities, we refer to the concept of Fair Value: the amount that would be obtained upon the sale of an asset, or the amount that would be paid to transfer a liability, in a structured transaction conducted between knowledgeable and willing market participants at a specified measurement date. (IFRS Foundation, 2011). Assets, therefore, have a value in the presence of a market where buyers and sellers agree on a specific price; using IFRS' words, "Fair value is a market-based measurement, not an entity-specific measurement." When approaching the problem of valuing a company, its shareholders should accept the principle that, in the absence of a buyer offering a price, their asset has no market value. Such a simple yet extreme concept is often rejected by start-up founders who have a significant emotional attachment to their entrepreneurial venture and seldom associate its value with the willingness of third parties to pay for a share of their company.

7.9.1 What the Market Tells Us About Valuations

Valuations are determined by market dynamics and fluctuate accordingly. Pitchbook US pre-money valuation data (PitchBook Report, 2024b) summarized in Fig. 7.69, show how every year the statistical distributions of values are wide and very asymmetric, with average values often above the top quartile, while year-on-year variations are significant (i.e., the highest valuations, top 10% in 2024 is five times higher than the respective top 10% of 2014).

These data confirm that the value recognized to the owners of an early-stage start-up company is significantly driven by market variables, rather than by intrinsic

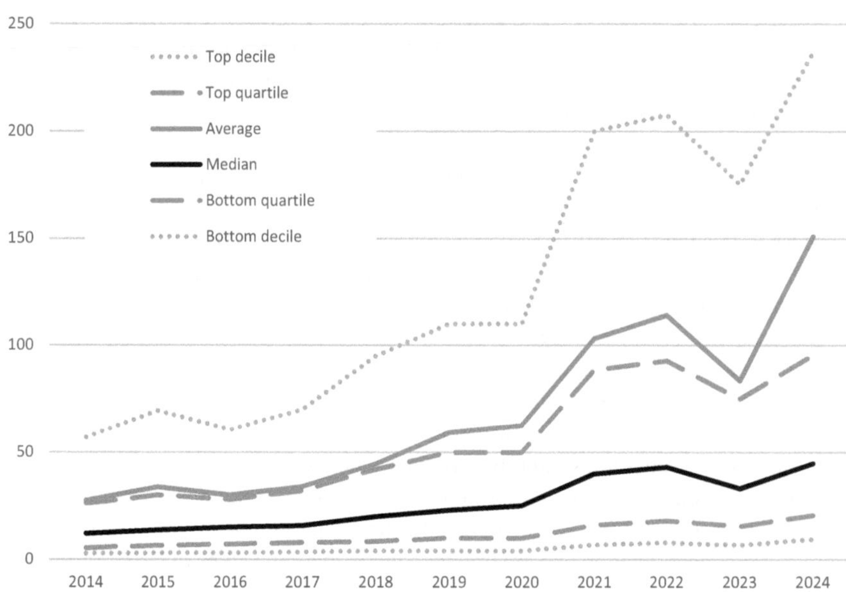

Fig. 7.69 Pitchbook US pre-money valuation data from 2014 to 2024

Table 7.20 Start-up companies have a lifecycle that can be divided into different phases, each determined by specific activities

Phase	Main activities
Pre-seed	Business idea validation, market research, and prototype development
Seed	Development of a core team, refining the product or service, and achieving early market entry
Early-stage	Complete initial market entry and product-market fit
Late-stage	Scale operations, expand markets, or prepare for exit
Venture growth	Additional scaling, new product launches, international expansion, acquisitions, or final IPO preparations

asset potential. As highlighted in Table 7.20 (Hillenbrand & Stafford, 2025; Kraemer-Eis & Lang, 2018), start-up companies have a lifecycle that can be divided into different phases, each determined by specific activities:

In the event of an investment, two valuations are to be considered: pre-money and post-money (Cyr, 2022). Pre-money is the price per share of a company multiplied by the total number of shares prior to the financing round. At the same time, post-money is the price per share of a company multiplied by the total number of shares after the financing round. A common rule of thumb calculation is that:

$$\text{Post} - \text{money} = \text{Pre} - \text{money} + \text{Investment Size}$$

Which leads to this simple assessment of the percentage of shares issued and acquired in the investment round:

$$\%\text{issued shares} = (\text{Investment Size}) / \text{Post} - \text{money}$$

Pre-money valuation of companies increases significantly, moving from one phase into the next one, as can be seen in Fig. 7.70, based on (PitchBook Report, 2024b) data; while the increase in median values from the earliest stage to the growth one is 42×, average values soar 151×.

A similar trend, both in terms of growth and variance between median and average values, can be noted when analyzing the size of the investments in each phase (Fig. 7.71).

Market data helps identify the typical amounts invested in the different phases of startups (Fig. 7.71) which are in constant need of capital to support their growth. Furthermore, it is not research and development activities that absorb the majority of financial resources; growing in global markets requires investments up to 10 times higher than those required for product engineering. Market data also suggest that the higher the cash need, the higher the company valuation, mitigating the dilution effect of founders and early investors.

In fact, while valuation soared for all phases between 2014 and 2024, so did the size of investment rounds, resulting in an almost stable, if not decreasing, percentage of shares acquired in each investment round. The graphs reported in Fig. 7.72 represent this phenomenon and are obtained using median values of both pre-money and investment size.

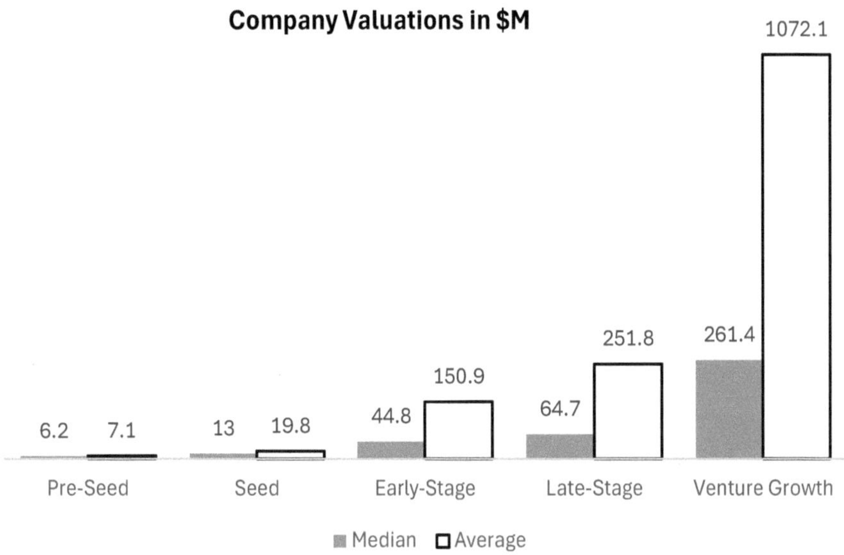

Fig. 7.70 Pre-money valuation of companies increases significantly, moving from one phase into the next one. Chart elaborated from (PitchBook Report, 2024b)

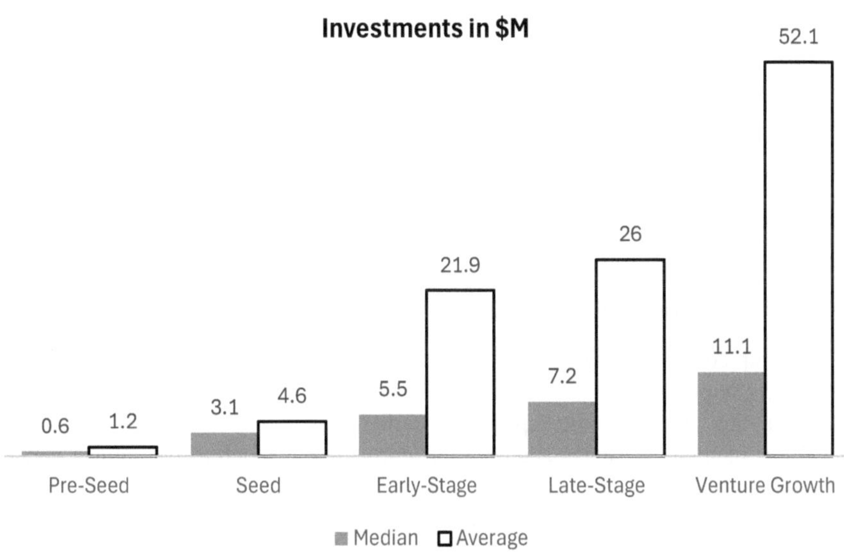

Fig. 7.71 Mean and median investment values in startups increase significantly, moving from one phase into the next one. Chart elaborated from (PitchBook Report, 2024b)

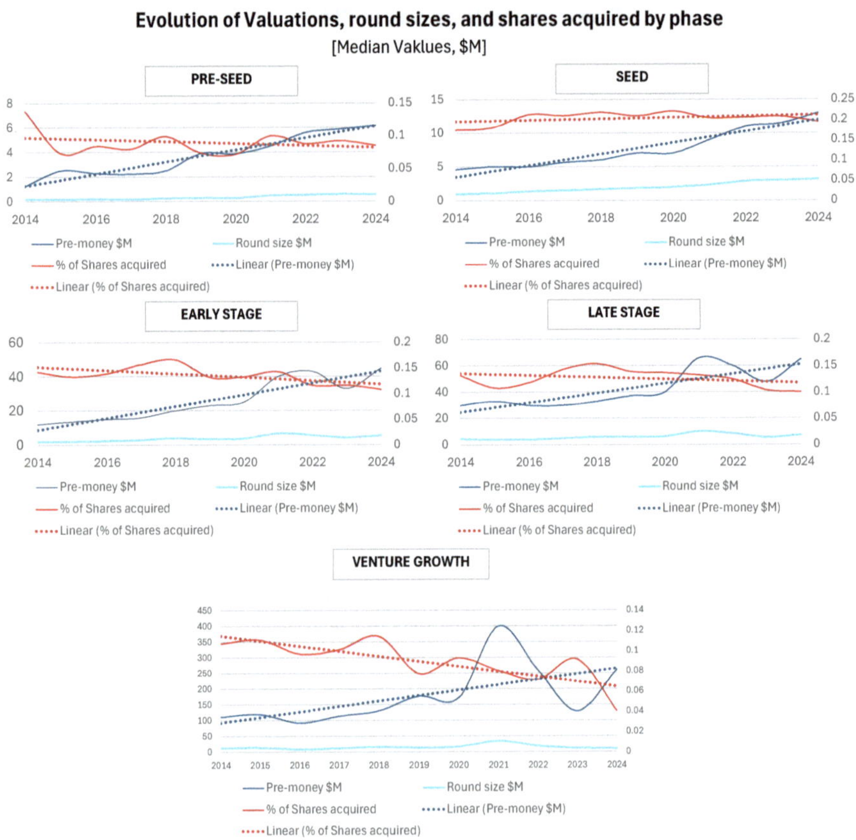

Fig. 7.72 Evolution of valuation, round sizes, and shares acquired by phase from 2014 to 2024

Analyzing these market data, and bearing in mind the definition of Fair Value, we can infer some principles:

- Start-up companies significantly increase their value while moving from earlier to later stages
- Later stages require more capital than earlier ones
- The percentage of shares acquired has not varied as much as pre-money valuations in the last 10 years
- The increase in round sizes mitigated the effect of pre-money valuations' growth on the percentage of shares acquired

7.9.2 Think About Value, Not Percentage

A common bias among entrepreneurs is to think in terms of percentage of shares and not in terms of value. This comes from the idea that the higher the percentage they own, the greater the expected return and the influence in the governance; both of these assumptions are biased. The reason for this is that shares are not equal (Ormsby, 2025): Founders and employees are usually entitled to "common shares" while investors require "preferred shares," divided into classes, each representing the specific investment round an investor joins. This concept is clarified in clearer terms by using an example: company Alfa is closing a Series B round, issuing B-series preferred stock in addition to common shares, Seed preferred shares, and A-series preferred shares. Each class of shares owes specific governance and financial rights following the market rule that the latest class of preferred shares has higher seniority over the other classes. One of the most common right every class of preferred shares has over the others is called "liquidation preference," meaning that related shareholders will be entitled to be liquidated in advance, up to a defined multiple of their investment (usually equivalent to 1 time, but in some cases well above it), following the same working principle of a senior repayment obligation in a loan agreement (Bartlett, 2025).

In this example of company Alfa, assuming the share types and price listed in Table 7.21.

- The distribution of proceeds will follow this scheme
- First, up to $15 M, to series B investors
- Second, up to $5 M of the remaining proceeds, to series A investors
- Third, up to $2.5 M of the remaining proceeds, to Seed investors
- Last, any remaining proceeds to be distributed proportionally to the percentage of shares owned by each shareholder.

Common shareholders will therefore receive a financial return only after $22.5 M have already been distributed to preferred shareholders in a proportion of 33%. In case Alfa is acquired entirely with a valuation of $100 M, common shareholders will receive 33% of $75 M ($100 M − $25 M) equal to $25 M; $8 M less than in case of absence of any liquidation preference clause.

The power of the effects of liquidation preference clauses with multiples exceeding 1× is as such that post-money face values are often not representative of the real value investors are recognizing to one company. A research performed on 15 unicorns (companies valued more than $1 billion in private transactions) demonstrated, assessing the pay out structure of the liquidation preference clauses, that the real fair value of such companies was almost half of their post-money value (Mirhaydari, 2017). In a broader and deeper analysis on 135 US unicorns, it was demonstrated that post-money valuations were on average 50% above fair value, when calculating

Table 7.21 An example of sequential emission and purchase of both common and preferred shares

Class of shares	Issued shares [$]	% of shares	Price per share[a] [$]
B-series preferred	10,000	33.33%	1500
A-series preferred	5000	16.67%	1000
Seed preferred	5000	16.67%	500
Common	10,000	33.33%	250
Total	30,000	100.00%	

[a]Paid by investors at the moment of their financing round

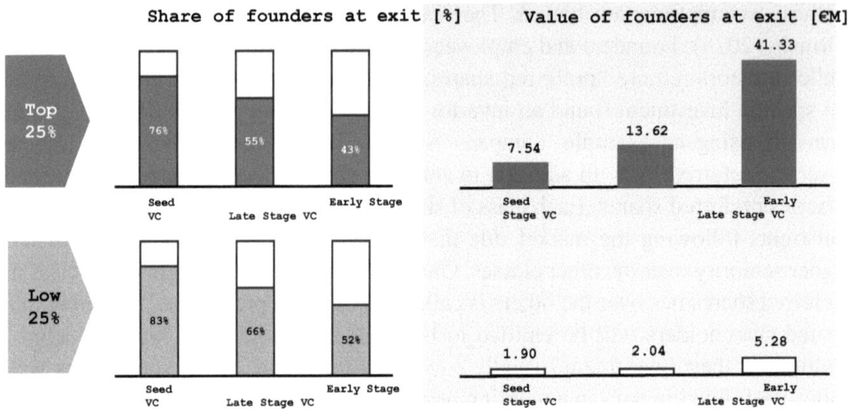

Fig. 7.73 Top quartile performers (those with the highest exit value) raised almost 3× the capital of bottom quartile ending up with lower shares in the company. Obtained from an analysis of (PitchBook Report, 2024a)

the value of each class of shares based on their different rights and seniority; as a result, 65 out 135 companies should not have been considered as unicorns (Gornall & Strebulaev, 2020).

While it is not the objective of this chapter to provide an exhaustive illustration of the mechanism of preferred stock rights, this example confirms the statement that founders and investors should think in terms of value, rather than the number of shares owned in the company. This is what was called as "The Founder's Dilemma" (Wasserman, 2008), as founders have to deal with a major trade-off: be "King," and keep the control of their companies, or get "Rich"; very few had the opportunity to be both of them. This study, performed on 212 American start-ups between late 1990s and early 2000s demonstrated that most founders left the control of their companies long before they became public. Analyzing European data (PitchBook Report, 2024a), we discovered (Fig. 7.73) that top quartile performers (those with the highest exit value) raised almost 3× the capital of bottom quartile (€76 M vs. €23 M) ending up with lower shares in the company (43% vs. 52%) which value was however more than 8× higher (€41 M vs. €5 M).

The Founder's Dilemma is well known among investors, and the market standard is not to excessively dilute founders in every round of financing. As we previously demonstrated in our analysis of median market data, the percentage of shares acquired by investors remained relatively stable between 2014 and 2024, despite a significant increase in both pre-money valuations and round sizes. Liquidation preference rights are therefore the balance investors use to mitigate the financial effects of valuation and size growth.

Market analysis, along with this brief explanation of preferred stock rights and the Founder's Dilemma, confirms that, according to the original definition of Fair Value, the value of a start-up company is the price of its equity assets. This value is determined not only by the firm's performance and potential but also by subjective value drivers linked to investment clauses and market dynamics.

7.9.3 Creating Value for Investors

An investment is the commitment of current resources in order to achieve higher returns in the future (Capital Market Authority, 2024). As the future always brings a degree of uncertainty, investing is an activity closely tied to the concept of risk. The goal of investors is to select the best opportunity among various investments, choosing the one that maximizes returns relative to risk.

This chapter is not intended to be an essay on venture capital principles; however, some basic information must be introduced to make the further dissertation on valuations easier to understand. Venture capital firms are financial intermediaries that manage capital on behalf of third parties and invest it in start-up companies. These intermediaries are called "General Partners" (GP), while "Limited Partners" (LP) are those who provide the capital to be invested; funds are organized as "Limited Partnerships." Venture capital funds do not offer any guarantee of return to their LPs and to have a proper alignment of interests, LPs provide GPs the opportunity to have a high percentage of fund's returns (usually in the range of 20% of total returns—called "Carried Interest" or "Carry"), in case of fund's financial performance exceeds a pre-defined threshold, called "Hurdle Rate" (Metrick & Yasuda, 2021). The idea of Limited Partnerships, with high financial incentives for their GP is not new, and was established for another very risky business in the early XIX century; that business was whaling. Hunting whales was a perilous venture, requiring highly skilled sailors to risk their lives while leaving their homes for months to navigate the most challenging waters. Industrial entrepreneurs, selling any by-product of whales, selected the most talented captains and their crew, provided them with all the required assets and financial support to run a whaling campaign, while offering a significant slice of the future returns (The Economist, 2015).

Creating value for investors means setting up a price for the shares in a company that provides a GP the opportunity to maximize its returns above the Hurdle Rate and access Carried Interest. So, how is Hurdle Rate determined? The market standard sees Hurdle Rate as an Internal Return Rate (IRR) of a fund equal to 8% (BVCA, 2024), where IRR is the discount rate of a series of cash inflows and outflows of an investment in time that equalizes their Net Present Value (NPV) to zero (Metrick & Yasuda, 2021). IRR is among the most commonly used metrics for comparing different investments with significantly different cash flows.

Let us assume two investments, A and B. Both require an initial cash outflow of \$10 M, while A returns \$30 M on year 3 and B returns \$50 M on year 5 (Fig. 7.74).

At first sight, B seems better than A, returning 5 times the initial investment, but when introducing the concept of IRR, time is of the essence: the longer an investment takes to provide a return, the higher its risk and, therefore, the lower is the actual value of future returns.

When assessing the value of an investment, the amount of future returns must be always considered together with their expected timing. In case of venture capital funds, the cash flows to be considered are not those generated by their target companies and projected in their cash flow statements; what only matters to investors

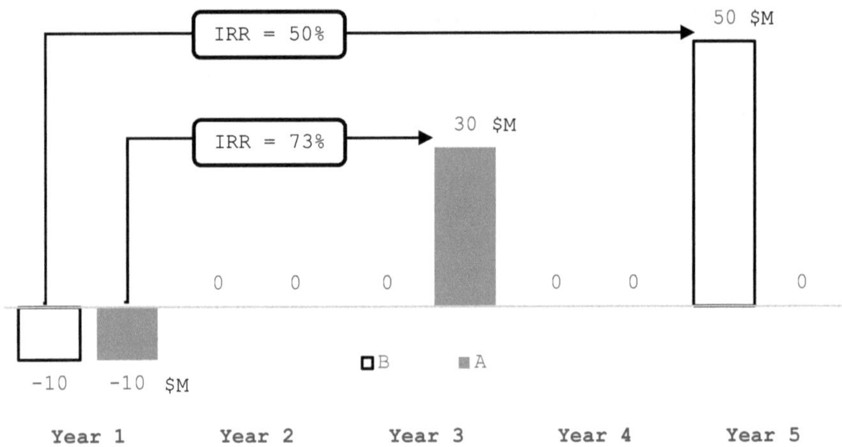

Fig. 7.74 Example of two companies receiving the same investment amount concurrently, but with different ending values at different times

are amounts and timings of their capital injections in a company and capital returns obtainable through de-investments (also known as "Exits"). This requires a significant shift in perspective when assessing a company's value.

7.9.4 Valuation Methodology

A variety of valuation methodologies are available; each has specific strengths and limitations, which should be known before applying it (IPEV, 2022). More broadly, valuating a company is a process, where the determination of the enterprise value using valuation techniques represents the first step only, to be followed by adjustments in value considering:

- Market perspectives such as surplus assets, excess liabilities, liquidity of transactions (i.e., private vs. public company)
- Seniority of share classes to be held and presence of other financial instruments that might dilute the fund's investment (i.e., stock option plans)
- Any other contractual and/or financial item affecting the value of shares acquired by the investor.

The most common valuation techniques used in corporate finance are based on present and/or prospective economic and financial performance of the target company.

Discounted Cash Flows (DCF) is the most used, often in isolation of other techniques (Koller et al., 2010). DCF discounts all cash flows generated by the firm to shareholders, using the Weighted Average Cost of Capital (WACC) as a discount factor. Mathematically, the value of a company's equity is defined as follows:

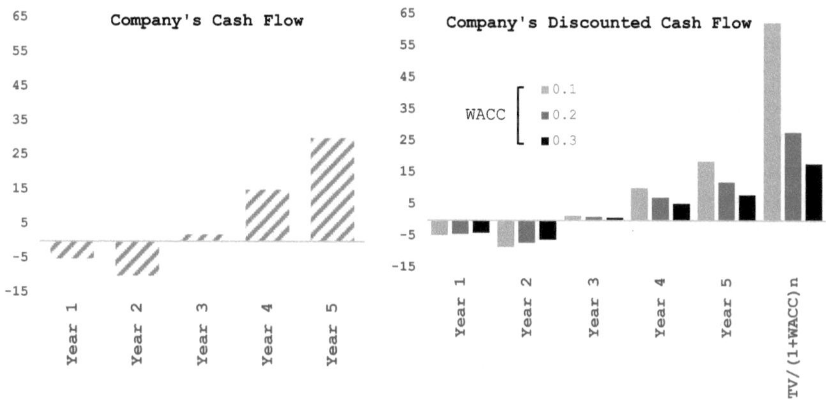

Fig. 7.75 Example of a company's cash flow (left side) modulated by three different levels of WACC (right side)

Table 7.22 Terminal values, together with WACC, represent the primary drivers of the company's valuation

WACC	Value ($M)	TV/Value (%)
10%	80.1	78%
20%	37.1	75%
30%	22.3	80%

$$\text{Value} = \sum_{t=1}^{n} \frac{\text{CF}_t}{(1+\text{WACC})^t} + \frac{\text{TV}}{(1+\text{WACC})^n}$$

where CF_t are the cash flows generated (if negative, absorbed) by the firm in every year "t" of the plan and TV is the Terminal Value (or Perpetuity) representing the sum of all future, discounted cash flows, from year "n" onwards, assuming a yearly growth factor of "g":

$$\text{TV} = \frac{\text{CF}_{t+1}}{(\text{WACC} - g)}$$

WACC is calculated based on the ratio between equity and debt financing of the firm, adjusted for the risk of the company.

A simple example might help understand how this technique works. The left chart of Fig. 7.75 describes the 5-year cash flow statement of a company (see left chart). While running a DCF analysis for three different scenarios of WACC (right chart) we can immediately appreciate the impact of higher WACC values with respect to the firm's value and the weight of TV in the final valuation (see right graph). Table 7.22 represents the results of these three scenarios.

We immediately notice two main implications of DCF:

1. Small variations of WACC have a significant negative relation with the value calculated (in our case, nearly 4 times lower value, increasing WACC from 10% to 30%).
2. TV drives most of the total value.

This final implication requires further analysis: it indicates that the value of a company, calculated using the DCF method, is between 75% and 80%, determined by TV, which is essentially a simple mathematical assumption based on last year's cash flow and a hypothetical perpetual growth rate of the company. As we have seen in previous chapters, there is a high level of uncertainty even in short-term assumptions of a firm's business plan; the way TV is calculated represents an even greater stretch of such uncertainty.

Besides being a common method, DCF remains a difficult one when it comes to reliability. "While this Valuation Technique may be applied to businesses going through a period of significant change, such as a rescue refinancing, turnaround, strategic repositioning, loss-making, or in its start-up phase, there is a considerable risk in using this Valuation Technique." (IPEV, 2022).

Multiples are another widespread valuation technique. The most commonly used multiples are those based on earnings (M_E = Value/EBITDA) and revenues (M_R = Value/Revenues). The concept behind multiples is that the value of a company in a specific market is determined by multiplying its revenues or earnings times a particular multiple, which is determined by the industry, size, and geography of the company. This technique values the company based on its current performance and assumes that such performance has been stable in recent years. Identifying the correct multiple for a specific firm is not an easy task. Multiples vary widely, industry by industry, as shown in the graph displaying EBITDA multiples across various US industries (Fig. 7.76). Such dispersion of values also occurs in both the healthcare and life science industries (Fig. 7.77).

When it comes to start-up companies, identifying the appropriate industry to benchmark against is also not straightforward. For example, which industry represents a new company developing a robotic platform for home rehabilitation? Would it be the one related to robotics or medical devices? Or even to healthcare services, in case the company's business model is to provide the service itself and not just the machine? A similar case is that of companies developing in silico drug discovery platforms that leverage AI and computational chemistry, or those engaged in AI-enhanced diagnostics.

Multiples are therefore not easy to identify and require further adjustments based on a variety of factors, such as (IPEV, 2022):

- Private vs. public companies
- Investments in tangible and intangible assets with consequent depreciations
- Firm's liquidity level
- Range of products and reference markets
- etc.

Finally, a key requirement for using multiples is that the company demonstrates stable and positive financial performance. Applying a revenue multiple to a pre-revenue company results in a value of zero, while using an earnings multiple for a company with negative EBITDA would produce a negative company value.

We highlighted how DCF and Multiples are valuation techniques with significant limitations; both are built on the economic and financial performance, both present

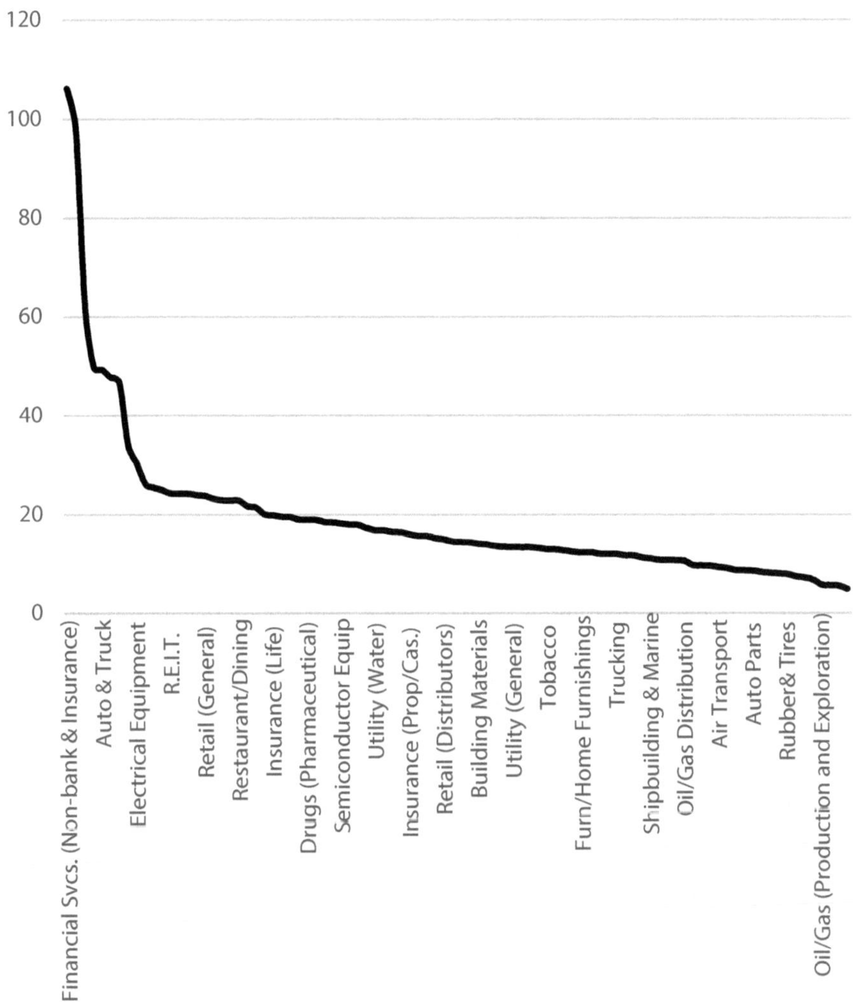

Fig. 7.76 Multiples vary widely across industries, as shown in this graph displaying EBITDA multiples for various US industries

and future, of the firm targeted for investment. DCF would mathematically fit for companies in very early stages and those with unstable cash flows, but the influence of WACC and the weight of TV recommend using it with extreme caution. Multiples do not apply for start-up companies and would mathematically lead to nonsensical valuations. In the previous paragraph, however, we highlighted how the value for an investor stays in the cash flows of the investment and exit phases. It is based on these cash flows that early-stage investors commonly use the "Venture Capital Method" (Metrick & Yasuda, 2021).

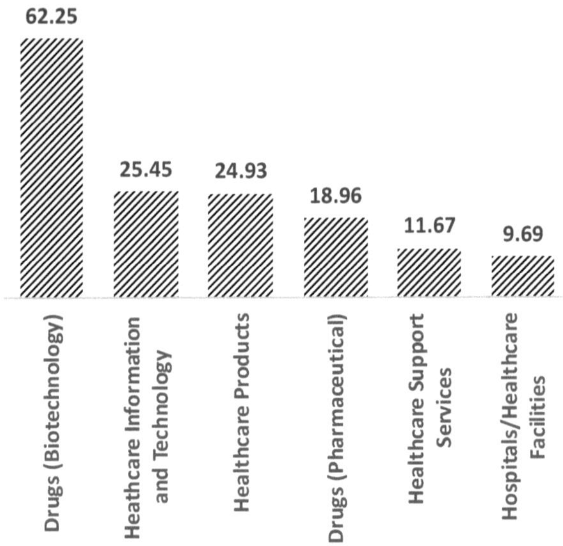

Fig. 7.77 Multiples vary significantly also across both the healthcare and life science industries

Venture Capital Method has a very simple principle: investors estimate both exit expected value and timing (positive cash flows), elaborate a target IRR and, based on the known investment rounds (negative cash flows), calculate the expected total value of their shares at the exit, the percentage of shares owned at the exit and therefore the price per share they are willing to pay while investing in the company. The following example illustrates the VC Method applied to a start-up company that has undergone only one round of investment. Company Alfa needs to raise \$10 M. Investors estimate an exit in 5 years with a valuation of \$144.64 M for 100% of the company and a target IRR of 50%. The target IRR is used to capitalize the amount invested up to the exit, Fig. 7.78 shows how capitalization works year by year:

$$\text{Expected Return} = \text{Investment}^{*}\left(1 + \text{Target IRR}\right)^{5}$$
$$= \$10\text{M} \times 1.5^{5} = \$50.625\text{M}$$

Therefore, if the Expected Exit Value is \$144.64 M, the percentage of shares the fund would need to have are:

$$\text{Shares\%@Exit} = \text{Expected Return} / \text{Expected Exit Value} = 35\%$$

If Alfa has a total of 6500 shares before the investment, in order to offer the investor an equivalent of 35% of the company after the investment, the firm issues 3500 new shares. The price per share paid upon the investment is:

$$\text{Price per share} = \text{Investment} / \text{New Shares} = \$2857.1429$$

Therefore, the value of the company before the investment (Pre-money) will be equal to:

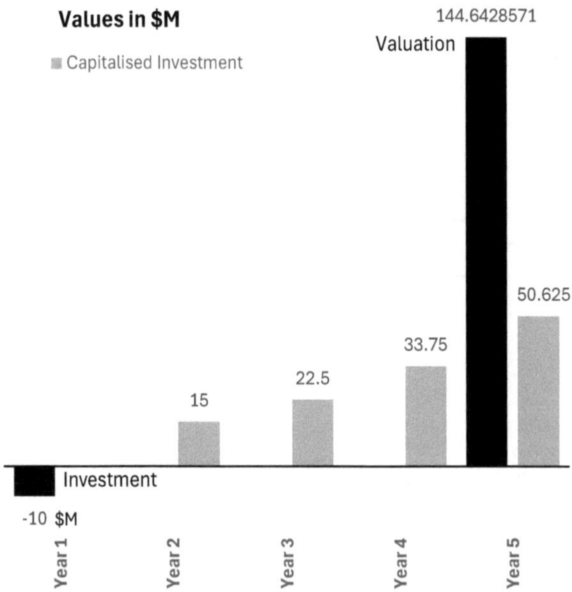

Fig. 7.78 This chart refers to the example about the VC method applied to a start-up company that has undergone only one round of investment

$$\text{Pre} - \text{money} = \text{Price per Share} \times \text{nr of Shares}\left(\text{before the new issue}\right) = \$18.57\text{M}$$

Two steps of this method require a dedicated review: (a) evaluating the exit value and (b) assessing the target IRR.

(a) Evaluating the Exit Value can be done mainly through two methodologies. If the company is expected to have consistent and stable revenues and margins at exit, multiples may be used. Otherwise, for instance, for drug development start-ups which usually exit at the end of Phase II clinical trials, value can be estimated by looking at comparable market transactions in the same therapeutic area. A reasonable approach would be identifying 5–10 comparable companies in the variety of sectors the company would fit, followed by choosing the median multiple value (Gray, 2025).
(b) IRR can be assessed based on a risk versus total portfolio target return. For instance, fund A has a total target IRR of 20%: assuming that the write-off rate of companies with a risk level close to the one in the current target is 50%, what should be the expected IRR of the target investment to ensure the sought fund's IRR of 20%?

Table 7.23 represents the hypothetical portfolio of Fund A: 10 companies have been invested in with $1 M each. After 5 years, just 50% of them will be positively exited. As the target IRR of Fund A is 20%, the total return to Fund A in 5 years should be:

Table 7.23 Hypothetical portfolio of Fund A in the current example

	Companies (nr)	Cash flow TgT ($M)	Cash flow TgT per surviving company ($M/nr)
Investment	10	(10)	(1)
Exit (5 years)	5	24.88	4.98

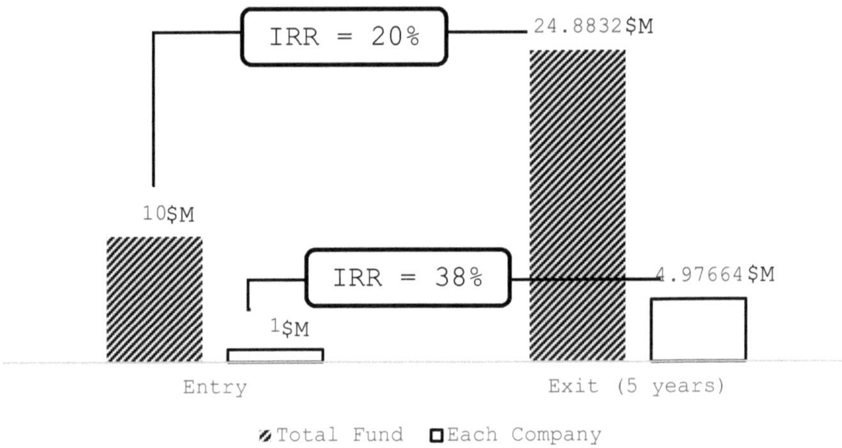

Fig. 7.79 The chart represents the overall fund's IRR and the average IRR of the 5 companies that could successfully exit ($M)

$$\text{Total Fund A Return} = \$10\text{M}^{*}\left(1+20\%\right)^{5} = \$24.88\text{M}$$

However, only five companies would be returning the investment, each with $4.98 M on average, representing a 38% IRR as described in Fig. 7.79.

The most complex assumption in calculating the target IRR of the target company is assessing the risk of default of that company. Once such a risk is clear, calculating the sought IRR, through a simple portfolio simulation (see figure above and the formula in the following paragraph), is straightforward and represents a reasonable trade-off between calculation complexity and reliability of the result.

7.9.5 Assessing the Risk of Default and Target IRR

Shareholders in any company run risks that are divided into two main categories: systematic and unsystematic risk, also known as firm-specific and market-specific risk (Berk, 2007). Firm-specific risk is driven by the company's performance and news (technical, economic, financial, ESG-related, reputational, etc.), while market-specific risk follows the fluctuations of market conditions (for public companies, this is mainly the fluctuations of the reference stock market). While public companies typically have a relatively long and consistent time series of data, both firm- and market-specific, and follow strict rules regarding transparency, private companies can seldom be analyzed using available secondary data. In the case of private start-up companies, even accessing their data rooms does not guarantee a significant set of quantitative information for risk calculation. K9 map represents a powerful tool for assessing the financial risk associated with a company and introducing it into the calculation of its value. More precisely, the first seven clusters of the map offer a compelling checklist for a qualitative, yet structured, risk valuation using a process similar to a Failure Mode and Effects Analysis (FMEA). FMEA was first introduced in 1963 by NASA as a design methodology to integrate reliability engineering into the development process. It is now widely used in several industries, including medical devices, engineering, and certification. During the last decades, its use for examining corporate failures has also been discussed by several studies (S. Balcaen, 2006) and recently proposed as a tool for financial audit (Askary et al., 2025). FMEA analyses risk through three dimensions:

1. The **severity** of a potential failure, defining the gravity of its impact in case it occurs
2. The expected **frequency** of such failure: the higher, the worst
3. The **ability to detect** the failure in advance and implement corrective actions in order to avoid it: the more it hides, the higher the risk

As “failure” we refer to something that produces a negative effect on any stakeholder involved inside or outside of the company, without prejudice on its relevance: its effects can span from a simple disappointment of a customer or employee to a threat for the health of a person or the continuity of the business. Using the first seven clusters of the K9 map and their detailed maps, entrepreneurs can identify all potential failures that would have an impact on the economic and financial performance of the firm. Once all potential failures are identified and listed, each of them is analyzed in terms of severity, frequency, and detectability. In a team exercise, entrepreneurs and their associates should provide a score from 1 to 10 to each dimension of the risk, where 1 is the lower risk score and 10 is the higher one. Once this exercise is completed, each failure is assigned a Risk Priority Number (RPN), defined as the product of the scores of the three risk dimensions:

$$\text{RPN}\left[1-1000\right] = \text{severity}\left[1-10\right] * \text{frequency}\left[1-10\right] * \text{detectability}\left[1-10\right]$$

Table 7.24 illustrates an example of the finished exercise to the first seven clusters of the K9 map, which represent firm-related and market-related risk variables. Besides its use for financial risk assessment, it represents a vital knowledge asset for the entrepreneurial team who becomes further aware of the potential impact on their business results of prospective failures. It defines a direction for implementing countermeasures. The very objective of an engineering FMEA is, in fact, to embed risk mitigation in the design phase of any complex product.

Once all seven clusters have been analyzed, the team should take the average value of the three highest RPNs for each cluster and use it as the cluster’s RPN. Later, the median value of all clusters’ RPNs is to be calculated (Table 7.25). While

Table 7.24 How to implement the Failure Mode and Effects Analysis using the K9 map

	Clusters	FMEA							
1	Market attractiveness and severity of unmet medical needs	Failure	Severity		Frequency		Detectability		RPN
		A1	Describe	4	Describe	3	Describe	1	12
		B1	Describe	2	Describe	5	Describe	3	30
		C1	Describe	7	Describe	9	Describe	9	567
2	Product strength and readiness	Failure	Severity		Frequency		Detectability		RPN
		A2	Describe	2	Describe	3	Describe	1	6
		B2	Describe	2	Describe	5	Describe	3	30
		C2	Describe	7	Describe	9	Describe	1	63
3	Value proposition attractiveness	Failure	Severity		Frequency		Detectability		RPN
		A3	Describe	4	Describe	3	Describe	1	12
		B3	Describe	4	Describe	10	Describe	3	120
		C3	Describe	5	Describe	5	Describe	5	225
4	Competitiveness	Failure	Severity		Frequency		Detectability		RPN
		A4	Describe	1	Describe	3	Describe	1	3
		B4	Describe	2	Describe	5	Describe	3	30
		C4	Describe	9	Describe	9	Describe	9	729
5	Suitability to the status quo	Failure	Severity		Frequency		Detectability		RPN
		A5	Describe	7	Describe	9	Describe	9	567
		B5	Describe	2	Describe	5	Describe	3	30
		C5	Describe	4	Describe	3	Describe	1	12
6	Attainability of the target segments.	Failure	Severity		Frequency		Detectability		RPN
		A6	Describe	2	Describe	5	Describe	3	30
		B6	Describe	2	Describe	10	Describe	3	60
		C6	Describe	7	Describe	9	Describe	9	567
7	Viability of monetization and commercialization	Failure	Severity		Frequency		Detectability		RPN
		A7	Describe	8	Describe	3	Describe	1	24
		B7	Describe	2	Describe	5	Describe	3	30
		C7	Describe	7	Describe	7	Describe	10	490

running this exercise, the team often identifies a multitude of failures in the first clusters, while later the exercise loses momentum and only the most relevant ones are considered. Selecting the top three RPNs for each cluster allows us not to dilute the real risk and underestimate the failure probability; the company's RPN will be calculated using the median value of the Clusters' RPNs in order to consider any asymmetrical distribution of risks.

Table 7.25 An example of how the results obtained implementing the Failure Mode and Effects Analysis using the K9 map

Clusters		Cluster RPN	Firm's RPN
1	Market attractiveness and severity of unmet medical needs	150	
2	Product strength and readiness	423	
3	Value proposition fit to needs	254	
4	Competitiveness	760	342 (median value of Clusters' RPNs)
5	Suitability to the status quo	125	
6	Attainability of the Market Segments	342	
7	Viability of monetization and commercialization	667	

Firm's RPN tells us the overall risk positioning of the company on a scale between 1 and 1000, where 1 refers to the lowest default risk level and 1000 represents the highest.

Medical start-ups are early-stage private companies; the lowest default rate for similar companies is represented by the lowest default rate in the portfolio of private equity funds. Private equity is the lowest-risk investment segment for private companies, representing firms that have stable and profitable economic and financial performances after successfully completing technological and market development. Still, they are private companies, with less liquid shares, lower regulatory compliance, and limited ability to issue financial instruments compared to public firms: using the default rate in publicly traded companies would therefore not be appropriate. Based on recent market data (Khalaf, 2024), 15% could be taken as a reasonable floor for the lowest risk level.

The highest default rate for start-up companies is 90% according to a study performed on multiple sources for both venture-backed and non-venture-backed new businesses (Howarth, 2025). This level doubles the default rate recorded in Europe following a research on 885 venture capital funds (Manigart, 2023), which, however, implies that it refers to a very exclusive sample of firms, already selected by investors and screened in terms of risk level. Accepting 90% as the default rate for the highest-risk level companies seems a reasonable assumption.

Based on the previous assumptions, we can build a simple linear model to calculate the probable default rate of a specific firm based on its RPN:

$$\text{Default Rate} = \text{MinDF} + \frac{\text{MaxDF} - \text{MinDF}}{\text{MaxRPN} - \text{MinRPN}} * (\text{RPN} - 1)$$

where:

MinDF = 15%

MaxDF = 90%

MinRPN = 1

MaxRPN = 1000

The Default Rate can therefore be expressed as follows:

$$\text{Default Rate} = 15\% + (75\% / 999) * (\text{RPN} - 1)$$

If we consider the target company analyzed in Table 7.25, with RPN = 342, its default can be calculated as:

$$\text{Default Rate} = 15\% + (75\% / 999) * (342 - 1) = 41\%$$

Figure 7.80 provides a graphical representation of the linear model introduced.

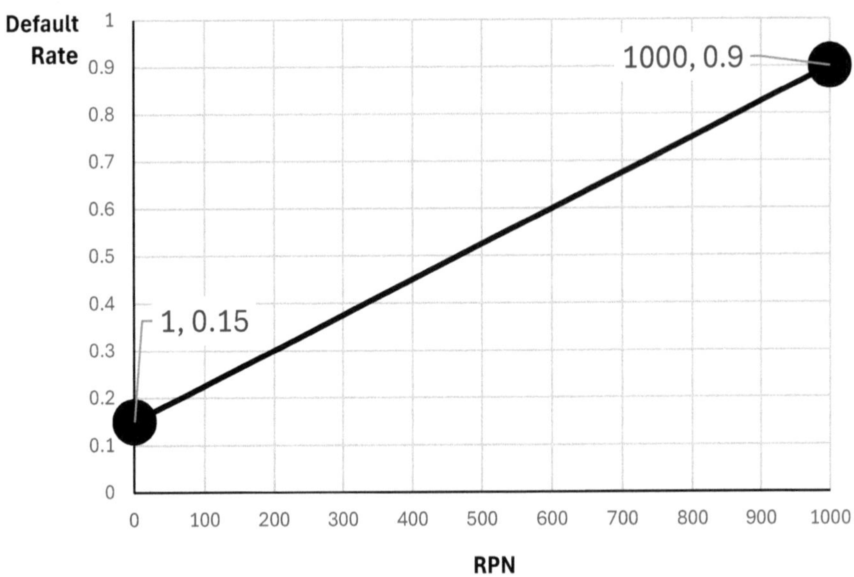

Fig. 7.80 The linear relationship between RMN and Default rates in the current example

Once the expected default rate is determined, we can use the portfolio simulation approach introduced in the previous paragraph to elaborate on the target IRR for this firm, based on the target IRR planned for the entire fund.

Let us define:

Val_E = Return to the fund while exiting the invested company

$TIRR_F$ = Target IRR of the fund, defined by the fund manager and proposed to its Limited Partners

$TIRR_C$ = Required IRR for the target company, enabling a total portfolio IRR equal to $TIRR_F$

Yrs = Expected average number of years between the investment and the exit for all portfolio companies

Inv = Amount of cash invested in a target company on average for the whole portfolio

N_C = Number of invested companies

DR = Default Rate, average on the portfolio

Based on the previous definitions: Val_E

$$\mathrm{TIRR}_C = \left(\mathrm{Val}_E / \mathrm{Inv}\right)^{(1-Yrs)} - 1$$

$$\mathrm{Val}_E = \frac{\left(\mathrm{Inv}\,\mathrm{x}\,N_C\right) * \left(1 + \mathrm{TIRR}_F\right)^{\mathrm{Yrs}}}{N_C \times \left(1 - \mathrm{DR}\right)}$$

For the sake of simplicity, if we assume a portfolio of companies where:

$$\mathrm{Inv} = 1\mathrm{M}, \mathrm{N_C} = 10, \mathrm{Yrs} = 5;$$

a firm with a 41% default rate should have a target IRR of 33% to provide a 20% IRR at the portfolio level.

We can now use this target IRR to calculate our company's valuation using the venture capital method. Table 7.26 summarizes the key steps of the venture capital valuation method.

Table 7.26 The key steps of the venture capital valuation method

Phase	Description	Steps
1	Risk assessment	• FMEA • RPN at firm level
2	Expected return	• Expected default rate • Target IRR
3	Exit value	• Identify comparable firms • Estimate multiples • Value at exit
4	% of shares to acquire	• Exit cash in for investors at exit • Expected % of shares at exit
5	Price per share	• Post money • Number of shares to be issued • Price per share
6	Adjustments	Price per share to be adjusted and negotiated with investors depending on liquidation preference and other rights of preferred shares issued in the investment round.

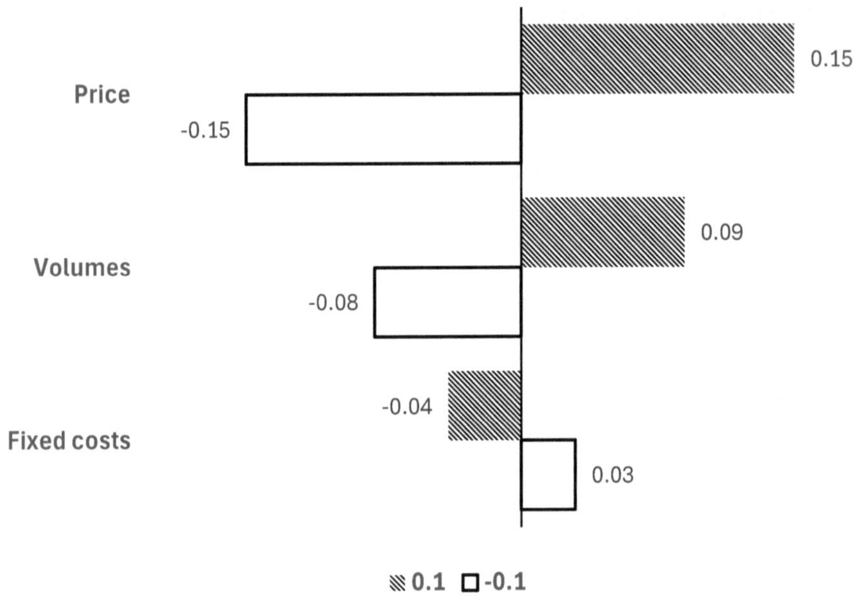

Fig. 7.81 An example of sensitivity analysis applied to the EBITDA

7.9.6 Dealing with Parallel Uncertainties: the Monte Carlo analysis

The previous paragraphs dealt with the process of company valuation in the case of early-stage, start-up companies. This process is based on the firm's risk assessment; however, it does not account for any variance in its financial model. In fact, the estimation of exit value using earnings or revenue multiples relies solely on the base scenario.

One standard methodology to make an initial "what-if scenario" is a sensitivity analysis. One input variable of the financial model is varied by a fixed percentage (usually −10% and +10%) with respect to the base scenario and the variation of an output variable in percentage is recorded. The result of such scenario analysis is usually plotted in a "tornado chart" (Fig. 7.81): input variables are sorted starting with the one with the highest variation on a selected output variable. In the example below, we plotted the percentage of variation of the year 5 EBITDA of Firm A's financial plan. A 10% increase in price leads to a 15% increase in year 5 EBITDA, and a 10% rise in product volumes sold leads to a 9% increase in EBITDA. Fixed costs also influence EBITDA, with a negative correlation: the higher the costs, the lower the EBITDA. An increase of 10% in fixed costs implies a decrease of 3% in year 5 EBITDA.

What-if scenarios are very useful for identifying input variables that significantly impact the output variable, allowing financial analysts and entrepreneurs to focus on

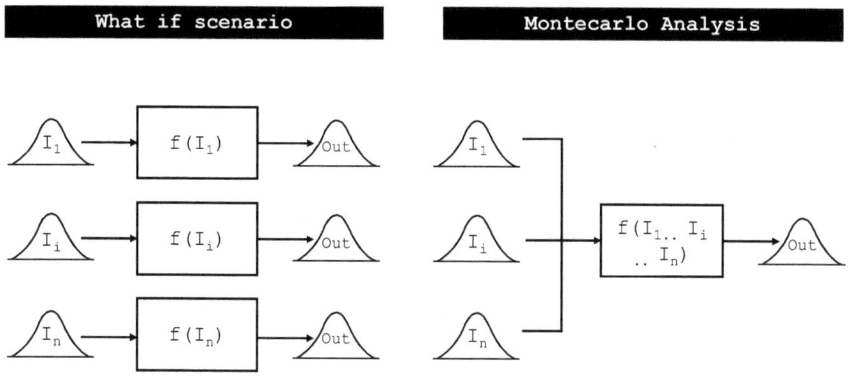

Fig. 7.82 The difference between the What if Scenario and the Monte Carlo Analysis

them and reduce related risks. However, what-if scenarios are based on the assumption that only one input variable varies at a time.

$$\text{Output} = \text{f}\left(\text{Input}_1\right)$$

The K9 map introduced a much wider and sophisticated model (Fig. 7.65), where multiple variables affect financial outputs, each one varying at the same time as the others:

$$\text{Output} = \text{f}\left(\text{Input}_{1,\ldots,}\text{Input}_{i,\ldots,}\text{Input}_n\right)$$

The statistical distribution of the output variable is determined by all possible scenarios where each input variable assumes a value following its own statistical distribution. This model is called "Monte Carlo Analysis" (Metrick & Yasuda, 2021). Figure 7.82 illustrates the difference between a what-if scenario and a Monte Carlo analysis.

Running a Monte Carlo simulation is a process itself, typically made up of the following steps.

1. **Identify the output variable(s)** to be analyzed. For example, if we want to assess the value of a company based on the venture capital method, where the exit value is calculated with earnings multiples, then EBITDA could be considered as our output variable.
2. **Build the firm's financial model**, ensuring that all input variables can be varied and affect EBITDA, our output one.
3. Following the K9 map, **link the financial model to an operative quantitative model** representing the most critical input variables. For example, revenues are a financial input variable that is determined by two main factors: price and volumes. Furthermore, such factors are also influenced by other variables, as described in detail in this chapter.
4. **Define**, based on available historical data or on (tested) assumptions, the **statistical distribution of input** variables.

5. Run a Monte Carlo analysis, creating a **consistent sample of random scenarios**, each of which represents a combination of input variables identified and follows their defined statistical distribution.
6. Plot and analyze the statistical distribution and the **cumulative probability graph of the output** variable.

Monte Carlo analysis is a compelling risk assessment method as it best represents a firm in its future daily operations: the company's performance is driven by a variety of factors, including time, which eventually influence economic and financial metrics (see example in Fig. 7.83). Focusing on the sole financial model of the firm and, in case, running a few what-if scenarios would provide a useful, yet limited view on the reliability of the business plan.

The role of time is often undervalued, but if we recall the definition of IRR, it is of the essence for an investment. In case a company delays by one year the launch of a new product, related revenues and profitability will be moved ahead by 12 months in the financial plan; still, the company will have to sustain fixed costs and investments for an additional year. The impact of this delay on the expected exit value of the company and, therefore, on the IRR for investors could be significant. Simple financial modeling provides limited opportunities to accurately assess the impact of delayed market entry, particularly when combined with

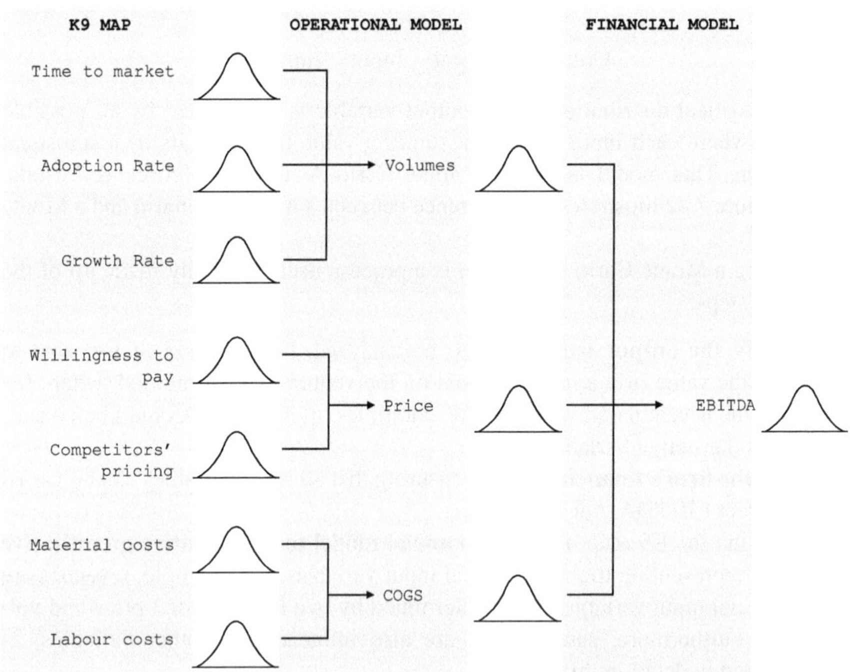

Fig. 7.83 The chart summarizes how some of the most relevant statistical distributions obtained by the K9 influence the distribution of the company's EBITDA

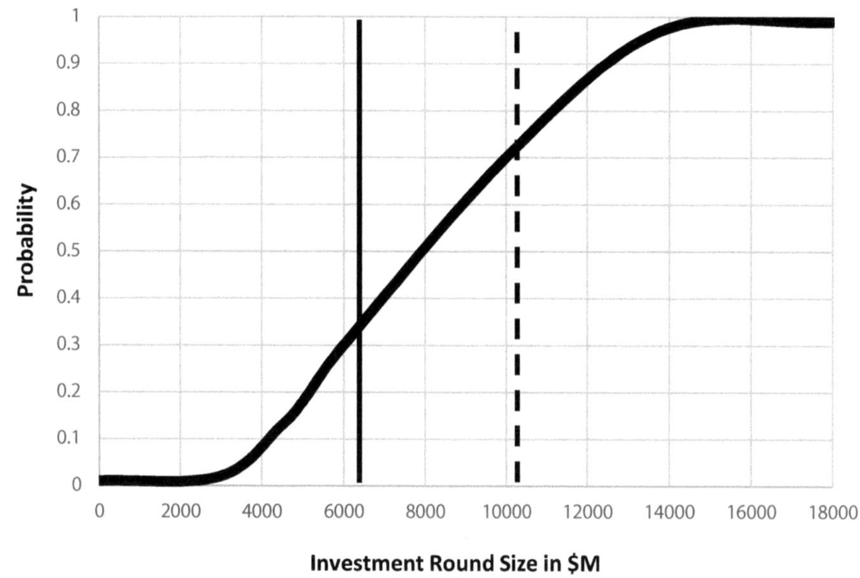

Fig. 7.84 The chart represents the cumulative probability that the total cash obtained from investors by a firm is not enough to operate without the need for further investments

other fluctuations in internal and external inputs. A Monte Carlo analysis on a financial model connected to an operational one based on the K9 map offers entrepreneurs the opportunity to mitigate the risks of internal and external variables, building up appropriate cash reserves and focusing on the critical path and variables of their venture.

The result of this analysis is the cumulative probability graph of output variables. This chart provides an immediate assessment of the statistical reliability of base scenarios and suggests mitigation actions. The example illustrated in Fig. 7.84 represents the cumulated probability of the total cash needs of a firm, and therefore the amount of capital that entrepreneurs should ask an investor. The x-axis represents the total investment needs in $ × 1000, while the y-axis represents the cumulative probability. The vertical line represents the base scenario, in which the company was supposed to look for approximately $6 M to expand its business. However, $6 M corresponds to a cumulated probability of 30%, meaning that there is a 70% chance that the firm would need additional capital. Based on this graph, an entrepreneur or an investor willing to have 100% certainty of running out of cash should invest approximately $14 M. Adding $8 M to the base capital request of $6 M represents a significant effort for both entrepreneurs and investors; thus, they would probably accept running a lower and shared risk. Let us assume that they consider satisfying 70% chances that the firm won't need additional cash; by following the chart of Fig. 7.84, they would agree on an investment size close to $10 M.

7.9.7 Eventually, Only the Market Will Be Right

This chapter highlights the unique characteristics of early-stage companies in evaluating their value. The most common methodologies used in corporate finance for mature companies do not apply. Entrepreneurs should therefore put themselves in the shoes of investors and understand how they would set up a price for acquiring shares of their firms and provide the capital needed for their growth. Venture Capital model of valuation does not focus on the firm's present performance, but looks forward and sets up today's price per share based on a target IRR and expected exit values.

Return on any investment should be proportional to the risk associated with it. The K9 map enables FMEA analysis and the assessment of the default probability of the company, allowing for the identification of a target IRR for an investment in that specific firm.

The K9 map is also fundamental for the broader risk assessment that entrepreneurs and financial analysts should run on a firm: the Monte Carlo analysis. This methodology simulates a company's future performance based on a quasi-real fluctuation of all main variables, including time-based ones; it provides selected critical outputs, such as EBITDA, cash needs, etc., in the form of statistical distributions and provides an effective tool to mitigate any risk through strategic choices, such as additional funding.

While all the methodologies we introduce provide quantitative tools, the overall very high level of uncertainty causes investors to add protective clauses, such as liquidation preference, and to reduce their investment risks. These clauses often result in premium or discount prices relative to the firm's calculated value. Additionally, market trends, positive or negative exit scenario outlooks, and other factors influence the price per share.

Besides all analyses and assessments, we close this paragraph recalling that the value of a company is the amount that would be obtained upon the sale of an asset, in a structured transaction conducted between knowledgeable and willing market participants at a specified measurement date (IFRS Foundation, 2011). If nobody is willing to pay anything for something, then the value of that something is considered to be zero. Whatever great a venture looks in the eyes of its founders, if no investors provide capital at an agreed price, then entrepreneurs should go back on the drawing board and rethink their project and their value expectations. The market is always right.

Bibliography

Abdullah, F., Lefferts, M., & Snyder, G. (2013). *The three rules in medical technology*. Deloitte University Press.

Alexander, A., McGill, M., Tarasova, A., Ferreira, C., & Zurkiya, D. (2019). Scanning the future of medical imaging. *Journal of the American College of Radiology, 16*(4), 501–507. https://doi.org/10.1016/j.jacr.2018.09.050

Alharbi, S. M., Alshaiti, J. H., Ghazwani, J. M., Ghazwani, A. M., Alshahrani, N. M., Banjar, M. Y., Alqahtani, H. S., Alkathiri, S. H., Alhamyani, R. A., Althumairy, A. I., & Amer, M. M. A. (2021). Epidemiology of hereditary anemias in Saudi Arabia. *International Journal of Community Medicine and Public Health, 8*(12), 6127–6131. https://doi.org/10.18203/2394-6040.ijcmph20214453

Arendt, K., & Segal, S. (2008). Why epidurals do not always work. *Reviews in Obstetrics & Gynecology, 1*(2), 49–55.

Askary, S., Askarany, D., & Ugras, Y. J. (2025). A financial auditing approach using failure effect mode analysis. *International Journal of Applied Decision Sciences, 18*(1), 1–23.

Auh, S., & Menguc, B. (2005). Balancing exploration and exploitation: The moderating role of competitive intensity. *Journal of Business Research, 58*(12), 1652–1661.

Aviles-Blanco, M. V. (2021). Economic evaluation of process utility: Elucidating preferences for a non-invasive procedure to treat restenosis. *Health Economics Review, 11*(1), 27.

Backhouse, M. E., Wonder, M., Hornby, E., Kilburg, A., Drummond, M., & Mayer, F. K. (2011). Early dialogue between the developers of new technologies and pricing and reimbursement agencies: A pilot study. *Value in Health, 14*(4), 608–615. https://doi.org/10.1016/j.jval.2010.11.011

Baig, A. A., Lopez, F. Y., DeMeester, R. H., Jia, J. L., Peek, M. E., & Vela, M. B. (2016). Addressing barriers to shared decision making among Latino LGBTQ patients and healthcare providers in clinical settings. *LGBT Health, 3*(5), 335–341.

Baines, R., Hoogendoorn, P., Stevens, S., Chatterjee, A., Ashall-Payne, L., Andrews, T., & Leigh, S. (2023). Navigating medical device certification: A qualitative exploration of barriers and enablers amongst innovators, notified bodies and other stakeholders. *Therapeutic Innovation & Regulatory Science, 57*, 238–250.

Bajwa, S. S., Wang, X., Duc, A. N., & Abrahamsson, P. (2017). 'Failures' to be celebrated: An analysis of major pivots of software startups. *Empirical Software Engineering, 22*(5), 2373–2408.

Ballantyne, D., Frow, P., Varey, R. J., & Payne, A. (2011). Value propositions as communication practice: Taking a wider view. *Industrial Marketing Management, 40*(2), 202–210.

Bardram, J. E. (2008). *Clinical proof-of-concept – A evaluation method for pervasive healthcare systems* (pp. 51–56). UbiComp.

Barnes, C., Blake, H., & Pinder, D. (2009). *Creating and delivering your value proposition: Managing customer experience for profit*. Kogan Page Limited.

Bartlett, R. (2025). Preferred stock liquidation preferences. *The Palgrave Encyclopedia of Private Equity*, 845–852.

Batalden, M., Batalden, P., Margolis, P., Seid, M., Armstrong, G., Opipari-Arrigan, L., & Hartung, H. (2016). Coproduction of healthcare service. *BMJ Quality & Safety, 25*(7), 509–517.

Baulia, S. (2020). Is household shock a boon or bane to the utilisation of preventive healthcare for children? Evidence from Uganda. *SSRN Electronic Journal*.

Behrns, K., Gingles, B., & Sarr, M. (2021). *Medical innovation*.

Berk, J. B. (2007). *Corporate finance, 5th Edition Berk and DeMarzo*. Pearson Education. https://perma.cc/5JDD-LW9Q

Bloom, D., Beetsch, J., Harker, M., Hesterlee, S., Moreira, P., Patrick-Lake, B., Selig, W., et al. (2018). The rules of engagement: CTTI recommendations for successful collaborations between sponsors and patient groups around clinical trials. *Therapeutic Innovation & Regulatory Science, 52*(2), 206–213.

Bobinac, A., van Exel, N. J. A., Rutten, F. F. H., & Brouwer, W. B. F. (2012). Get more, pay more? An elaborate test of construct validity of willingness to pay per QALY estimates obtained through contingent valuation. *Journal of Health Economics, 31*(1), 158–168.

Bonella, F., Wijsenbeek, M., Molina-Molina, M., Duck, A., Mele, R., Geissler, K., & Wuyts, W. (2016). European IPF patient charter: Unmet needs and a call to action for healthcare policymakers. *European Respiratory Journal, 47*(2), 597–606.

Bratic, W., Blok, J. R., & Gostola, M. M. (2014). Valuation of early-stage companies in the biotechnology industry. *Journal of Commercial Biotechnology, 20*(2), 51–58.

Bruhn, W. E., Fracica, E. A., & Makary, M. A. (2018). Group purchasing organizations health care costs and drug shortages. *JAMA, 320*(18), 1859. https://doi.org/10.1001/jama.2018.13604
Burns, L. R., Andrew Lee, J., Bradlow, E. T., & Antonacci, A. (2007). Surgeon evaluation of suture and endo-mechanical products. *Journal of Surgical Research, 141*(2), 220–233.
Burns, L. R., Bradlow, E. T., Andrew Lee, J., & Antonacci, A. C. (2007). Assessment of medical devices: How to conduct comparative technology evaluations of product performance. *International Journal of Technology Assessment in Health Care, 23*(4), 455–463.
BVCA. (2024). *Performance measurement survey 2024*. BVCA. https://perma.cc/3Z7K-HBEH
Caffery, B., Petris, R., Hammitt, K. M., Montecchi-Palmer, M., Haque, S., Malkowski, J.-P., & Barabino, S. (2023). Patient perspectives on dry eye disease and chronic ocular surface pain: Insights from a virtual community-moderated dialogue. *European Journal of Ophthalmology, 33*(2), 856–864.
Cameron, D., Ubels, J., & Norström, F. (2018). On what basis are medical cost-effectiveness thresholds set? Clashing opinions and an absence of data: A systematic review. *Global Health Action, 11*(1), 1447828.
Camuffo, A., Gambardella, A., Messinese, D., Novelli, E., Paolucci, E., & Spina, C. (2024a). A scientific approach to entrepreneurial decision-making: Large-scale replication and extension. *Strategic Management Journal, 45*(6), 1209–1237.
Camuffo, A., Gambardella, A., & Pignataro, A. (2024b). Theory-driven strategic management decisions. *Strategic Science, 9*(4), 382–396.
Canavan, C., West, J., & Card, T. (2014). Review article: The economic impact of the irritable bowel syndrome. *Alimentary Pharmacology & Therapeutics, 40*(9), 1023–1034.
Capital Market Authority. (2024). *Investment*. Capital Market Authority. Riyadh.
Carayon, P., & Wooldridge, A. R. (2020). *Improving patient safety in the patient journey: Contributions from human factors engineering*. 275–299.
Carman, K. L., Dardess, P., Maurer, M., Sofaer, S., Adams, K., Bechtel, C., & Sweeney, J. (2013). Patient and family engagement: A framework for understanding the elements and developing interventions and policies. *Health Affairs, 32*(2), 223–231.
Cendron, M., Hitchcock, D., & Hampshire, N. (1999). Primary nocturnal enuresis: Current concepts. *American Family Physician*.
Chamova, J. (2017). *Mapping of HTA national organisations, programmes and processes in EU and Norway*.
Chan, M. S., Kowalik, J. C., Ashfield, T., & Pearson-Stuttard, J. (2023). Improving health evaluations to capture wider value of therapeutics and incentivise innovation. *Frontiers in Public Health* 11 (March).
Chen, F., Long, Y., Yang, J., Zhong, K., & Liu, B. (2023). Economic evaluation of Serplulimab plus chemotherapy as the first-line treatment of oesophageal squamous cell carcinoma in China. *BMJ Open, 13*(12), e078924.
Choi, P. P. (2015). Patient advocacy: The role of the nurse. *Nursing Standard, 29*(41), 52–58.
Cohen, S., Fleishaker, J. C., Madani, S., McLeod, J. F., Musser, B., & Williams, S. A. (2010). Proof of concept: A PhRMA position paper with recommendations for best practice. *Clinical Pharmacology & Therapeutics, 87*(3), 278–285.
Constantino, Jr. S. K., & Waisel, D. B. (2018). The history of the endotracheal tube, from concept to modern design. *Journal of Anesthesia History*.
Cook, A. (2006). *Forecasting for the pharmaceutical industry*.
Coster, J. E., Turner, J. K., Bradbury, D., & Cantrell, A. (2017). Why do people choose emergency and urgent care services? A rapid review utilizing a systematic literature search and narrative synthesis. *Academic Emergency Medicine, 24*(9), 1137–1149.
Cyr, L. A. (2022). *Note on pre-money and post-money valuation (A) | Harvard Business Impact Education*. HBR. https://hbsp.harvard.edu/product/802179-PDF-ENG
D'Aveni, R. A. (2007). *Mapping your competitive position*. HBR. https://perma.cc/W2Z7-JNTM
Davey, S. M., Brennan, M., Meenan, B. J., Mcadam, R., Girling, A., Chapman, A., & Liliford, R. L. (2011). A framework to manage the early value proposition of emerging healthcare technologies. *Irish Journal of Management, 31*(1), 59–75.

Deloitte & AdvaMed. (2017). *A framework for comprehensive assessment of medical technologies: Defining value in the new health care ecosystem.* https://perma.cc/T8DQ-Y3FAh

DeMeester, R. H., Lopez, F. Y., Moore, J. E., Cook, S. C., & Chin, M. H. (2016). A Model of organizational context and shared decision making: Application to LGBT racial and ethnic minority patients. *Journal of General Internal Medicine, 31*(6), 651–662.

Dodds, W. (1973). An application of the bass model in long-term new product forecasting. *Journal of Marketing Research, 10*(3), 308–311.

Ecocibi, M. L. (2017). The use of Ansoff Matrix in the field of business. *Analele Universităţii Constantin Brâncuşi Din Târgu Jiu: Seria Economie, 2*(special issue), 141–149.

Ehrman, C. M., & Shugan, S. M. (1995). The forecaster's dilemma. *Marketing Science, 14*(2), 123–147.

Eliashberg, J., Elberse, A., Mark, A. A. M., & Leenders. (2006). The motion picture industry: Critical issues in practice, current research, and new research directions. *Marketing Science, 25*(6), 638–661.

Elliott, J., Forbes, D., Chesworth, B. M., Ceci, C., & Stolee, P. (2014). Information sharing with rural family caregivers during care transitions of hip fracture patients. *International Journal of Integrated Care, 14*(2).

Dean, E. B., Pierre, R., Carter, S., & Bond, A. M. (2024). Role of supply chain intermediaries in steering hospital product choice: Group purchasing organizations and biosimilars. *Health Affairs Scholar, 2*(6), qxae067. https://doi.org/10.1093/haschl/qxae067

European Medicines Agency. (2019). *Defining unmet medical need EMA payers community meeting.*

FDA. (2017). *Software as a Medical Device (SAMD): Clinical evaluation.* https://perma.cc/J6Q8-VUNV

Fischer, M., Leeflang, P. S. H., & Verhoef, P. C. (2010). Drivers of peak sales for pharmaceutical brands. *Quantitative Marketing and Economics, 8*(4), 429–460.

Fontrier, A.-M., Visintin, E., Kanavos, P., & Fontrier, A.-M. (2024). *Similarities and differences in health technology assessment systems and implications for coverage decisions: Evidence from 32 countries*. PharmacoEconomics – Open.

Friedman, A., & Miles, S. (2006). *Stakeholders theory and practice.*

Frow, P., & Payne, A. (2011). A stakeholder perspective of the value proposition concept. *European Journal of Marketing, 45*(1), 223–240.

Gamova, E., Zakirova, O., & Rodionova, E. (2020). Leasing of hardware and software solutions in health care systems: Potential benefits and perspectives. In *Proceedings of the "New Silk Road: Business Cooperation and Prospective of Economic Development" (NSRBCPED 2019).* Atlantis Press.

Gao, Y., Li, H., & Luo, Y. (2015). An empirical study of wearable technology acceptance in healthcare. *Industrial Management and Data Systems*. Emerald Group Holdings.

Ghodeswar, B. M. (2008). Organisational adoption of medical technology in healthcare sector. *Journal of Services Research, 7*(2).

Gipson, D. S., Selewski, D. T., Massengill, S. F., Modes, M. M., Desmond, H., Lee, L., Kamil, E., et al. (2018). NephCure Accelerating Cures Institute: A multidisciplinary consortium to improve care for nephrotic syndrome. *Kidney International Reports, 3*(2), 439–446.

Girling, A., Young, T., Brown, C., & Lilford, R. (2010). Early-stage valuation of medical devices: The role of developmental uncertainty. *Value in Health, 13*(5), 585–591.

Girling, A., Lilford, R., Cole, A., & Young, T. (2015). Headroom approach to device development: Current and future directions. *International Journal of Technology Assessment in Health Care, 31*(5), 331–338.

Globe Newswire. (2022). *Global surgical procedures market to generate 372,837.* https://perma.cc/QQ2E-AAYF

Gornall, W., & Strebulaev, I. A. (2020). Squaring venture capital valuations with reality. *Journal of Financial Economics, 135*(1), 120–143.

Gospe, M. (2011). *The marketing high ground: The essential playbook for B2B marketing practitioners everywhere* (Vol. 1). Createspace. ASIN: B0056AE7JO.

Goyat, S. (2011). The basis of market segmentation: A critical review of literature. *European Journal of Business and Management, 3*(9), 45–55.

Gray, D. (2025). *How to use comparables effectively in startup valuation | Equidam*. Equidam. https://perma.cc/NQM5-8WPY

Green, C. G., Krause, D. O., & Wylie, J. L. (2006). Spatial analysis of campylobacter infection in the Canadian Province of Manitoba. *International Journal of Health Geographics, 5*(1), 2.

Grimaldi, D., Shalla, K., Fontanals, I., & Carrasco-Farré, C. (2022). From smart city to data-driven city. *Implementing Data-Driven Strategies in Smart Cities: A Roadmap for Urban Transformation*, January, 1–45.

Harrison, R., Walton, M., Chauhan, A., Manias, E., Chitkara, U., Latanik, M., & Leone, D. (2019). What is the role of cultural competence in ethnic minority consumer engagement? An analysis in community healthcare. *International Journal for Equity in Health, 18*(1), 191.

Heidari, B. (2011). Rheumatoid arthritis: Early diagnosis and treatment outcomes. *Caspian Journal of Internal Medicine, 2*(1), 161–170.

Herzlinger, R. E. (2006). Why innovation in health care is so hard. *Harvard Business Review, 1*(May).

Hillenbrand, S., & Stafford, E. (2025). *Venture capital as portfolios of compound options*. Harvard Business School. https://perma.cc/FE2C-DXDW

Hmieleski, K., & Baron, R. (2017). Entrepreneurs' optimism and new venture performance: A social cognitive perspective. *Academy of Management Journal, 52*(3), 473–488.

Hofdijk, J., Weber, P., Mantas, J., Mihalas, G., & Masic, I. (2014). A short factography about IMIA and EFMI. *Acta Informatica Medica, 22*(1), 71.

Hook-Barnard, India, Sheena M. Posey Norris, Joe Alper, Board on health sciences policy, board on life sciences, Institute of Medicine, and National Research Council. (2014). Technology Readiness Levels in the Department of Defense. *National Academy of Sciences*, December. https://perma.cc/4SJT-2JRK

Howarth, J. (2025). *Startup failure rate statistics (2025)*. ExplodingTopics.Com. https://perma.cc/U8YP-EV7A

Hracs, L., Windsor, J. W., Gorospe, J., Cummings, M., Coward, S., Buie, M. J., Quan, J., Goddard, Q., Caplan, L., Markovinović, A., Williamson, T., Abbey, Y., Abdullah, M., Abreu, M. T., Ahuja, V., Ali, R. A. R., Altuwaijri, M., Balderramo, D., Banerjee, R., Benchimol, E. I., Bernstein, C. N., Brunet-Mas, E., Burisch, J., Chong, V. H., Dotan, I., Dutta, U., El Ouali, S., Forbes, A., Forss, A., Gearry, R., Dao, V. H., Hartono, J. L., Hilmi, I., Hodges, P., Jones, G.-R., Juliao-Baños, F., Kaibullayeva, J., Kelly, P., Kobayashi, T., Kotze, P. G., Lakatos, P. L., Lees, C. W., Limsrivilai, J., Lo, B., Loftus Jr, E. V., Ludvigsson, J. F., Mak, J. W. Y., Miao, Y. L., Ng, K. K., Okabayashi, S., Olén, O., Panaccione, R., Paudel, M. S., Quaresma, A. B., Rubin, D. T., Simadibrata, M., Sun, Y., Suzuki, H., Toro, M., Turner, D., Iade, B., Wei, S. C., Yamamoto-Furusho, J. K., Yang, S.-K., Ng, S. C., & Kaplan, G. G. (2025). Global evolution of inflammatory bowel disease across epidemiologic stages. *Nature, 642*(8067), 458–466. https://doi.org/10.1038/s41586-025-08940-0

Hu, F. B., & Willett, W. C. (2002). Optimal diets for prevention of coronary heart disease. *Journal of the American Medical Association, 288*(20), 2569–2578.

Hu, Q., & Schwarz, L. B. (2011). Controversial role of GPOs in healthcare-product supply chains. *Production and Operations Management, 20*(1), 1–15. https://doi.org/10.1111/j.1937-5956.2010.01121.x

Hussain, S., Khattak, J., Rizwan, A., & Latif, M. (2013). ANSOFF matrix, environment, and growth– An interactive triangle. *Management and Administrative Sciences Review, 2*(2), 196–206. https://perma.cc/VFB3-GFU4

ICER. (2018). *Emicizumab for Hemophilia A with inhibitors: Effectiveness and value*. Final Report. Vol. April 16. https://perma.cc/78NT-YNQ7

IFRS Foundation. (2011). *Fair value measurement*. IFRS Foundation. https://perma.cc/TL67-JCS9

Ijzerman, M. J., & Steuten, L. M. G. (2011). Early assessment of medical technologies to inform product development and market access: A review of methods and applications. *Applied Health Economics and Health Policy, 9*(5), 331–347.

IPEV. (2022). *IPEV: Valuation guidelines*. International Private Equity and Venture Capital Valuation Guidelines. London. https://perma.cc/K9VT-PQ72v

Jambulingam, T. (2018). The R&D marketing interface in biopharma and medtech. *Journal of Commercial Biotechnology, 24*(1), 48–55.

Jazzar, M. (2023). Financial leasing in Palestine 'Prospects & Obstacle'. *Indian Journal of Economics and Finance, 3*(2), 1–7.

Joseph, J. J., Gray, D. M., Williams, A., Zhao, S., McKoy, A., Odei, J. B., Brock, G., et al. (2023). Addressing non-medical health-related social needs through a community-based lifestyle intervention during the COVID-19 pandemic: The black impact program. *PLoS ONE, 18*(March 9th).

Jufriyanto, M., Kurniati, N., & Supriatna, A. (2018). Preventive maintenance policy on leasing by considering the usage rate. *MATEC Web of Conferences, 204*(September), 02016.

Kapferer, J.-N. (2012). *The new strategic brand management: Advanced insights and strategic thinking*. Kogan Page. ISBN-10:9780749465162.

Kaplan, G. G., & Windsor, J. W. (2021). The four epidemiological stages in the global evolution of inflammatory bowel disease. *Nature Reviews Gastroenterology & Hepatology, 18*(1), 56–66. https://doi.org/10.1038/s41575-020-00360-x. Epub 2020 Oct 8. PMID: 33033392; PMCID: PMC7542092.

Kaplan, G., et al. (2025) Global evolution of inflammatory bowel disease across epidemiologic stages Nature 642(8067) 458–466. https://doi.org/10.1038/s41586-025-08940-0.

Kasi, S. S. (2023). A new strategy for rejuvenating primary healthcare: The government-society partnership model. *Journal of Primary Care & Community Health, 14*. https://doi.org/10.1177/21501319231200763

Khalaf, R. (2024). Private equity skirts losses to the economy's cost. *Financial Times*, October.

Kim, H.-S., & Yoon, K.-H. (2020). Lessons from use of continuous glucose monitoring systems in digital healthcare. *Endocrinology and Metabolism, 35*(3), 541–548.

Knight, S. R., Aujla, R., & Biswas, S. P. (2011). Total hip arthroplasty-over 100 years of operative history. *Orthopedic Reviews*.

Koller, T., Goedhart, M. H., & Wessels, D. (2010). *Valuation: Measuring and managing the value of companies*. Wiley. https://perma.cc/WXM3-TDRZ

Konishi, T., & Yasunaga, H. (2023). Current status of health technology assessment research in Japan. *Annals of Clinical Epidemiology, 5*(1), 1–4. https://doi.org/10.37737/ace.23001

Kraemer-Eis, H., & Lang, S. D. (2018). *The European venture capital landscape: An EIF perspective*. European Investment Fund.

Kugeler, K. J., Farley, G. M., Forrester, J. D., & Mead, P. S. (2015). Geographic distribution and expansion of human lyme disease, United States. *Emerging Infectious Diseases, 21*(8), 1455–1457.

Lambert, W. J. (2010). Considerations in developing a target product profile for parenteral pharmaceutical products. *AAPS PharmSciTech, 11*(3), 1476–1481.

Lamptey, A. A., Nsiah-Boateng, E., Agyemang, S. A., & Aikins, M. (2017). National health insurance accreditation pattern among private healthcare providers in Ghana. *Archives of Public Health, 75*(1), 36.

Laurisz, N., Ćwiklicki, M., Żabiński, M., Canestrino, R., & Magliocca, P. (2023). The stakeholders' involvement in healthcare 4.0 services provision: The perspective of co-creation. *International Journal of Environmental Research and Public Health, 20*(3), 2416.

Lauvergeon, S., Burnand, B., & Peytremann-Bridevaux, I. (2012). Chronic disease management: A qualitative study investigating the barriers, facilitators and incentives perceived by Swiss healthcare stakeholders. *BMC Health Services Research, 12*(1).

Lee, S.-S., & Salole, E. (2014). Medical device reimbursement coverage and pricing rules in Korea: Current practice and issues with access to innovation. *Value in Health, 17*(4), 476–481. https://doi.org/10.1016/j.jval.2014.03.1719

Lee, C. C., Langdo, J., Hwang, D., Marques, V., & Hwang, P. (2023). Impacts of distributors and group purchasing organizations on hospital efficiency and profitability: A bilateral data envelopment analysis model. *International Transactions in Operational Research, 30*(1), 476–502.

Lewinski, A. A., & Simmons, L. A. (2018). Nurse knowledge and engagement in health policy making: Findings from a pilot study. *The Journal of Continuing Education in Nursing, 49*(9), 407–415.

Liang, M. H. (2006). Translating outcomes measurement in experimental therapeutics of systemic rheumatic disease to patient care. *Rheumatic Disease Clinics of North America, 32*(1), 1–8.

Lopatina, E., Miller, J. L., Teare, S. R., Marlett, N. J., Patel, J., Barber, C. E. H., Mosher, D. P., Wasylak, T., Woodhouse, L. J., & Marshall, D. A. (2019). The voice of patients in system redesign: A case study of redesigning a centralized system for intake of referrals from primary care to rheumatologists for patients with suspected rheumatoid arthritis. *Health Expectations, 22*(3), 348–363.

Lukas, P., Ladislav, H., Maresova, P., Augustynek, M., & Penhaker, M. (2020). Medical devices: Regulation, risk classification, and open innovation. *Journal of Open Innovation: Technology, Market, and Complexity, 6*, 42.

Makower, J. (2010). *FDA impact on US medical technology innovation.*

Malkin, R., & von Oldenburg Beer, K. (2013). Diffusion of novel healthcare technologies to resource poor settings. *Annals of Biomedical Engineering, 41*(9), 1841–1850. https://doi.org/10.1007/s10439-013-0750-5

Mangen, M.-J. J., Rozenbaum, M. H., Huijts, S. M., van Werkhoven, C. H., Postma, D. F., Atwood, M., van Deursen, A. M. M., et al. (2015). Cost-effectiveness of adult pneumococcal conjugate vaccination in the Netherlands. *European Respiratory Journal, 46*(5), 1407–1416.

Mangen, M.-J. J., Huijts, S. M., Bonten, M. J. M., & Ardine de Wit, G. (2017). The impact of community-acquired pneumonia on the health-related quality-of-life in elderly. *BMC Infectious Diseases, 17*(1), 208.

Manigart, S. (2023). *45% of European VC investments fail or do not secure high returns.* Entrepreneurship Academy. https://perma.cc/S8AU-YUF3

Man-Son-Hing, M., Laupacis, A., O'rourke, K., Molnar, F. J., Mahon, J., Chan, K. B. Y., & Wells, G. (2002). Determinants of the clinical importance of study results. *Journal of General Internal Medicine, 17*, 469–476.

Markovitch, D. G., Steckel, J. H., Michaut, A., Philip, D., & Tracy, W. M. (2015). Behavioral reasons for new product failure: Does overconfidence induce overforecasts? *Journal of Product Innovation Management, 32*(5), 825–841.

Maron, D., & Harrington, R. (2019). *Stents, bypass surgery show no benefit in heart disease mortality rates among stable patients.* Stanford Medicine. https://perma.cc/YBJ2-3FFH

MaRs Discovery District. (2011). *Defining you target product profile: Medical device products.*

Martin, J. L., Clark, D. J., Morgan, S. P., Crowe, J. A., & Murphy, E. (2012). A user-centred approach to requirements elicitation in medical device development: A case study from an industry perspective. *Applied Ergonomics, 43*(1), 184–190.

Masters, R., Anwar, E., Collins, B., Cookson, R., & Capewell, S. (2017). Return on investment of public health interventions: A systematic review. *Journal of Epidemiology and Community Health, 71*(8), 827–834.

Mayfield, D. L. (2016). Medical patents and how new instruments or medications might be patented. *Missouri Medical Journal, 113*(6), 456–462.

McCarthy, S., O'Raghallaigh, P., Woodworth, S., Lim, Y. L., Kenny, L. C., & Adam, F. (2016). An integrated patient journey mapping tool for embedding quality in healthcare service reform. *Journal of Decision Systems, 25*(sup1), 354–368.

McDermott, A. M., & Pedersen, A. R. (2016). Conceptions of patients and their roles in healthcare. *Journal of Health Organization and Management, 30*(2), 194–206.

McNiff, M. M., Hawkins, S., Haase, B., Bullivant, J., McIver, T., Mitelman, O., Emery, N., Tasca, G., Voermans, N., & Diaz-Manera, J. (2024). Facioscapulohumeral muscular dystrophy European patient survey: Assessing patient reported disease burden and preferences in clinical trial participation. *Journal of Neuromuscular Diseases, 11*(2), 459–472.

MDDI. (1997). *Determining the size of a medical device market.* https://perma.cc/V38Z-YC9B

Memish, Z. A., Owaidah, T. M., & Saeedi, M. Y. (2011). Marked regional variations in the prevalence of sickle cell disease and β-thalassemia in Saudi Arabia: Findings from the premarital

screening and genetic counseling program. *Journal of Epidemiology and Global Health, 1*(1), 61–68. https://doi.org/10.1016/j.jegh.2011.06.002
Metrick, A., & Yasuda, A. (2021). *Venture capital & the finance of innovation* (3rd ed.). Wiley. https://perma.cc/9HEV-7NYF
Mirhaydari, A. (2017). *Why Unicorns are overvalued (and the industry knows it) – PitchBook.* PitchBook. November 27, 2017. https://perma.cc/M63H-HLA6
Mulligan, K., Paneva, D., Ryan, M., & Schaeffer Center Staff. (2020). *Health Technology Assessment for the U.S. Healthcare System.*
Mun, J. (2004). *Applied risk analysis: Moving beyond uncertainty in business.* Wiley.
Mushkin, S. J. (1958). Toward a definition of health economics. *Public Health Reports (1896–1970), 73*(9), 785–793.
Naidu, A. (2009). Factors affecting patient satisfaction and healthcare quality. *International Journal of Health Care Quality Assurance, 22*(4), 366–381.
NASA. (2025). *Technology readiness level definitions.* https://perma.cc/7N5B-ETEL
NCI. (2024a). *Definition of primary endpoint.* National Cancer Institute. https://perma.cc/ZV5M-JPWK
NCI. (2024b). *Dictionary of cancer terms.* NCI. https://perma.cc/586H-3PG8
NIMH. (2022). *What is prevalence?* National Institute of Mental Health. https://perma.cc/L3SF-388P
Novignon, J., Olakojo, S. A., & Nonvignon, J. (2012). The effects of public and private health care expenditure on health status in sub-Saharan Africa: New evidence from panel data analysis. *Health Economics Review, 2*(1), 22.
Oben, P. (2020). Understanding the patient experience: A conceptual framework. *Journal of Patient Experience, 7*(6), 906–910.
Ochwo, S., VanderWaal, K., Munsey, A., Ndekezi, C., Mwebe, R., Okurut, A. R. A., Nantima, N., & Mwiine, F. N. (2018). Spatial and temporal distribution of lumpy skin disease outbreaks in Uganda (2002–2016). *BMC Veterinary Research, 14*(1), 174.
Ormsby, D. (2025). *Understanding share classes in venture capital: A guide for founders and investors | DWF.* DWF Group. March 13, 2025. https://perma.cc/FK5Y-KSMR.
Ostasevičiute, R., & Šliburyte, L. (2008). Theoretical aspects of product positioning in the market. *Engineering Economics, 56*(1), 97–103.
Parker, W., & Iqbal, J. (2020). Comparison of contemporary drug-eluting coronary stents – Is any stent better than the others? *Heart International, 14*(1), 34.
Patel, S., Huang, M., & Miliara, S. (2025). Understanding treatment adherence in chronic diseases: Challenges consequences and strategies for improvement. *Journal of Clinical Medicine, 14*(17), 6034. https://doi.org/10.3390/jcm14176034
Pietzsch, J. B., Zanchi, M. G., & Linehan, J. H. (2013). Medical device innovators and the 510(k) regulatory pathway: Implications of a survey-based assessment of industry experience—Part 2: Medical device ecosystem and policy. *Journal of Medical Devices, Transactions of the ASME, 7*(2), 1–5.
PitchBook Report. (2024a). *2024 Annual European Venture Report.* PitchBook. https://perma.cc/4APV-578U
PitchBook Report. (2024b). *2024 Annual US VC Valuations Report.* PitchBook. https://perma.cc/X75E-H4KT
Pollack, M. M. (2016). Severity of illness confusion. *Pediatric Critical Care Medicine: A Journal of the Society of Critical Care Medicine and the World Federation of Pediatric Intensive and Critical Care Societies, 17*(6), 583.
Popa, A. L., Ţarcă, N. N., Sasu, D. V., Bodog, S. A., Roşca, R. D., & Tarcza, T. M. (2022). Exploring marketing insights for healthcare: Trends and perspectives based on literature investigation. *Sustainability, 14*(17), 10499.
Porter, M. E., & Canada, M. M. (1985). *Competitive advantage: Creating and sustaining superior performance.*

Preko, M., & Budu, J. (2024). The impact of digital health technologies on the quality of service delivery: A pre and post assessment of the healthcare consumer's experiences in Ghana. *The Electronic Journal of Information Systems in Developing Countries, 90*(4).
Qermane, K., & Mancha, R. (2021). WHOOP, Inc.: Digital entrepreneurship during the covid-19 pandemic. *Entrepreneurship Education and Pedagogy, 4*(3), 500–514.
Racchi, M., Govoni, S., Lucchelli, A., Capone, L., & Giovagnoni, E. (2016). Insights into the definition of terms in European medical device regulation. *Expert Review of Medical Devices*. Taylor and Francis.
Raghupathi, V., & Raghupathi, W. (2020). Healthcare expenditure and economic performance: Insights from the United States data. *Frontiers in Public Health, 8*(May).
Rahmani, K., Karimi, S., Rezayatmand, R., & Raeisi, A. R. (2021). Value-based procurement for medical devices: A scoping review. *Medical Journal of the Islamic Republic of Iran*, October.
Raja, J. M., Elsakr, C., Roman, S., Cave, B., Pour-Ghaz, I., Nanda, A., Maturana, M., & Khouzam, R. N. (2019). Apple watch, wearables, and heart rhythm: Where do we stand? *Annals of Translational Medicine, 7*(17), 417–417.
Ramanujam, M., & Tacke, G. (2016). *Monetizing innovation: How smart companies design the product around the price*.
Rettig, R. A. (2017). Medical innovation duels cost containment. *Health Affairs, 13*(3), 7–27. https://doi.org/10.1377/Hlthaff.13.3.7
Richardson, M., & Evans, C. (2007). Strategy in action applying Ansoff's Matrix. *Manager*, no. SUMMER.
Ringel, D. M. (2023). Multimarket membership mapping. *Journal of Marketing Research, 60*(2), 237–262.
Roback, K., Gäddlin, P.-O., Nelson, N., & Persson, J. (2007). Adoption of medical devices: Perspectives of professionals in Swedish Neonatal Intensive Care. *Technology and Health Care, 15*(3), 157–179.
Ronald, W. R. (1975). A protection motivation theory of fear appeals and attitude change. *Journal of Psychology Preview Publication Details, 91*, 93–114.
Ronnebaum, E. D., & Schmer, C. (2015). Patient advocacy and the affordable care act: The growing need for nurses to be culturally aware. *Open Journal of Nursing, 05*(03), 237–245.
Russo, G., Tartaglione, A. M., & Cavacece, Y. (2019). Empowering patients to co-create a sustainable healthcare value. *Sustainability, 11*(5), 1315.
Ryen, L., & Svensson, M. (2015). The willingness to pay for a quality adjusted life year: A review of the empirical literature. *Health Economics, 24*(10), 1289–1301.
Saha, R. L., Seidmann, A., & Tilson, V. (2010). A research agenda for emerging roles of healthcare GPOs and their evolution from group purchasing to information sharing to strategic consulting. In *2010 43rd Hawaii international conference on system sciences* (pp. 1–9). IEEE.
Saha, R. L., Seidmann, A., & Tilson, V. (2019). The impact of custom contracting and the 'infomediary' role of healthcare GPOs. *Production and Operations Management, 28*(3), 650–667.
Sahanaa, C., Niranjan, R., Pradeep, K., Gopinath, S., Dhanasekar, E., Vendhan, S., Maniradjou, V., Konduru, R. K., & Phalsalkar, M. (2023). Innovative AETCOM session on health care as a right: Experience at the Medical College in Puducherry. *Journal of Education and Health Promotion, 12*(1).
Saing, S., van der Linden, N., Hayward, C., & Goodall, S. (2019). Why is there discordance between the reimbursement of high-cost 'life-extending' pharmaceuticals and medical devices? The funding of ventricular assist devices in Australia. *Applied Health Economics and Health Policy, 17*(4), 421–431. https://doi.org/10.1007/s40258-019-00470-x
Salazar, G., & Natalia Russi-Vigoya, M. (2021). Technology readiness level as the foundation of human readiness level. *Ergonomics in Design, 29*(4), 25–29.
Salikhanov, I., Katapodi, M. C., Kunirova, G., & Crape, B. L. (2023). Improving palliative care outcomes in remote and rural areas of LMICs through family caregivers: Lessons from Kazakhstan. *Frontiers in Public Health, 11*(August).
Scheier, M. F., & Carver, C. S. (2018). Dispositional optimism and physical health: A long look back, a quick look forward. *The American Psychologist, 73*(9), 1082.

Scuffham, P. A., Whitty, J. A., Mitchell, A., & Viney, R. (2008). The use of QALY weights for QALY calculations. *PharmacoEconomics, 26*(4), 297–310.

Selgas, R., Rodriguez, L., Julian, J. C., Remon, C., Prieto-Velasco, M., Perez-Contreras, J., & Fontan, M. P. (2017). How a bottom-up multi-stakeholder initiative helped transform the renal replacement therapy landscape in Spain. *Applied Health Economics and Health Policy, 15*(6), 755–762.

Sendyona, S., Odeyemi, I., & Maman, K. (2016). Perceptions and factors affecting pharmaceutical market access: Results from a literature review and survey of stakeholders in different settings. *Journal of Market Access & Health Policy, 4*(1), 31660.

Seva, R. R., Tan, A. L. S., Tejero, L. M. S., Dorothy, M. L., & Salvacion, S. (2023). Multidimensional readiness assessment of medical devices. *Theoretical Issues in Ergonomics Science, 24*(2), 189–205.

Shiel, W. (2020). *Definition of standard of care*. RxList. 2020. https://perma.cc/LQD8-V287

Sillup, G. (1992). Forecasting the adoption of new medical technology using the Bass model. *Journal of Health Care Marketing, 12*, 42–51.

Silva-Pinto, A. C., Costa, F. F., Gualandro, S. F. M., Fonseca, P. B. B., Grindler, C. M., Souza Filho, H. C. R., Bueno, C. T., & Cançado, R. D. (2022). Economic burden of sickle cell disease in Brazil. *PLoS One, 17*(6), e0269703.

Smith, M. (2021). Scope of health economics. *Journal of Health & Medical Economics, 7*(5), 63.

Smith, S., & Connolly, S. (2020). Re-thinking unmet need for health care: Introducing a dynamic perspective. *Health Economics, Policy and Law, 15*(4), 440–457.

Sofie, B., & Ooghe, H. (2006). 35 years of studies on business failure: An overview of the classic statistical methodologies and their related problems. *The British Accounting Review, 38*(1), 63–93. https://doi.org/10.1016/j.bar.2005.09.001

Sorenson, C., Drummond, M., & Burns, L. R. (2013). Evolving reimbursement and pricing policies for devices in Europe and the United States should encourage greater value. *Health Affairs, 32*(4), 788–796. https://doi.org/10.1377/hlthaff.2012.1210

Stein, J., Pabst, A., Weyerer, S., Werle, J., Maier, W., Heilmann, K., Scherer, M., et al. (2016). The assessment of met and unmet care needs in the oldest old with and without depression using the Camberwell Assessment of Need for the Elderly (CANE): Results of the AgeMooDe Study. *Journal of Affective Disorders, 193*(March), 309–317.

Stern, A. D. (2017). Innovation under regulatory uncertainty: Evidence from medical technology. *Journal of Public Economics, 145*, 181–200.

Stremersch, S., & Van Dyck, W. (2009). Marketing of the life sciences: A new framework and research agenda for a nascent field. *Journal of Marketing, 73*(4), 4–30.

Sun, D., Gao, W., Hongxiang, H., & Zhou, S. (2022). Why 90% of clinical drug development fails and how to improve it? *Acta Pharmaceutica Sinica B, 12*(7), 3049.

Symmank, C., Krause, J., & Gurtner, S. (2015). Stakeholder variety in healthcare and their integration in the medical device development process. In *Challenges and opportunities in health care management* (pp. 247–258). Springer.

Tamura, M., Nakano, S., & Sugahara, T. (2019). Reimbursement pricing for new medical devices in Japan: Is the evaluation of innovation appropriate? *The International Journal of Health Planning and Management, 34*(2), 583–593. https://doi.org/10.1002/hpm.2719

Taylor, P. C., Moore, A., Vasilescu, R., Alvir, J., & Tarallo, M. (2016). A structured literature review of the burden of illness and unmet needs in patients with rheumatoid arthritis: A current perspective. *Rheumatology International*. Springer.

The Economist. (2015). *Fin-Tech*. The Economist. https://perma.cc/85PM-EAUP

The Orphan Disease Collective. (2025). *The orphan disease collective*. https://perma.cc/UM59-L7YT

Thomas, R. K. (2008). *Introduction to healthcare*. Stanford Medicine.

Tilling, C. J., Krol, M., Attema, A. E., Tsuchiya, A., Brazier, J., van Exel, J., & Brouwer, W. (2014). Exploring a new method for deriving the monetary value of a QALY. *SSRN Electronic Journal*.

Toloo, G.-S., Aitken, P., Crilly, J., & FitzGerald, G. (2016). Agreement between triage category and patient's perception of priority in emergency departments. *Scandinavian Journal of Trauma, Resuscitation and Emergency Medicine, 24*(1), 126.

Tsai, H.-Y., Huang, Y.-W., Chang, S.-Y., Huang, L.-Y., Lin, C. J., & Lee, P.-C. (2022). The reimbursement coverage decisions and pricing rules for medical devices in Taiwan. *GMS Health Innovation and Technologies, 16*, Doc02. https://doi.org/10.3205/hta000134

Tynan, A. C., & Drayton, J. (1987). Market segmentation. *Journal of Marketing Management, 2*(3), 301–335.

Vallejo-Torres, L., García-Lorenzo, B., & Serrano-Aguilar, P. (2017). Estimating a cost-effectiveness threshold for the Spanish NHS. *Health Economics, 27*(4), 746–761.

van Angeren, J., Vroom, G., McCann, B. T., Podoynitsyna, K., & Langerak, F. (2022). Optimal distinctiveness across revenue models: Performance effects of differentiation of paid and free products in a mobile app market. *Strategic Management Journal, 43*(10), 2066–2100.

Venkatesh, V., Thong, J. Y. L., & Xin, X. (2012). Consumer acceptance and use of information technology: Extending the unified theory of acceptance and use of technology. *MIS Quarterly: Management Information Systems, 36*(1), 157–178.

Villela, lucia maria aversa. (2013). Ten digital health tensions driving innovation in MedTech. *Journal of Chemical Information and Modeling, 53*(9), 1689–1699.

Vreman, R. A., Inkatuuli Heikkinen, A., Schuurman, C. S., Garcia, J. L., Hedberg, N., Athanasiou, D., Grueger, J., Leufkens, H. G. M., & Goettsch, W. G. (2019). Unmet medical need: An introduction to definitions and stakeholder perceptions. *Value in Health, 22*(11), 1275–1282.

Wasserman, N. (2008). The Founder's dilemma. *Harvard Business Review*. https://perma.cc/UV39-B8MM

WHO. (2020). *Target product profiles*, 126–143. https://perma.cc/HD8F-GNQW

WHO. (2015). WHO | Health Technology Assessment. *WHO*. World Health Organization. https://perma.cc/MT67-Q29N

Wiklund, M., Kendler, J., & Yale, A. S. (2011). *Usability testing of medical devices*.

Willis, W., Allison, B. A., Evanoff, M. L., Criswell, S. R., & Racette, B. A. (2010). Geographic and ethnic variation in Parkinson disease: A population-based study of US Medicare beneficiaries. *Neuroepidemiology, 34*(3), 143–151.

Wu, L. T., Glass, G. F., Chew, E. Y. H., Joo, E., et al. (2023). *Developing a theory of change to guide the design and implementation of a caregiver-centric support service*, September.

Yoon, Y. S., Jung, B., Kim, D., & Ha, I. H. (2019). Factors underlying unmet medical needs: A cross-sectional study. *International Journal of Environmental Research and Public Health, 16*(13).

Yu, H., Yang, J., & Zhou, T. (2024). The optimization of the pay-per-use model for the healthcare supply chain. *Journal of Global Economy Business and Finance, 6*(8), 70–74. ISSN: 2141–5595.

Zhou, L., Deng, Y., Li, N., Zheng, Y., Tian, T., Zhai, Z., Yang, S., et al. (2019). Global, regional, and national burden of Hodgkin Lymphoma from 1990 to 2017: Estimates from the 2017 Global Burden of Disease Study. *Journal of Hematology & Oncology, 12*(1), 107.

Zoltners, A. A., Sinha, P., & Lorimer, S. E. (2004). *Sales force design for strategic advantage*. Palgrave Macmillan UK.

Prioritizing Decision-Related Problems

8

Abstract

This chapter presents a rational framework for prioritizing decision-related problems in medical technology ventures. After mapping the venture's full decision landscape, entrepreneurs must allocate limited resources among competing priorities. This chapter introduces a method based on two fundamental elements: financial uncertainty—defined as the product of a problem's potential economic impact and the uncertainty of its future scenarios. Problems with the greatest combined impact and uncertainty deserve precedence. The approach is visualized through bi-dimensional mapping and exemplified for medical startups. It emphasizes that prioritization is a continuous, iterative process, constantly refined as information quality improves. This chapter also applies this logic to higher-level K9 map clusters, combining quantitative reasoning with managerial consensus to ensure that fundamental, high-impact issues—such as value proposition, market assumptions, or commercialization readiness—receive timely, collective, and evidence-based attention.

8.1 Principles of Decision Problem Prioritization

Once the mapping process is thoroughly completed, the medical entrepreneur can finally refer to a complete chart in which important problems related to decisions and assumptions are visually linked and grouped in clusters. Understandably, the numerous, interrelated, and competing decision-related problems can disorient any decision-maker. The constrained resources of early-stage medical ventures do not allow a concurrent approach to multiple decision problems; therefore, it becomes necessary to prioritize and target all mapped problems.

L. Negri, *De-Risking Investments in Medical Technology Innovation*, Springer Series in Healthcare Management and Innovation,
https://doi.org/10.1007/978-3-032-15007-3_8

In decision-making, resources are inherently finite. It is impossible to pursue every objective—or even a selection of them—in every conceivable way. This limitation requires us to establish priorities. Because rational prioritization is essential for effective decision-making, addressing and mitigating the lack of prioritization should itself be treated as one of our highest priorities (Kovic, 2018). In other words, the entrepreneur must strive to make rational and explicit decisions about which problems to prioritize at any given time.

We have observed several medical startups through this process, and we have learned that it is essential to recall that this prioritization across decision-related problems:

- It is not a one-time, initial duty but a recurring process throughout the startup lifecycle.
- It should not be intended to exclude any problem from the map but to allow, over time, the identification of the ones that are currently most critical.

Sooner or later, all the K9 map's decision-related problems will be confronted by the venture's management throughout the venture's path to commercialization. They will acquire a different priority level in various steps of the venture's development. For instance, while the assumptions about the validity of the value proposition constitute a key issue that must be addressed in the early stages of the startup, fine-tuning decisions about price per market segment usually gains relevance in the later days. Most assumptions about the actual market size usually hold high priority in the earlier development phases, while problems concerning the commercial approach and organization gain higher relevance in later stages.

Therefore, the entrepreneur will find that specific decision-related problems show higher degrees of priority at the beginning of the venture's life. In contrast, other issues, initially less relevant, gain importance in the later stages. Prioritization of decision-related problems represents an ongoing, cyclical, and necessary effort to ensure that relevant business decisions and assumptions are promptly addressed and that the venture's constrained resources are rationally allocated. In contexts of high uncertainty, such as those faced by medical startups, practitioners may be drawn to fast, instinctual prioritization approaches and overlook the discipline required by a rational prioritization approach.

Field experience has taught us that a rational approach is more necessary in contexts of great degrees of uncertainty. A constant tension toward rational decision-making remains necessary even when entrepreneurial perceptions of uncertainty suggest the use of the so called "common sense" approaches. This chapter adapts well-known prioritization principles to the problem of prioritizing decision-related problems in the entrepreneurial context of medical technology innovation. It suggests a rational approach to decision-related problem prioritization, which has been successfully tested in medical startups employing the scientific approach to decision-making.

8.2 The Theory Behind the Practice

The suggested process prioritizes decision-related problems by focusing on the monetary risk of not resolving them. Not resolving a decision-related problem means not making a decision or not making a business assumption. For instance, what risk is associated with not making a tested assumption about the product's competitive standing until the market launch? What is the risk associated with not explicitly deciding which markets will be targeted until the venture is very close to market entrance and commercialization?

Under the suggested method, the decision-maker seeks to prioritize problems based on the financial uncertainty associated with each of them. The value of this uncertainty is high when the financial impact of the decision-related problem is high and when the uncertainty about the problem's future scenarios is high.

Risk is often defined as a function of an event's potential impact (or severity) and its likelihood (or probability) of occurring. It is often represented as follows:

$$\text{Risk} = \text{Likelihood} \times \text{Impact}$$

Suppose we assume that a risk's likelihood can be estimated by the probability of occurrence, p, and its impact can be expressed as a financial value, v. In that case, the financial risk can be expressed as follows:

$$\text{Financial Risk} = p \times v = f\left(p,v\right)$$

However, the decision-maker wants to prioritize decision-related problems, not risks. Indeed, if the decision-maker knew the value of p for all problems, those decision-related problems would already be very close to being solved. The predicted value of p is most often unknown before actually working on a decision-related problem and weighing its future scenarios. While looking at a group of decision-related problems to be prioritized, the first thing the decision-maker may be able to estimate about the respective future scenarios is their uncertainty, a value that we identify with the function $U(p)$. Therefore, the decision-maker can attempt to estimate the financial uncertainty of each problem as follows:

$$\text{Problem's Financial Uncertainty} = U\left(p\right) \times v = f\left(U\left(p\right),v\right)$$

The problem's financial uncertainty can therefore be approximated as a function of the potential financial impact of the decision-related problem and the degree of uncertainty of its respective future scenarios. This prioritization approach suggests focusing first on problems bearing the highest financial impact and the highest predictive uncertainty (Table 8.1).

Figure 8.1 graphically summarizes these prioritization criteria.

The following represents a simple example of how to apply these criteria to a simplified context of binary decision-related problems and assess its financial risk explicitly.

Table 8.1 A summary of the main criteria for the prioritization of decision-related problems

Criterion	Goal
Financial impact	Helps prioritize problems that have an essential impact on value compared to those that are less impactful
Degree of uncertainty	Helps focus decision-making resources on problems that deserve more attention due to a lack of high-quality signals

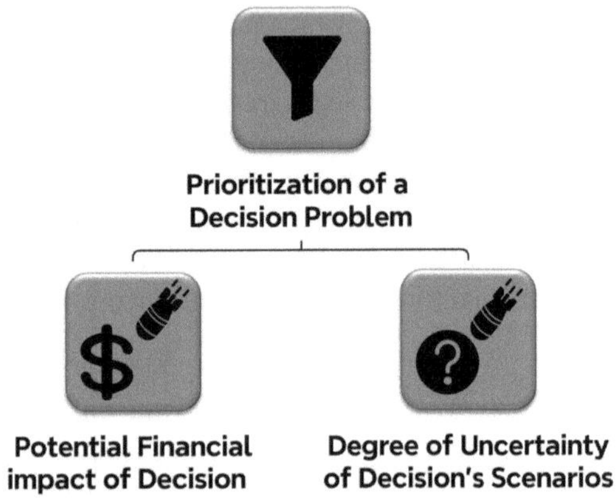

Fig. 8.1 A graphical representation of the two main criteria for the prioritization of decision-related problems

8.2.1 The Financial Impact Criterion

The financial impact of a binary decision-related problem can be estimated by calculating the difference between the highest and the lowest monetary values among all the scenario–action pairs. For instance, in an invest-exit decision, the most rewarding value corresponds to the value of investing in a high scenario, while the most penalizing value corresponds to the value of investing in a low scenario (type 1 error).

The value of the problem described in Table 8.2 can therefore be expressed as follows:

$$\text{Problem's Value} = v = \text{Value}\left(\text{Best Case}\right) - \text{Value}\left(\text{Worst Case}\right)$$

In other terms, when using the criteria of financial impact, the decision-maker prioritizes problems characterized by the highest financial value gap between a right investment decision and a Type 1 error decision. By substituting the respective variables r and z it is simple to obtain the following expression:

$$v = r + z$$

In general, in a group of n problems, the value of the problem "i" can be expressed as follows:

$$v_i = r_i + z_i \ \text{with}\, i = 1\ldots, n.$$

Table 8.2 Scenario–action table that helps calculating the value of a decision problem

	H P(H)=p	L P(L)=1-p
Invest	r	-z
Exit	0	0

	H	L
Invest	Right decision	Type 1 error
Exit	Type 2 error	Right decision

On a Cartesian axis, the potential financial value of problem "*i*" can therefore be placed on the interval from zero to v_i. Each of the decision-related problems being prioritized can consequently be placed on the same axis.

Suppose the decision-maker wants to place all problems on the same axis, on an interval ranging from zero to four. In that case, all values must be normalized to four, obtaining:

$$\hat{v}_i = 4 * (r_i + z_i) / \max(v_i) \text{ with } \hat{v}_i \in [0,4]$$

8.2.2 The Uncertainty Criterion

As discussed in Section 5.1, in a binary decision-related problem, the degree of uncertainty is a function of the probability of the high scenario *p*.

$$\text{Degree of Uncertainty} = U = U(p)$$

In a two-scenario decision-related problem, the degree of uncertainty is higher when it is more difficult for the decision-maker to predict the likelihood of the expected future scenario. This happens when the probability p is close to 50%. Therefore,

$$\lim_{p \to 50\%} U(p) = \max(U(p))$$

Indeed, in this case, the decision-maker does not dispose of information (signals) helpful to predict the likelihood of either scenario.

On the contrary, the degree of uncertainty decreases when decision-makers possess accurate predictive information about possible future scenarios. It happens when *p* is close to either 0% or 100%:

$$\lim_{p \to 0\%} U(p) = \min(U(p)) \text{ or } \lim_{p \to 100\%} U(p) = \min(U(p))$$

Indeed, in this case, the decision-maker knows almost with certainty which scenario the venture will encounter. In synthesis, *U*(*p*) is a function that shows its maximum value when $p = 50\%$ and its minimum values when $p = 0\ \%$ or $p = 100\%$.

In a practical attempt to map *U*(*p*), the decision-maker can approximate it using a simple function. One of the possible approximations can be written as follows:

$$U(p) = |0.5 - p| \text{ with } U \in [0,0.5]$$

When expressed for a generic decision-related problem “*i*”, this expression becomes as follows:

$$\hat{u}_i = |0.5 - \mathrm{p_i}| \text{ with } \hat{u}_i \in [0,0.5]$$

The variable u_i can be normalized to the interval [0,4], as expressed below:

$$\hat{u}_i = 8 * |0.5 - p_i| \text{ with } \hat{u}_i \in [0,4]$$

This function returns normalized values that can be easily plotted.

8.2.3 Mapping Decision-Related Problems

At this point it is possible to map a group of decision-related problems on a bi-dimensional chart, using the normalized values of v_i and u_i

$$\hat{u}_i = 8 * |0.5 - p_i| \text{ with } \hat{u}_i \in [0,4]$$
$$\hat{v}_i = 4 * (r_i + z_i) / \max(v_i) \text{ with } \hat{v}_i \in [0,4]$$

Here, $\mathrm{Max}(\hat{v}_i)$ represents the highest decision value among all the decision problems. The normalization of the values of v_i and p_i allows for easy placement on a prioritization map such as the one shown in Fig. 8.2. If the decision-maker assumes that the highest priority position corresponds to the point with coordinates $v_i = 4$ and $u_i = 4$, then the relative priority of the different decision-related problems is represented by their respective distance from the highest priority point. Therefore, iso-priority lines can be drawn as arcs of circles centered in (4,4); these confines facilitate the visual identification of the priority levels. In the example Fig. 8.2, the final prioritization sequence (from high to low) of decision problems (DP) is the following: DP9, DP3, DP7, DP1, DP4, DP8, DP6, DP5, and DP2.

8.2.4 Applying the Theory: Example of Decision-Related Problems Prioritization

The medical entrepreneur must decide which of four problems to prioritize. The four pending problems have the characteristics described in the tables below (Table 8.3).

It is necessary to calculate the respective coordinates (v_i, u_i) using the formulas discussed above in order to position all four decision-related problems on the map. Table 8.4 summarizes both the partial and the final calculation results.

Figure 8.3 shows the respective positions of the four problems: the relative mapped positions, compared with the iso-priority lines, suggest the following prioritization sequence: DP2, DP4, DP1, and DP3.

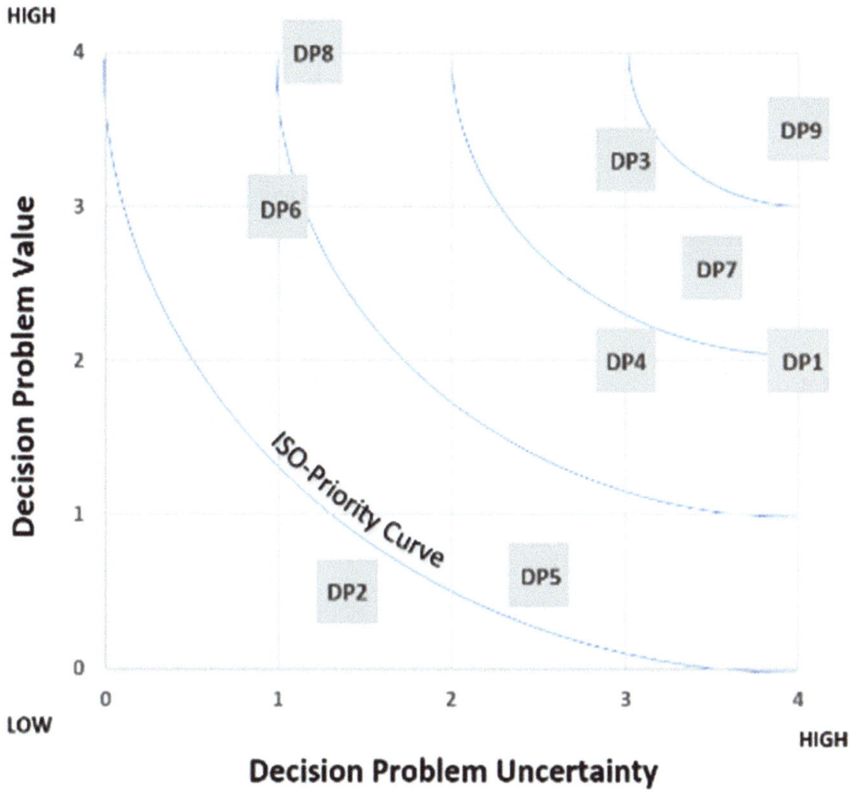

Fig. 8.2 An example of graphic decision problems' prioritization in a chart contains iso-priority curves

Table 8.3 The scenario–action tables of the four decision problems that need to be prioritized in this practical case

	Decision-Related Problem 1	
	H P(H) = 55%	L P(L) = 45%
Invest	US$5 million	-US$2 million
Exit	0	0

	Decision-Related Problem 2	
	H P(H) = 75%	L P(L) = 25%
Invest	US$5 million	-US$2 million
Exit	0	0

	Decision-Related Problem 3	
	H P(H) = 25%	L P(L) = 75%
Invest	US$2.5 million	-US$0.5 million
Exit	0	0

	Decision-Related Problem 4	
	H P(H) = 25%	L P(L) = 75%
Invest	US$2.5 million	-US$0.5 million
Exit	0	0

Table 8.4 The summary of the values relevant to the prioritization of the four problems

	p_i	$r_i + z_i$	u_i	v_i
DP1	55%	US$7 M	**0.4**	**4.00**
DP2	75%	US$7 M	**2**	**4.00**
DP3	60%	US$3 M	**0.8**	**1.71**
DP4	80%	US$3 M	**2.4**	**1.71**

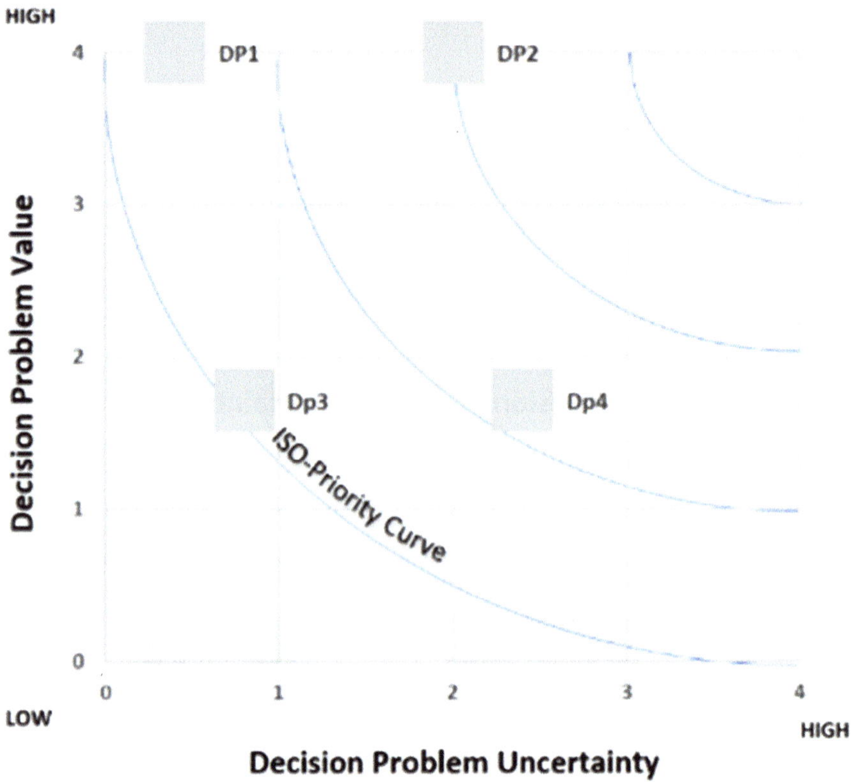

Fig. 8.3 The graphic prioritization of the four problems

8.3 The Practice

Understanding the fundamental mechanisms outlined in the above theory provides valuable guidance for prioritizing decision-related problems. In real-life practice, a disciplined, precise application of this prioritization theory is very challenging. Indeed, not all decision-related problems are characterized by binary decisions, and many problems involve more than two future scenarios. Moreover, for certain types of decision-related problems, such as scenario decisions, it is challenging to approximate the values of each scenario–action pair. The application of this approach, therefore, involves expert judgment for assigning qualitative scores to both financial impact and uncertainty. Such assessments may still utilize the bi-dimensional charts already discussed: plotting decision-related problems based on their uncertainty and financial implications, facilitating visualization.

The practical mechanisms consist of two steps:

- Decision-related problems are placed on a bi-dimensional map (such as the one shown in Fig. 8.3) based on the best quantitative approximation of their expected financial impact and degree of uncertainty.
- Priorities are assigned using the iso-priority curves, by observing problems' and iso-curves' relative positions and distances.

A robust process for implementing this theory consists of sharing its basic principles with all members of the venture's executive team and asking each of them to suggest a mapped prioritization. Then, in a dedicated consensus meeting, all members can share and evaluate one another's views to reach a consensus on prioritization for the venture. Even if it is not applied to the entire depth of the mathematical and statistical calculations, the illustrated theory's fundamentals help practitioners avoid instinctive jumps into prioritization conclusions. These instinctive prioritization practices may even inadvertently cause the undesired exclusion of specific key decision-related problems. As mentioned already, the prioritization process is a recursive exercise that must be repeated regularly throughout the life of a medical venture. Management will realize that the constant refinement of the available information on each decision-related problem diminishes its degree of uncertainty and fine tunes the estimation of its value. Moreover, when decision-related problems are finally approached, their degree of uncertainty inevitably diminishes, causing a shift in the bi-dimensional map. An example of practical problem prioritization is illustrated by the prioritization of top-level K9 map clusters, discussed in the following section.

8.4 Prioritizing the Top Clusters of the K9 Map

Before applying the discussed decision-ranking process to a few individual decision-related problems, it is feasible to start prioritizing at the higher levels of decision clusters. Once a high-priority, high-level cluster has been identified, the decision-maker can gradually start prioritizing those at lower levels within the same cluster. The highest possible level of prioritization is among the clusters of the K9 map (Fig. 8.4) and it requires the consideration of two prioritization criteria.

8.4.1 First Prioritization Criterion

One of the two criteria illustrated for prioritizing individual problems, such as financial value, cannot be safely used in such a high-level case. Indeed, the monetary

Fig. 8.4 The K9 map

value of each cluster is challenging to estimate: Each of the nine mapped clusters contains several lower-level clusters, which, in turn, contain individual decision-related problems holding their respective monetary values. Moreover, prioritization at the level of the K9 map is more commonly performed (or reviewed) during the early phases of the venture's life, when all of the nine clusters may be considered to hold a high estimated value concurrently. In summary, the criterion based on the decision's monetary value is hardly applicable while prioritizing high-level clusters.

While looking for a substitute criterion, we found that the sequence provided by the K9 map contains critical information relevant to prioritization. Using the sequence of the K9 map means encouraging the decision-maker to prioritize problem clusters located at the beginning of the sequential roadmap while deprioritizing those in downstream positions. In this way, critical, "primordial" questions that are potentially life-changing for the venture can be addressed in the earlier stages.

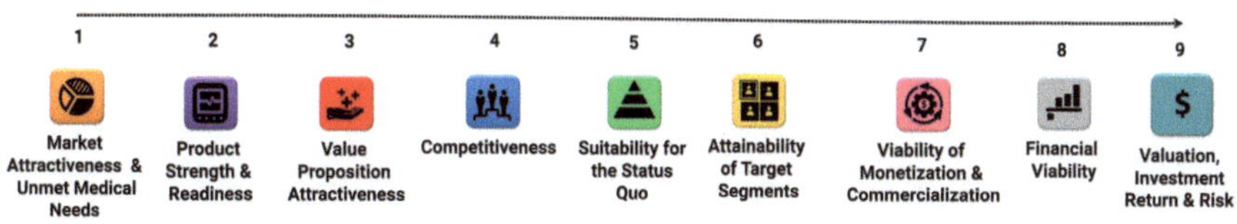

Fig. 8.5 The sequential method for the prioritization of the top-level K9 clusters

This approach helps avoid focusing on problems whose relevance and solution may depend on upstream decisions that have not yet been addressed (Fig. 8.5). For instance, if early assumptions about expectations of a "high" market value are rejected, then the subsequent problems become significantly less interesting or even worthless.

For instance, understanding the reimbursement context is helpful at any point in time for a medical startup. However, drilling into the reimbursement problem and attempting to crystallize the related business assumptions before having a clear position on the competitiveness of the value proposition may constitute an inefficient way of allocating resources.

On the other hand, it is advisable to include the position-on-map criterion with care. For instance, it is essential to remember that this approach is not intended to suggest to the entrepreneur that it is acceptable to overlook or ignore any of the downstream problems. Instead, this criterion helps avoid in-depth, resource-draining resolutions of downstream issues, while fundamental, upstream problems are only superficially approached or not approached at all.

How can this criterion be transformed into mappable coordinates? The related numeric index can be obtained from the relative positions of the various clusters, $i = 1, 2, \ldots, 9$. The respective index r_i can be identified with the series:

$$r_i = 9 - i \text{ with } i \in [0,9] \text{ and } r_i \in [0,8]$$

Or, it can be expressed on a normalized scale [0,4]:

$$\hat{r}_i = r_i * \left(\frac{4}{8}\right) \text{ ensuring that} : \hat{r}_i \in [0,4]$$

For instance, cluster #1 holds priority 9–1 = 0, cluster #2 has priority (9–2)/2 = 3.5, and cluster #9 holds priority 0. The values of the first prioritization indicator for each cluster are represented in Table 8.5.

8.4.2 Second Prioritization Criterion

The second criterion already presented for prioritizing individual decision problems refers to the degree of uncertainty of the problem's scenarios. In the case of high-level clusters, such as those in the K9 map, it is difficult to assess specific values of p and to use them as an estimator of uncertainty. Therefore, to determine the degree of a cluster's uncertainty, the decision-maker must rely on broader assessments that

Table 8.5 The priority indexes (normalized to the interval [0–4]) of the top-level K9 clusters

Cluster	i	r_i	$\hat{r}_i$
Cluster 1	1	8	4.0
Cluster 2	2	7	3.5
Cluster 3	3	6	3.0
Cluster 4	4	5	2.5
Cluster 5	5	4	2.0
Cluster 6	6	3	1.5
Cluster 7	7	2	1.0
Cluster 8	8	1	0.5
Cluster 9	9	0	0.0

require more extensive use of managerial judgment (both individual and collective). A practical approach involves identifying the uncertainty drivers for each cluster and assessing them individually. For instance, the degree of uncertainty of the cluster "Market Attractiveness and Unmet Medical Needs" strongly depends on the uncertainty related to the following sub-clusters:

1. The estimation of the social perception of the unmet medical needs addressed by the new diagnostic technology.
2. The precise identification of diagnostic indications, and the estimation of the respective patient group sizes.
3. The estimation of the value of the potential market size.
4. The identification of accessible global markets and the estimation of the respective sizes.

Each member of the venture's management team can quantify the level of uncertainty $\hat{u}_i$ associated with each of the four listed drivers using a scale of [0,4]. The obtained values are, by definition, already normalized on the desired scale; therefore, they do not need further normalization. In an internal consensus meeting, each team member can present, justify, and support their estimations of the uncertainty for each driver, comparing them with the results given by other members.

Later, the venture can find consensus on the degree of uncertainty for each cluster or, if consensus is not reached, the team can use the averages of the presented values. The following section provides a practical example of prioritizing the K9 map clusters.

8.4.3 Example of K9 Map Prioritization

A medical startup ready to commercialize an AI system for diagnosing diabetic retinopathy during regular oculistic checks was highly concerned about designing and investing in a viable international commercial process and organization. Before diving into the decisions related to this aspect, the venture's management organizes a review of the K9 map clusters' prioritization. The two selected prioritization criteria are the sequential position on the K9 map and the degree of uncertainty associated with each cluster.

The first prioritization indicator can, therefore, be expressed on a normalized scale [0,4] as follows:

$$\hat{r}_i = (9 - \mathrm{i}) * \left(\frac{4}{9}\right) \text{ ensuring that: } \hat{r}_i \in [0,4]$$

The first cluster's prioritization indicator is therefore the one already calculated and represented in Table 8.5.

The second criterion requires assessing each cluster's degree of uncertainty on a scale [0,4].

The exercise was perceived as difficult by management, as assessing each cluster's uncertainty at such a high level was considered very challenging. Therefore, to make the individual assessments more approachable and the group effort more rewarding, the venture's management associated four different uncertainty assessments with each cluster. Practically, each cluster-level question was subdivided into four easier questions. For instance, the question about the first cluster's degree of uncertainty was split into agreement questions about four separate aspects of the cluster:

- The clarity on the identification of patients' and stakeholders' needs.
- The accuracy in the determination of the market size.
- The accuracy of the estimation of current regional spending on the same diagnosis.
- The degree of accuracy of the assessed market rivalry.

Therefore, for this first cluster, the team members were asked to quantitatively express their degree of agreement (Table 8.6) about each of the statements listed in Table 8.7. Form their answers it is simple to calculate the respective degree of uncertainty as shown in the third column of the same Table.

By extending the same approach to the remaining eight clusters and after a few discussions and consensus rounds, the venture's management reached an agreement on each of the four uncertainty-related questions, for each of the nine respective clusters: This led to the uncertainty assessments shown in Table 8.8.

The concurrent consideration of the two prioritization indicators ($\hat{r}_i$ and $\hat{u}_i$) led to the prioritization map shown in Fig. 8.6, which drew management's attention from the monetization and commercialization cluster toward the value proposition cluster. In summary, while the startup's management initially rushed and focused on the decisional cluster of monetization and commercialization models, it later realized that more "primordial," fundamental, unresolved critical questions remained open regarding the attractiveness of the value proposition. As a consequence, in a second stage, the venture's management agreed on the need to give more explicit and ground-breaking answers to questions such as the following:

- How much is a sleeker and cheaper automatic diagnostic system for diabetic retinopathy worth to patients or payers?
- What is the value for the specialist in ophthalmology?

Table 8.6 An example of a numeric scale which is helpful to quantify the degree of agreement (disagreement) of the company's management over selected statements

Statement	Degree of Agreement $\hat{u}_i$
I strongly disagree	0
I disagree	1
I am neutral	2
I agree	3
I strongly agree	4

Table 8.7 Example of assessment of the degree of uncertainty of the first cluster: market attractiveness and unmet medical needs

Assessment of the degree of Uncertainty of the First Cluster: Market Attractiveness & Unmet Medical Needs		
Statements	Degree of Agreement with the statement: u_i **0 = I strongly disagree** **1 = I disagree** **2 = I am neutral** **3 = I agree** **4 = I strongly agree**	Degree of Problem's Uncertainty: $\hat{u}_i$ $\hat{u}_i = 4 - u_i$
Patients' and stakeholders' needs are precisely identified.	$u_1 = 4$	$\hat{u}_1 = 0$
Market size is Precisely Known.	$u_2 = 4$	$\hat{u}_1 = 0$
The estimation of current regional spending on the same diagnosis is precisely identified.	$u_3 = 4$	$\hat{u}_1 = 0$
The degree of accuracy of the assessed market rivalry	$u_4 = 3$	$\hat{u}_1 = 1$
Averages	$u = 3.7$	$\hat{u} = 0.3$

– What is the cost of a later diagnosis of retinopathy? How many patients actually become blind due to the lack of an earlier diagnosis? What is the personal and the social cost of missing this diagnosis?

On the one hand, applying these suggested prioritization methods might feel excessively complicated and time-consuming; on the other hand, a rational, transparent, and fact-based assessment of clusters' prioritization represents a recurring duty for medical startup management. This kind of approach can save the venture from overlooking or leaving indefinitely untouched time-critical, life-or-death, business-defining problems.

Table 8.8 An example of the typical statements to be used—for each of the 9 top-level clusters—to estimate the respective management's degrees of agreement, over their respective degrees of uncertainty

Decision cluster	Assessing decision cluster's degree of uncertainty								Scenario uncertainty
	Statement 1	u1	Statement 2	u2	Statement 3	u3	Statement 4	u4	
Market attractiveness and unmet medical needs	Patient's and stakeholder needs are precisely identified	4	Market size is precisely known	4	Current regional spending on the same diagnosis	4	The degree of market rivalry is fairly assessed	3	0.3
Product strength and readiness	TPP is detailed and complete	4	The product is ready to be commercialized globally	2	Product has been awarded global regulatory clearance	3	Scientific evidence of product accountabilities is fully available	3	1.0
Value proposition attractiveness	Value proposition is precisely spelled	1	Value proposition addresses precisely all stakeholders	2	Unmet medical needs have been ranked	3	The ability to add value vs. existing therapeutic alternatives is defined	2	2.0
Competitiveness	Superiority vs. CSC is defined clearly	1	Superiority vs. existing competition is defined clearly	2	Superiority vs. emerging competition is defined clearly	2	Competitiveness protection is fully set	1	2.5
Suitability to status quo	Reimbursement challenges are clear	2	Regulatory challenges are clear	3	Buyer's alignment is clear	3	Other stakeholders' positions is clear	4	1.0
Attainability of target segments	Target geographies are prioritized and identified	2	Target in-region segments are prioritized and identified	3	Positioning statements are defined for each target segment	4	Positioning claims are all supported by data	3	1.0
Viability of monetization and commercialization	Business model is completely set in each of the target regions	3	Commercial organization is designed for each region	4	Commercial capabilities are set, and the implementation plan is ready	3	Margin splits and profit degradation are defined across regions	2	1.0
Financial viability	Cost of product is known and the volume/price trend is known	1	Local currency prices are identified	1	Cost of sales (commissions and margins) and SG&A are set globally	1	Venture's NPV and IRR were calculated	1	3.0
Valuation, investment Return and risk	Key product development milestones are identified	1	Potential strategic investors are identified	1	Potential financial investors were identified	1	Key exit alternatives were explored	1	3.0

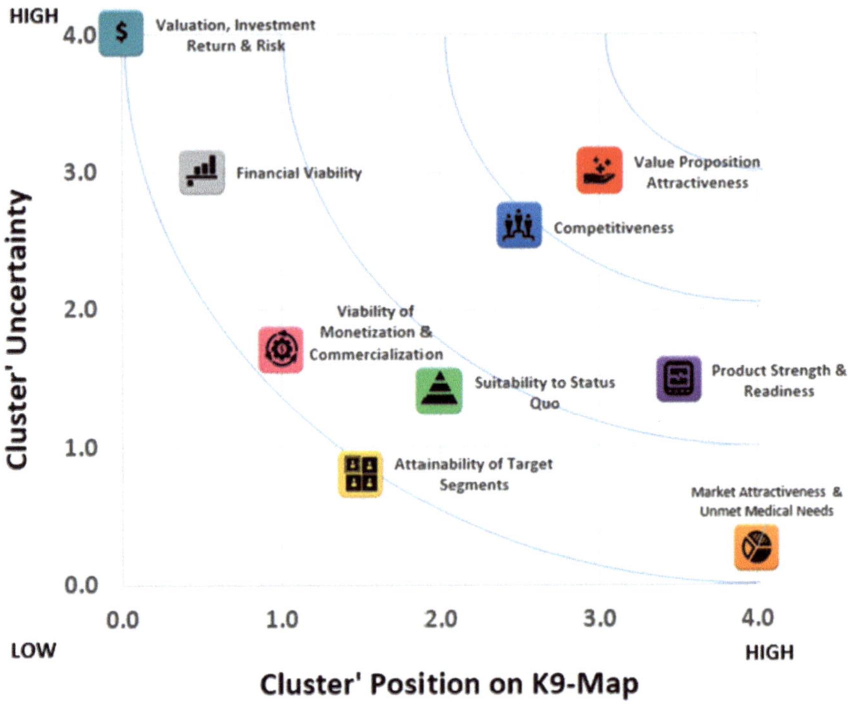

Fig. 8.6 A graphical example of the prioritization of the top-level K9 map clusters

Bibliography

Kovic, M. (2018). *The general problem of prioritization ZIPAR discussion paper series.*

Correction to: Decision-Making Using a Scientific Approach

Correction to:
Chapter 5 in: L. Negri, *De-Risking Investments in Medical Technology Innovation*, Springer Series in Healthcare Management and Innovation,
https://doi.org/10.1007/978-3-032-15007-3_5

The original version of the chapter was inadvertently published with the incorrect figure for Fig. 5.18, which has been updated now.

The updated version of this chapter can be found at
https://doi.org/10.1007/978-3-032-15007-3_5

L. Negri, *De-Risking Investments in Medical Technology Innovation*, Springer Series in Healthcare Management and Innovation,
https://doi.org/10.1007/978-3-032-15007-3_9

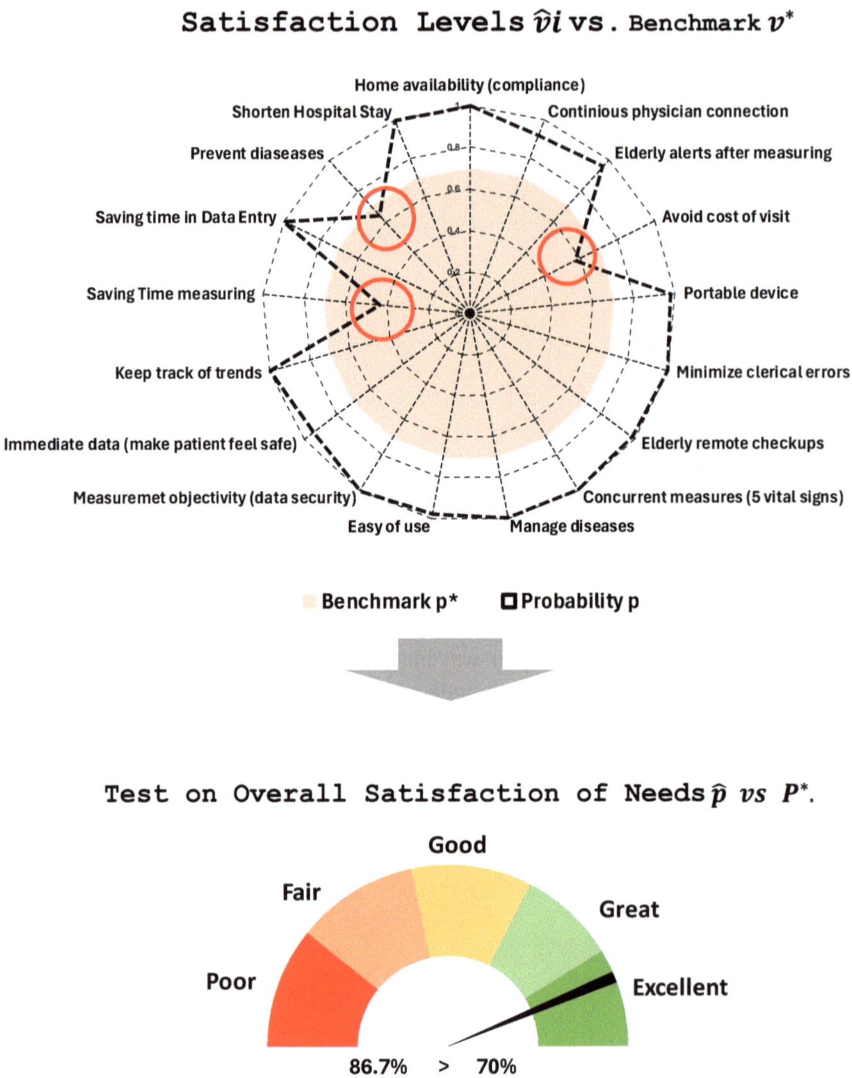

Fig. 5.18 Plotted values of pi representing the probability that the satisfaction of need "" falls above the benchmark of 70%. It also shows that the cross-needs average $\hat{p}$ is higher than the benchmark

MIX
Papier aus verantwortungsvollen Quellen
Paper from responsible sources
FSC® C105338

If you have any concerns about our products,
you can contact us on
ProductSafety@springernature.com

In case Publisher is established outside the EU,
the EU authorized representative is:
Springer Nature Customer Service Center GmbH
Europaplatz 3, 69115 Heidelberg, Germany

Printed by Libri Plureos GmbH
in Hamburg, Germany